THE 100 BEST BIRDWATCHING SITES IN INDIA

Contributing Photographers

Aidan Fonseca
Alka Vaidya
Amano Samarpan
Amit Sharma
Anand Arya
Arka Sarkar
Arpit Bansal
Arpit Deomurari
Avinash Khemka
Barnita Newar
Biswapriya Rahut
Chinmay Agnihotri
Clement M Francis
Deboshree Gogoi
Dhritiman Chatterjee
Dhairya Jhaveri
Dushyant Prashar
Falguna Shah
Ganesh Adhikari
Garima Bhatia
Gopinath Kollur
Goutam Mohapatra
Gururaj Moorching

Ingo Waschaies
James Eaton
Khushboo Sharma
Kintoo Dhawan
Koshy Koshy
Kunan Naik
M. V. Shreeram
Manjula Mathur
Mohit Mishra
Natalia Parkina
Nikhil Devasar
Niranjan Sant
Nitin Bhardwaj
Nitin Srinivasamurthy
P. B. Biju
P. S. Anand
Panchami Manoo Ukil
Parth Satvalekar
Prasanna Parab
Pia Sethi
Purushottam Lad
Rahul Sharma
Ramki Sreenivasan

Raju Kasambe
Rathika Ramasamy
SarwanDeep Singh
Satinder Sharma
Satya Singh
Savio Fonseca
Shahnawaz Khan
Shashank Dalvi
Shyam Ghate
Sonu Anand
Sugata Goswami
Subhoranjan Sen
Sujan Chatterjee
Sumit K Sen
Supriyo Samanta
Swati Kulkarni
Tapas Misra
Tejas Naik
Tripta Sood
Uma & Ganesh
Urmi Nath
Vaibhav Deshmukh
Vaidehi Gunjal

Bikram Grewal & Bhanu Singh

PRAKASH BOOKS

This edition of *The 100 Best Birdwatching Sites in India* is published and distributed in India by Prakash Books India Pvt. Ltd., 113A Ansari Road, Daryaganj, New Delhi-110002, India, by arrangement with John Beaufoy Publishing Ltd.

10 9 8 7 6 5 4 3 2 1

Copyright © 2020 John Beaufoy Publishing Limited
Copyright in text © 2020 Bikram Grewal and Bhanu Singh
Copyright in maps © 2020 John Beaufoy Publishing Limited
Copyright in photographs, see individual photos except as below

Photo Credits
Front cover: Great Hornbill © Clement M Francis.
Back cover: clockwise from top left: Orange Bullfinch © Dhritiman Mukherjee; Rufous Sibia © Clement M Francis; Satyr Tragopan © Nikhil Devasar; Pale-capped Pigeon © Panchami Manoo Ukil; Sunda Teal © Clement M Francis; Yellow-throated Bulbul © Garima Bhatia; Eurasian Oystercatcher © Kunan Naik; Great Indian Bustard © Garima Bhatia.
Spine: Common Kingfisher © Avinash Khemka

All rights reserved. No part of this publication may be reproduced, stored in a retrieval system or transmitted in any form or by any means, electronic, mechanical, photocopying, recording or otherwise, without the prior written permission of the publishers.

ISBN 978-93-8971-705-1

Edited by Krystyna Mayer
Cartography and design by Samiha Grewal Mishra, Alpana Khare Graphic Design, New Delhi
Project management by Rosemary Wilkinson
Printed and bound in Malaysia by Times Offset (M) Sdn. Bhd.

CONTENTS

Introduction	4
India's Climate	5
India's Geography	7
Bird Habitats	10
Threats to Birds	13
Glossary of Terms	15

Andaman & Nicobar Isands	16	Maharashtra	164
Andhra Pradesh	26	Manipur	176
Arunachal Pradesh	40	Mizoram	180
Assam	56	Meghalaya	184
Goa	72	Nagaland	188
Gujarat	84	Odisha	196
Himachal Pradesh	100	Punjab	206
Jammu & Kashmir	110	Rajasthan	210
Karnataka	120	Sikkim	222
Kerala	134	Tamil Nadu	232
Lakshadweep	144	Uttar Pradesh	242
Delhi and NCR	148	Uttarakhand	250
Madhya Pradesh	154	West Bengal	260

Recommended Reading & Acknowledgements	272
Conservation & Birdwatching Organizations	273
Checklist of Birds of India	274
Index	291

Introduction

India is one of the 17 megadiversity regions of the world and contains two of the world's 34 biodiversity hotspots. It is a country with a myriad landscapes, great heritage and culture, and unrivalled birdlife. Of more than 9,000 of the world's birds, the Indian subcontinent contains about 1,300 species, or more than 13 per cent. One of the main reasons for India's high avian diversity is the presence of diverse habitats, from the arid cold desert of Ladakh and Sikkim and the steamy, tangled jungles of the Sunderbans, to the wet, moist forests of the Western Ghats and Arunachal Pradesh.

The Indian region exhibits an enormous variation in climate, geomorphology and surface topography. On the basis of physiography it is divisible into three zones, namely the Himalayan region, the Indo-Gangetic Plains and the Indian Peninsula. Each of these zones can be further classified based on relief features, and patterns of soil and water regimes. The resulting variations in habitats have manifested in a great deal of floral and faunal diversity, and interesting patterns of vegetation formations.

The region is home to many bird groups such as pheasants, laughingthrushes, drongos, leafbirds, pittas, parrotbills and flowerpeckers. With so many birds to be found and such mind-boggling levels of biological and cultural diversity, India ought to be on the radar for every serious birdwatcher. However, many birdwatchers planning to visit India quickly encounter two major impediments. First, because the region's economies are developing so rapidly and its population is soaring well past 1.3 billion, forests are being cleared, rivers dammed and swamps drained, and in their places spring factories, cities, plantations and roads. Second, historically, there has been a very limited selection of good bird books that you can take with you on a birdwatching trip to this part of the world. Although this is rapidly changing, with many new guides and mobile apps appearing every year, it is still quite rare to find a book that provides a broad overview of the region's distinctive birdlife or its key birdwatching sites.

The 100 Best Birdwatching Sites in India was conceived to fill the gaps in information in book form, and is the result of numerous conversations with our peers working in the region, whether as birdwatching guides, bird-tour leaders, researchers or conservationists.

We are grateful to the individuals who have shared their knowledge or images of the places featured here. We hope that the book will not only intrigue armchair naturalists with the remarkable birdlife of India, but also encourage birdwatchers and general ecotourists to visit the region, and in so doing create the economic incentives needed to protect its incredible biodiversity. While the book is not intended to be a photographic guide to the birdlife of India, it includes images of more than 300 bird species, including many of the region's rarest and most elusive.

Organization of the Book

The book is divided into three sections, namely the introduction, the systematic section and the appendices. The introductory section provides a broad overview of India's geography, climate and habitats, and an indication of key birds found in different regions. The systematic section provides details on each of the 100 birdwatching sites across the region. The profiles describe the nearest major towns to the birdwatching sites, travel options, and key bird species and other wildlife, and advise on the best time of year to visit. This information is by no means exhaustive, and some details may have changed by the time of the book's publication. A list of the key bird species for each site is included, although the lists do not capture every species that occurs at the site. For consistency in names, the nomenclature and taxonomy recommended by the International Ornithological Union's checklist has been adopted. The appendices contain useful information for birdwatchers and naturalists planning visits to the region, including a checklist of India's birds and their status.

Himalayan Monal

India's Climate

India is a vast country with varied climatic conditions. It includes an extraordinary variety of climatic regions, ranging from tropical in the south, to temperate and alpine in the Himalayan north. The nation's climate is strongly influenced by the Himalayas and the Thar Desert. There are three climatic seasons in a year:

1. Monsoons (June–September, south-west monsoon; October–November, north-east monsoon).
2. Summer (April–July).
3. Winter (October–March).

The winter in South India is not as cold as it is in North India. It is marked by clear skies, hot days and cool nights. This kind of weather prevails from September to March. The south-west monsoon sets in over Kerala in June and progresses towards the north, enveloping the entire country by the end of July. The eastern coastal regions, the coasts of Andhra Pradesh and Tamil Nadu, experience the north-east monsoon in October–November. Along the east coast, this period is marked by cyclones due to severe atmospheric depression in the Bay of Bengal and the Indian Ocean that moves towards the mainland at a high speed, causing widespread destruction to life and property. The west coast rarely experiences such cyclonic effects.

The annual average rainfall in India varies from a low of 50mm in the extreme western parts of Jaisalmer bordering Pakistan, to a high of 11,000mm in the Cherrapunjee region of Meghalaya. Similarly, the temperature is highly variable, ranging from more than 50 °C in the Thar Desert to minus 50 °C at Siachen in Jammu and Kashmir.

India has been broadly divided into the groups below based on the different climatic conditions encountered in each region.

Tropical Humid

The tropical humid group experiences tropical wet, tropical monsoon, and tropical wet and dry types of climate. The most humid is the tropical wet and monsoon climate that covers a strip of the southwestern lowlands, the Western Ghats and southern Assam. India's two island territories, Lakshadweep, and the Andaman and Nicobar Islands, are also subject to this climate. The tropical wet and dry climate is most common in India. Noticeably drier than areas with a tropical monsoon climate, it prevails over most of inland peninsular India except for a semi-arid rain shadow east of the Western Ghats.

Tropical forest

Tropical humid areas are characterized by moderate to high year-round temperatures, and seasonal but heavy rainfall. The months of May–November experience the most rainfall, and the rain received during this period is sufficient to support vegetation throughout the year.

Tropical Dry
The tropical dry climate group is divided into three subdivisions:

1. Tropical semi-arid (steppe) climate.
2. Subtropical arid (desert) climate.
3. Subtropical semi-arid (steppe) climate.

Karnataka, central Maharashtra, some parts of Tamil Nadu and Andhra Pradesh experience the tropical semi-arid (steppe) climate. Rainfall is very unreliable in this type of climate, and the hot and dry summers occur in March–May. With scanty and erratic rainfall and extreme summers, western Rajasthan experiences the subtropical arid (desert) climate. The subtropical semi-arid (steppe) climate occurs in the areas of the tropical desert that run from the regions of Punjab and Haryana to Kathiawar.

Subtropical Humid
This climate occurs in most of north and north-east India. The summers are very hot, while in winter the temperature can plunge to as low as 0 °C. In most of this region there is very little precipitation during the winter, due to powerful anticyclonic and katabatic (downwards-flowing) winds from Central Asia.

Humid subtropical regions are subject to pronounced dry winters. Most summer rainfall occurs during powerful thunderstorms associated with the south-west summer monsoon; occasional tropical cyclones also contribute. Annual rainfall ranges from less than 1,000mm in the west to more than 2,500mm in parts of the north-east. As most of this region is far from the ocean, the wide temperature swings more characteristic of a continental climate predominate; the swings are wider than those in tropical wet regions, ranging from 24 °C in north-central India to 27 °C in the east.

Mountain
The Himalayan regions of India's northernmost areas are subject to a montane, or alpine, climate. For every 100m rise in altitude in the Himalayas, the temperature falls by 0.6 °C, resulting in a number of climates, from tropical to tundra. The trans-Himalayan belt, which is the northern side of the western Himalayas, is cold, arid and windswept. There is less rain in the leeward side of the mountains, whereas heavy rainfall is received by the well-exposed slopes. The heaviest snowfall occurs in December–February.

The mountain areas of Uttarakhand experience a tundra type of mountain climate, with an average temperature of 0–10 °C. Higher areas of Jammu and Kashmir, and Himachal Pradesh, experience extreme cold with a polar-type climate. In these regions the temperature during the warmest month is 0–10 °C, and precipitation mainly occurs in the form of snow.

The Himalayas

India's Geography

India is situated entirely on the Indian Plate, a major tectonic plate that was formed when it split off from the ancient continent Gondwanaland. The land mass of India is bounded in the south-west by the Arabian Sea and in the south-east by the Bay of Bengal. To the north and north-east lies the mighty Himalayan range. To the west lies Pakistan, and to the east, Bangladesh and Myanmar. In the north, China (Tibet), Nepal and Bhutan share the international boundary with India. To the south, Sri Lanka shares the maritime boundary and is separated from India by a narrow channel of the Bay of Bengal.

India is one of the largest countries in the world and covers an area of about 3,287,263km^2. It measures 3,214km from north to south and 2,933km from east to west, and has a land frontier of 15,200km and a coastline of 7,516km. The mountain ranges such as the Himalayas in the north, the Aravallis in the west, the central highlands of the Vindhyas and Satpuras, and the Eastern and Western Ghats in the east and west, comprise several submontane tracts of varied lengths and heights that support a great variety of birdlife in the region.

Himalayas & Trans-Himalayan Region

The Himalayan region is spread over an area of about 210,626km^2 within India. This region as a whole is regarded as an important biodiversity hotspot. It supports a wide range of vegetation ranging from tropical to alpine types. It is home to more than 8,000 species of flowering plant and nearly 10,000 species of lower plant.

The Indian Trans-Himalayas, also known as the Indian cold desert, support very sparse vegetation in general. The region is poorer in floral as well as bird diversity compared with the moist alpine meadows of the Greater Himalayas. A small portion of the Indian Trans-Himalayas is represented in the Central Himalayas (Sikkim), which are relatively higher in terms of species diversity compared to the northwestern region.

Indian Desert

Most of the Indian desert region is located in western India, and it covers nearly 12 per cent of the overall land. Biogeographically, the Thar Desert is the eastwards extension of the Sahara-Arabian desert system that spreads through Iran, Afghanistan and Baluchistan. The Indian desert supports about 500 species

Indian Desert

of vascular plant. This low species diversity compared to that of the other biogeographic zones is due to harsh climatic factors coupled with intense biotic pressure (anthropogenic factors). Despite uniformly adverse climatic conditions, the Indian desert region supports a remarkable and distinctive birdlife.

Western Ghats

The Western Ghats, a chain of ancient mountains parallel to the west coast of the Indian peninsula, occupy only 5 per cent of India's land area, yet they harbour nearly 27 per cent of its total flora. With a latitudinal range of more than 10 degrees, they lie more or less parallel to the west coast of India. Their forests are one of the best representatives of non-equatorial tropical forests in the world. Wet evergreen forests are mostly confined to the windward side of the Western Ghats, where the rainfall exceeds 2,000mm. Dry deciduous forests are confined to the rain-

Zanskar Range

Deccan landscape

shadow areas, especially towards the northern parts of the Nilgiris, Palnis, and areas bordering the Mysore and Karnataka plateau. Areas above 1,800m are dominated by natural grassland and adjacent pockets of montane evergreen forests, frequently termed the Shola-Grassland Complex. The Western Ghats are home to numerous endemic species of flora and fauna that are found nowhere else in the world. This whole region is a hotspot for birding.

Deccan Peninsula

The Deccan Peninsula is one of the largest biogeographic zones of India with extensive forested tracts. The zone is relatively homogenous and supports various vegetation types, ranging from tropical thorn forests, to tropical dry and moist deciduous forests. There are five biogeographic provinces within this zone: Central Highlands comprising the Vindhya and Satpura hill ranges, Chota Nagpur Plateau, Eastern Ghats, Tamil Nadu Plains and Karnataka Plateau.

Gangetic Plains

The Gangetic Plains include the areas adjacent to the Terai-Bhabar tracts in Uttar Pradesh, Bihar and West Bengal. This area is dominated by grassland and savannah woodland. It is strongly influenced by frequent fires and floods, which deposit silt from the Himalayan foothills. Some of the plant communities within the grassland, for example *Imperata cylindrica* (commonly known as cogongrass), are also reported to be the habitat of the threatened Bengal Florican.

North-east India

North-east India, including the Assam Valley and the adjacent hill ranges, exhibits a complex mosaic of vegetation types, ranging from northern tropical wet evergreen to montane and wet temperate types. The north-east is considered as the 'biological gateway' for much

North-east forests

Andaman coral reefs

of India's fauna and flora, as Gondwanaland first touched this region during the Tertiary period. It represents the transition zone between the Indian, Indo-Malayan and Indo-Chinese biogeographic regions.

The Islands
There are 1,208 islands (including uninhabited ones) in India. The total area under 'islands' within the Indian territory covers approximately 8,358km^2, of which the Andaman and Nicobar Islands occupy about 8,249km^2 and the Lakshadweep Islands 109km^2. These islands have about 2,200 species of higher plant, including around 200 strict endemics. Roughly 1,300 species are not found elsewhere in India.

Rivers & Lakes
A number of major rivers, with their sources in either the Himalayas or the eastern Tibetan Plateau, flow through India. Some of the important ones are the Ganga, Godavari, Kavery, Brahmaputra, Krishna, Narmada, Mahanadi, Tapti and Yamuna. The Ganga is considered to be the holy river. Its source is at Gaumukh in Uttarakhand state, where the mighty river emerges from the depths of the Gangotri glacier. The Godavari is the only river in India that flows from western to southern India, and is considered to be one of the big river basins in India. The Brahmaputra River is one of the major rivers of the world. It originates in Mansarovar near Mt Kailash in the Himalayas, and flows via Tibet, China, India and Bangladesh to the Bay of Bengal. The total length it travels from the Himalayas to the Bay of Bengal is about 2,900km.

Along with the strong river system in India, there are many important lakes. Wular Lake, one of the largest freshwater lakes in Asia and the largest in India, is located in Jammu and Kashmir. Dal Lake is one of the most beautiful lakes in India, and is the second largest lake in Jammu and Kashmir. Chilika Lake is a brackish-water lake and is the largest coastal lake in India; it is situated in Odisha (formerly Orissa), and is Asia's largest inland salt-water lagoon. Pulicat Lake, an important waterbody of southern India, is a saline backwater lake lying along the Tamil Nadu and Andhra Pradesh coast.

Bird Habitats

High-altitude Himalayan Region
The Himalayan region and Trans-Himalayan region collectively form one of the most important bird habitats in India. Nearly 6.41 per cent of the total area of India consists of the Himalayan mountain ranges. There are extreme temperature and rainfall variations within the western and eastern Himalayas, with the western Himalayas being colder and drier than the eastern Himalayas.

As the western Himalayas merge with the Hindu Kush and then the mountains of Central Asia, and the eastern Himalayas merge with the Indo-Chinese and south-east Asian forests, there are not many endemic birds confined to the Himalayas, but this is the centre of species radiation of pheasants. The important pheasant species found in the Indian Himalayan region are Western, Satyr, Blyth's and Temminck's Tragopans, Monal Pheasant, Sclater's Monal, Tibetan Eared, Cheer, Blood, Kalij and Koklass Pheasants, and Red Junglefowl. Important birding habitats of the western Himalayas are primarily located in Jammu and Kashmir, Himachal and Uttarakhand. In the eastern Himalayas, Sikkim, Arunachal, Meghalaya, Assam and Nagaland offer some great birding habitats.

The Trans-Himalayas, consisting of Ladakh in Jammu and Kashmir, Lahaul-Spiti in Himachal Pradesh and a small area of Sikkim, comprise part of a much larger Tibetan plateau of Tibet and China. Many major rivers, for example the Brahmaputra, Sutlej and Indus, start from this region, but much of the area has an internal drainage system where the rivers end in vast lakes. Such lakes, and marshes, mostly saline, are important breeding grounds for birds such as the Black-necked Crane, Bar-headed Goose and Great Crested Grebe, while the flat plains provide habitat for the Tibetan Sandgrouse, Horned Lark and various species of wheatear. Tibetan and Himalayan Snowcocks can be seen on the treeless mountains, with both species sometimes occurring in the same area.

Indian Desert
The Thar Desert occupies nearly 10 per cent of India's geographical area, and covers 208,751km^2 in Rajasthan alone. To the north, it extends into Punjab, and in the north-east it joins the desert areas of Haryana. The Aravalli Mountains, starting from North Gujarat

Indian Desert

and extending up to Delhi, form the eastern boundary of the Thar.

Despite its comparatively small area, the Thar Desert has a high avian diversity, due to its location on the crossroads of the Palaearctic and Oriental biogeographic regions. The grassland of this desert region is the major habitat of the highly endangered Great Indian Bustard, and the winter migrant MacQueen's Bustard. Other important desert species are the Cream-coloured Courser, Greater Hoopoe Lark, and various sandgrouse, raptor, wheatear, lark, pipit and munia species. Both Greater and Lesser Flamingos breed in the Rann of Kutch of Gujarat.

Semi-arid Region
This region with a rainfall of 400–1,000mm is dominated by grass and shrub species. It includes high avian numbers, especially birds such as finches, munias, larks, doves and pigeons. It has dry deciduous forests and extensive tracts of grassland on the Deccan

High-altitude Himalayas

Western Ghats

Gangetic Plains

Plateau in central India, Malwa Plateau in north-west India, and Saurashtra region in Gujarat. The semi-arid region merges with the desert on the western side and with the Gangetic Plains in the north.

Some of the important species of this region are Greater Spotted and Eastern Imperial Eagles, Long-billed Vulture, Great Indian Bustard, Sarus Crane, Lesser Kestrel, Lesser Florican, Sociable Lapwing, Green Avadavat and Stoliczka's Bushchat.

Western Ghats

The Western Ghats on the north-west coast of India extend for about 1,600km from the Tapti River in the north to Kanyakumari in the south. The highest peak is Anamudi (2,700m). The Western Ghats is one of the most diverse regions of India. More than 600 bird species have been reported from here, including 16 endemics found nowhere else in the world.

Some of the restricted range species of the Western Ghats are the Nilgiri Wood Pigeon, Malabar Grey Hornbill, Malabar Parakeet, Nilgiri Pipit, Grey-headed Bulbul, Broad-tailed Grassbird, Wayanad Laughingthrush, Black-and-Orange Flycatcher and White-bellied Treepie.

Eastern Ghats

The Eastern Ghats are spread through Odisha, Andhra Pradesh and Tamil Nadu. They comprise about 75,000km^2, with an average width of 200km in the north and 100km in the south. They extend over a length of 1,750km between the Mahanadi River in the north and the Vaigai River in the south, along the east coast of India. The Eastern Ghats are important from the avifaunal distribution point of view. Two identified IBAs in the Eastern Ghats, Sri Lankamalleswara and Sri Peninsula Narasimha Wildlife Sanctuaries, provide refuges for Jerdon's Courser, an endangered and poorly known species. The area is also important for other species like the Spot-billed Pelican, Lesser Adjutant, Long-billed Vulture, Greater Spotted Eagle, Lesser Kestrel, Lesser Florican, Forest Owlet and Pale-capped Pigeon.

Gangetic Plains

The Gangetic Plains are one of the most fertile areas of the world, with a nearly 3,000-year history of human occupation. They are also one of the most densely populated areas of the world. The twin combination of a long history of human occupation, and a dense and still growing human population, has resulted in the almost complete conversion of the original vegetation into cropland and human settlements. The Gangetic Plains are drained by numerous rivers and streams, the most famous being the Ganga River.

This region is famous for its floodplain wetlands – the results of copious rainfall in the Gangetic Plains and also in the Himalayas, where most of the rivers originate. Large areas are annually flooded, and when the floods recede they leave low-lying areas under water. These wetlands are very productive in terms

North-east Hills

Andaman Islands

of vegetation biomass and avian diversity. The Gangetic Plains Biogeographic Zone is also important for many Near Threatened species, especially the Oriental Darter, Painted and Black-necked Storks, Black-headed Ibis, Ferruginous Duck and Black-bellied Tern. A few identified IBAs for the conservation of the endangered Bengal Florican are also located in this region.

North-east India

North-east India is one of the biodiversity hotspots of the world, and one of the most important birding habitats of north-east India is the Assam Plains Endemic Area. This area in Assam, the lowlands of Sikkim, northern West Bengal, parts of Arunachal Pradesh, Nagaland, Manipur and Meghalaya, basically consists of the floodplains of the mighty Brahmaputra River and its tributaries. The main vegetation of the Assam Plains is floodplain forest and grassland, with adjacent strips of undulating land at the bases of the foothills that are marshy, with tall elephant grass and forest. The Assam Plains support some of the most threatened bird (and mammal) species of the world. They also support some of the restricted range birds associated with the remaining grassland and wetland habitats found below 1,000m elevation.

This area is home to a globally threatened species, the White-winged Duck. Some of the other important species here are Lesser and Greater Adjutants, Baikal and Marbled Duck, Baer's Pochard, Pallas's Fish Eagle, Swamp Francolin, Chestnut-breasted Hill Partridge, Blyth's Tragopan, Mrs Hume's Pheasant, Sclater's Monal, Bengal Florican, Hodgson's Bushchat, Marsh and Jerdon's Babblers, and Beautiful Nuthatch.

The Islands

The Andaman and Nicobar Islands, consisting of more than 560 islands and rocks, are the peaks of a submerged mountain range stretching from Myanmar to Sumatra. The vegetation is mainly tropical evergreen, with some grassland in inland areas. The coastline of 1,962km^2 is mainly covered by mangroves. About 350 bird species and subspecies have been recorded in the islands, of which about 40 per cent are endemic species and subspecies.

While the islands form only 0.25 per cent of the land mass of the subcontinent, they contain 12 per cent of the endemic avifauna of the region, making them priority areas for conservation. The highest conservation priority species are the Nicobar Megapode, Edible-nest Swiftlet and Narcondam Hornbill. Other important and restricted range species here are the Andaman Serpent Eagle, Andaman Crake, Andaman Wood Pigeon, Andaman Cuckoo Dove, Andaman Hawk Owl, Brown Coucal, Andaman Woodpecker, White-headed Starling and Andaman Drongo.

Threats to Birds

India's birds are confronted by a multitude of threats. As the human population in the region soars, demand for forest products, timber and agricultural land continues to rise at the expense of the wilderness. Climate change is one of the major concerns across the globe. In some areas the climate is becoming wetter, while in others it is plunged into drought, and bouts of severe weather are likely to occur more often. The effects will be profound in almost every region, and birdlife is going to be affected significantly. In addition to climatic conditions there are also many man-made threats – for example, the pet-bird trade and human consumption place enormous pressures on the populations of some species.

Deforestation

Habitat Loss & Degradation

Across India, extensive conversion of natural habitat has been a real threat in recent times. While factors such as a rapidly changing demography and changes in traditional natural resource-management practices have been areas of concern, large-scale development projects are a major emerging threat to biodiversity in the region. These include expansion of traditional sectors such as commercial plantations, and oil and coal mining, as well as an increased focus on sectors such as large hydroelectric power projects and communications infrastructure. As per the current development plans for India, a number of these development projects (dams, mines, industries and so on) are proposed in and around biodiversity-rich areas, including protected areas and wildlife corridors. It has been observed that

Illegal timber logging

Nagaland deforestation

Threats to Birds

Parakeets trapped for trade

Owls are used for rituals

environmental and social concerns are not being appropriately addressed in the planning and decision making of many of these projects. Moreover, communication with citizens (local people in particular) about the impacts of projects, and their involvement in the whole clearance process, is minimal.

Pet-bird Trade
Among the several wildlife species and their derivatives in commerce, trade in birds is the most extensive in terms of its widespread nature, species diversity, numbers traded and value. The Indian Wildlife (Protection) Act, has completely banned the export and domestic trade in Indian birds since 1990–1991. Despite this law, the bird business is still prevalent in many Indian cities, small towns and villages. More than 450 of the approximately 1,300 Indian species have been documented in the international and domestic bird trade.

The top ten traded species in terms of numbers are Rose-ringed, Plum-headed and Alexandrine Parakeets, followed by passerines, namely Black-headed, Scaly-breasted Munia, Red Avadavat, Indian Silverbill, Baya Weaver, Red-headed Bunting and Common Pigeon. These ten species contribute to nearly 75 per cent of the indigenous bird trade. The rest of the trade comprises waders, ducks, larks, pipits and mynas.

Hunting of Birds for Food
Human consumption of wild birds remains rampant across some parts of India, especially in northeastern remote regions. As well as being hunted for subsistence, birds are trapped for sale in food markets and for religious release at worrying levels. While large birds such as game birds and waterfowl are usually preferred, smaller songbirds are also taken for consumption when local populations of larger birds have collapsed. Though strict regulations are enforced on the mainland, there are still many places in remote corners of India where bundles of pheasants, flycatchers, bulbuls, leafbirds and other songbirds are openly sold in markets for consumption.

Glossary of Terms

aberrant Abnormal or unusual.
accidental Vagrant.
adult Mature, capable of breeding.
aerial Making use of the open sky.
altitudinal Refers to migrant birds moving between high mountains and lower foothills.
arboreal Living in trees.
bheel Shallow lake or wetland.
biosphere The part of the Earth that is inhabited by life.
biotope Area of uniform environment, flora and fauna.
bund Man-made mud embankment.
canopy Leafy foliage of treetops.
colonial Roosting or nesting in groups.
crepuscular Active at dusk (twilight) and dawn.
deciduous Refers to trees that shed their leaves annually.
duars Forested areas south of eastern Himalayas.
endemic Indigenous and confined to a place.
endangered Facing high risk of extinction.
Eurasian Of both Europe and Asia.
extinct No longer in existence.
extirpated Locally extinct.
extralimital Not existing in the region.
fallow land Cultivated land after cultivation and before ploughing.
family Specified group of genera.
game birds Pheasants, partridges and allied species.
genus (plural genera) Group of related species.
Ghats Hills parallel to the east and west coasts of India.
hunting party Refers to group of birds usually of different species, seeking food.
hybridization Cross-breeding between two different species.
immature Plumage phases before adult.
jheel Shallow lake or wetland.
jizz Essence or striking characteristics of a species.
juvenile Refers to immature bird immediately after leaving nest.
Lantana Invasive aggressive shrub introduced to a region.
local Unevenly distributed within a region.
mangroves Coastal salt-resistant trees or bushes.
migration Seasonal movement between distant places.
monsoon Rainy season in India.
montane Pertaining to mountains.
nocturnal Active at night.
nomadic Refers to bird species without specific territory except when breeding.
nullah Dry or wet stream bed or ditch.
order Group of related families.
paddy fields Rice fields that are often flooded.
Palaeartic Old World and Arctic zone.
pelagic Ocean-going.
plumage Feathers of a bird.
race Subspecies.
range Geographical area or areas inhabited by a species.
raptors Birds of prey and vultures, excluding owls.
resident Non-migratory and breeding in same place.
riparian Occurring along creeks, streams, rivers and waterways.
salt pans Shallow expanses of ground with naturally drying out salt, usually found in deserts. Shallow, artificial salt-evaporation pans are designed to extract salts from sea or other briny water.
savannah Open flat land with grass and scattered bushes.
storey Level of a tree or forest.
submontane Hills below highest mountains.
subspecies Distinct form beneath species that does not have specific status.
tank Water reservoir.
taxonomy Science of classifying organisms.
teak Dominant tree of South Indian forests.
terai Alluvial stretch of land south of the Himalayas.
vagrant Accidental, irregular.
waders Shorebirds; usually, small, long-legged waterbirds.
wildfowl Ducks and geese.

ANDAMAN & NICOBAR ISLANDS

The Andaman Islands, situated in the Bay of Bengal, are one of the nine union territories of India. These clusters of large and small islands are actually the peaks of a submerged mountain chain. The Great Andaman, lying to the north, is the largest contiguous block of islands and is administratively divided into the North, Middle and South Andaman. The Little Andaman is squeezed in between the Great Andaman and Nicobar cluster of islands. Port Blair, located in South Andaman, is the capital city of the group.

The majority of these islands are uninhabited and covered in pristine tropical rainforest. The Andaman and Nicobar Islands constitute a globally important biodiversity hotspot. Because they are off the mainland and isolated, endemicity is very high in all taxa. The proximity to the Myanmar coast has also resulted in the presence of many species with Southeast Asian affinities, with the fauna resembling Malayan and Myanmarese species. The endangered Narcondam Hornbill, the dream species for every birdwatcher, is endemic to this region.

Climate
The climate of the islands is humid tropical. The recorded average annual rainfall varies from 1,400 to 3,000mm and the average annual temperature is 24–28 °C, with relative humidity up to 80 per cent. Tropical wet evergreen forests occur throughout the islands on higher altitudes, and moist deciduous forests are found on the slopes. Southern hilltop evergreen forests occur on hilltops and steep slopes, and are usually exposed to high winds. Semi-evergreen forests also constitute an important part of the vegetation of the islands, and include both deciduous and evergreen species.

Access, Transportation & Logistics
Port Blair, the capital of the islands, is the usual

1. Mount Harriet National Park
2. Chidiya Tapu
3. Mahatma Gandhi Marine National Park, Wandoor
4. Saddle Peak National Park

port of arrival by air or sea. Port Blair is directly connected by air from Kolkata and Chennai. Sea passage takes up to three or four days and can be taken from Kolkata and Chennai. There are regular taxis and a local bus from the airport to the rest of the city.

There is a complete range of accommodation options available in and around Port Blair. Andaman Beach Resort, Hotel Landmark and Seashell are some of the popular hotels, and there is also a youth hostel, which is a cheaper option.

Health & Safety
There are no major health concerns in the Andaman Islands. The area is a low- to no-risk area of India, but when it comes to malaria, do take precautions, especially after the monsoons. Visit a GP or travel clinic 6–8 weeks before departure to make sure you are up to date with all vaccinations. Avoid roadside food and drink only bottled water. During the summer months keep well hydrated and carry

> **KEY BIRDS**
> Top 10 Birds
> 1. Andaman Teal
> 2. Andaman Hawk Owl
> 3. Andaman Crake
> 4. Beach Thick-knee
> 5. Pied Imperial Pigeon
> 6. Andaman Nightjar
> 7. Andaman Bulbul
> 8. White-breasted Woodswallow
> 9. Long-tailed Parakeet
> 10. Andaman Treepie

Andaman forest

Beach Thick-knee

a stole, scarf or cap to protect yourself from direct sunlight. Afternoons are really hot, so be especially careful at this time. It is also advisable to carry sunblock and tanning lotions, a pair of sunglasses and hats. Wear a life jacket during boat rides, and always listen to the instructions of your guide and lifeguard while birding along the shores.

Birdwatching Highlights

The Andaman and Nicobar Islands constitute a globally important biodiversity hotspot. The avifauna of the region is very rich and rather different from that of the rest of India, having closer affinities with Southeast Asia. More than 360 different bird species have been recorded from the region, including many endemics. The Narcondam Hornbill, one of the star birds, is endemic only to Narcondam Island, which is difficult to access. Apart from that, there are other endemics like the Andaman Wood Pigeon, Andaman Cuckoo Dove, Andaman Nightjar, Andaman Crake, Andaman Serpent Eagle, Andaman Shama, Andaman Bulbul, Andaman Barn Owl, Andaman Scops Owl, Andaman Treepie and White-headed Starling.

There are categorized Important Birding Areas (IBAs) and well-known birdwatching sights in the Andaman Islands. Along with the IBAs, Port Blair and its surrounding areas are good for birding. Corbyn's Cove marsh and Sippighat are two such locations where many species of bird can be seen, including some of the endemic species.

Andaman landscape

Mount Harriet National Park

Andaman Teal

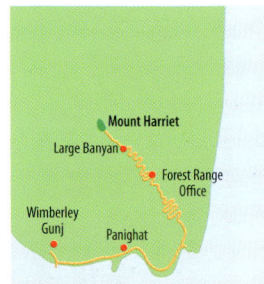

KEY FACTS

Nearest Major Towns
Port Blair and Hope Town

Habitats
Tropical wet evergreen, tropical semi-evergreen and littoral forests

Key Species
Andaman Crake, Andaman Serpent Eagle, Andaman Wood Pigeon, Andaman Cuckoo Dove, Andaman Scops Owl, Andaman Hawk Owl, Andaman Woodpecker, Andaman Drongo, Andaman Treepie, Andaman Coucal, White-headed Starling, Scarlet Minivet, Plain Flowerpecker, Large Cuckooshrike, Andaman Shama, Green Imperial Pigeon, Olive-backed Sunbird, Brown Shrike, Red-whiskered Bulbul, Greater Racket-tailed Drongo, Spot-breasted Pied Woodpecker

Other Specialities
Andaman Gecko, Andaman Spiny Shrew, Andaman Horseshoe Bat, Andaman Rat

Best Time to Visit
October to March

This park is situated in Ferrargunj Tehsil of the Andaman district, about 38km from Port Blair. The park is named in commemoration of Harriet Tytler, the second wife of Robert Christopher Tytler, a British army officer, administrator, naturalist and photographer, who was appointed superintendent of the convict settlement at Port Blair in the Andamans.

The entire park forms a major chunk of the hill ranges in the eastern part of the south Andaman Islands. The eastern face of the park has steep slopes, and the beaches here are also formed of rocks, interspersed with small sandy areas. The park is drained by many streams, which rise in the hills and flow into the sea on the east. The avifauna of the park is very rich and diverse due to dense forests, the presence of many varieties of wild fruit plant, and open seashore on the eastern side. The park is also a butterfly hotspot.

Birdwatching Sites

Mount Harriet National Park is categorized as an IBA. More than 100 bird species have been recorded in this area. The swamp area and a small pond near the base of Mt Harriet are good places to look for the Slaty-breasted Rail and Ruddy-breasted Crake. There have also been records of the Andaman Crake from this area, although this is quite an elusive species and sightings are difficult.

Long-tailed Parakeet

Pied Imperial Pigeons

Andaman Hawk Owl

Andaman Serpent Eagle

The scrubby region of higher ground is good for Thick-billed Warblers, Andaman Serpent Eagles and Andaman Woodpeckers. Further up the road, the forests become quite dense and are good habitats for Andaman rainforest species like the Vernal Hanging Parrot, Red-breasted and Long-tailed Parakeets, Violet Cuckoo, Dollarbird, Andaman Coucal, Greater Racket-tailed Drongo, Asian Fairy-bluebird and Black-naped Monarch. It is worth spending some time after sunset to look for some nocturnal bird species like Andaman and Hume's Hawk Owls, and Oriental and Andaman Scops Owls.

Access & Accommodation

The quickest way to reach Mount Harriet National Park is to take the ferry from Phoenix Bay to Bamboo Flat. There are regular ferries, and it is recommended that you check the latest ferry schedules with the Marine Department office. There are also car ferries to Bamboo Flat. From here there is a metallized road that goes up to Mt Harriet.

There is a forest rest house at Mount Harriet National Park. The accommodation facilities provided in the guest house can be availed of with prior permission of the park authorities. Foreign nationals need to obtain permission from the park authorities to visit and stay in the area.

Conservation

One of the main conservation issues of this park is encroachment on forest land. There is an urgent need for a buffer zone around the park. Some of the other conservation issues are grazing, poaching of wild animals and birds, illegal cutting of trees and firewood collection.

Andaman Cuckoo Dove

Chidiya Tapu

Chidiya Tapu

Chidiya Tapu (sometimes written as Chiriya Tapu) means 'bird island'. It lies at the southeastern tip of South Andaman. The island is situated 30km from Port Blair. It is one of the most famous tourist destinations of the Andaman Islands for its wide variety of birds and brilliant sunsets. The island is a treasure trove of thick forests and stunning views of the ocean, with almost every turn providing a mesmerizing view of dense mangrove cover and azure sea waters.

Birdwatching Sites

There is reasonably good forest cover both inland and along the coastline that harbours several bird species. A motorable road goes to almost all birding spots and offers a comfortable birding experience. This is one of the best places to see one of the star

KEY FACTS

Nearest Major Towns
Port Blair

Habitats
Thick forests and dense mangroves

Key Species
Andaman Coucal, Green Imperial Pigeon, Indian Cuckoo, White-breasted Waterhen, Crested Serpent Eagle, Collared Kingfisher, Black-naped Oriole, White-headed Starling, Andaman Treepie, Andaman Drongo, Changeable Hawk Eagle, White-bellied Sea Eagle, Stork-billed Kingfisher, Chestnut-headed Bee-eater, Blue-tailed Bee-eater, Greater Racket-tailed Drongo, Asian Fairy-bluebird, Oriental Magpie Robin, Asian Glossy Starling, Andaman Flowerpecker, Small Minivet, Long-tailed Parakeet, Oriental White-eye, Brown-backed Needletail, Dollarbird, Spot-breasted Pied Woodpecker, Barn Swallow, Alexandrine Parakeet

Other Specialities
Crab-eating Macaque, Andaman Masked Civet

Best Time to Visit
October to March

Andaman Crake

Andaman Scops Owl

birds of the Andaman Islands – the Pale or Andaman Serpent Eagle. Other bird species that can be seen in this region are the Black Baza, Changeable Hawk Eagle, Green Imperial Pigeon, Asian Emerald Dove, Asian Glossy and White-headed Starlings, and Chestnut-headed Bee-eater.

Access & Accommodation
Chidiya Tapu is just 30km from Port Blair and can be reached by bus or private taxi. A private vehicle is recommended as buses have scheduled arrival and departure timings and may not provide sufficient time for a good birding experience. There is a forest rest house and a couple of homestay options in Chidiya Tapu. There are also a couple of newly constructed resorts in the area. For a comfortable stay book a hotel in Port Blair, which is easily accessible from this place.

Conservation
There are no major conservation threats and issues. Some of the other issues are illegal use of land for agriculture, illegal tree cutting and grazing of livestock. There are small villages at the fringes of this area and the main occupation of the villagers is agriculture. To meet their various needs, they enter the forest illegally and collect local forest produce.

Andaman Woodpecker

Andaman Treepie

Hume's Hawk Owl

Mahatma Gandhi Marine National Park, Wandoor

Andaman Cellular Jail

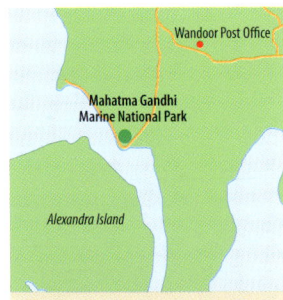

KEY FACTS

Nearest Major Towns
Port Blair

Habitats
Tropical wet evergreen, tropical semi-evergreen and littoral forests

Key Species
Andaman Serpent Eagle, Andaman Crake, Andaman Wood Pigeon, Andaman Cuckoo Dove, Andaman Coucal, Andaman Scops Owl, Andaman Woodpecker, Andaman Hawk Owl, Andaman Treepie, Nicobar Sparrowhawk, Nicobar Megapode, Nicobar Bulbul

Other Specialities
Andaman Lesser Short-nose Fruit Bat, Jungle Cat, Indian Chevrotain, Andaman Wild Pig, Indian Pipistrelle

Best Time to Visit
October to March

The Mahatma Gandhi Marine National Park, previously called the Wandoor National Park, stretches over 15 islands and islets in the Labyrinth Island group. Its boundaries run across the coast as well as inland. The park encompasses a stretch of marine waters, with lushly vegetated islands, vast coral reefs and beaches.

The flora and fauna of the park are extremely diverse. The park includes stretches of protected mangrove forest that are among the largest in India. The rich flora and fauna make it an ideal destination for nature lovers. The abundant marine life consists of a variety of coral reefs, colourful fish, molluscs, shells, starfish, turtles and Saltwater Crocodiles.

Birdwatching Sites

The park is situated about 30km from Port Blair, and is located on the southwestern coast of South Andaman, in the Bay of Bengal. It is home to birds like the White-bellied Sea Eagle, parakeets, Andaman Teal, herons, terns, waders and swifts. A scientific checklist of the birds of the park is not available, but restricted range (endemic) species are present. According to various reports, there are more than 271 species of bird inhabiting the idyllic landscape of the park. The Pacific Reef Egret, Japanese Sparrowhawk, Slaty-breasted Rail, Greater and Lesser Sand Plovers, Terek Sandpiper, Andaman Wood Pigeon, Brown-backed Needletail, Andaman Drongo and Dollarbird are some of the key species found in the area.

Access & Accommodation

The nearest airport to reach the park is Port Blair Airport, which is about 30km away. Port Blair is well connected to the park by

Andaman Cuckooshrike

a road network. A number of government and privately operated vehicles go to the park at frequent intervals. The Tourist Department of Andaman runs a few mini-bus trips each day. The buses start from Andaman Teal House and can be boarded from the office of the Director of Information and Publicity and Tourism, Port Blair.

A forest guest house with limited facilities and a private lodge are situated at Wandoor. The Directorate of Tourism provides cozy and comfortable accommodation at affordable prices at Hornbill Nest at Port Blair. The Andaman and Nicobar Islands Integrated Development Corporation runs the Megapode Tourist Home Complex at Port Blair. More varied accommodation options are available at Port Blair.

Conservation

Urbanization and uncontrolled tourism are the main challenges of this park. Due to the proximity of Port Blair, both Indian and foreign tourists visit this place quite frequently, attracting the concentration of settlements along its borders. This has resulted in disturbance to both the birdlife and marine life. Siltation caused by inland forestry operations is also affecting the coral reefs.

Rahul & Khushboo Sharma

Andaman Barn Owls

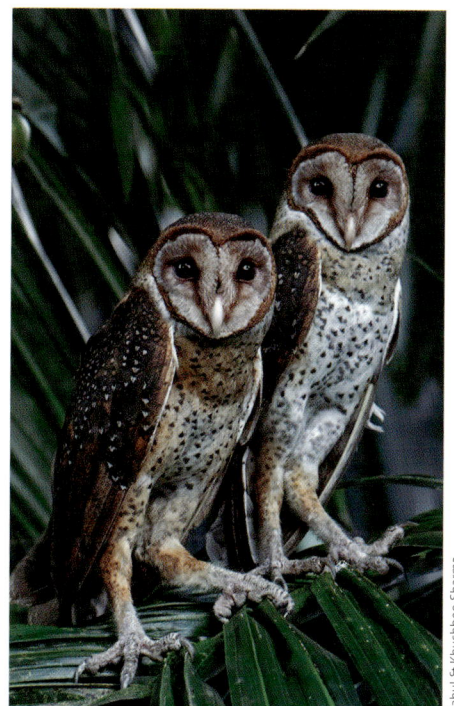

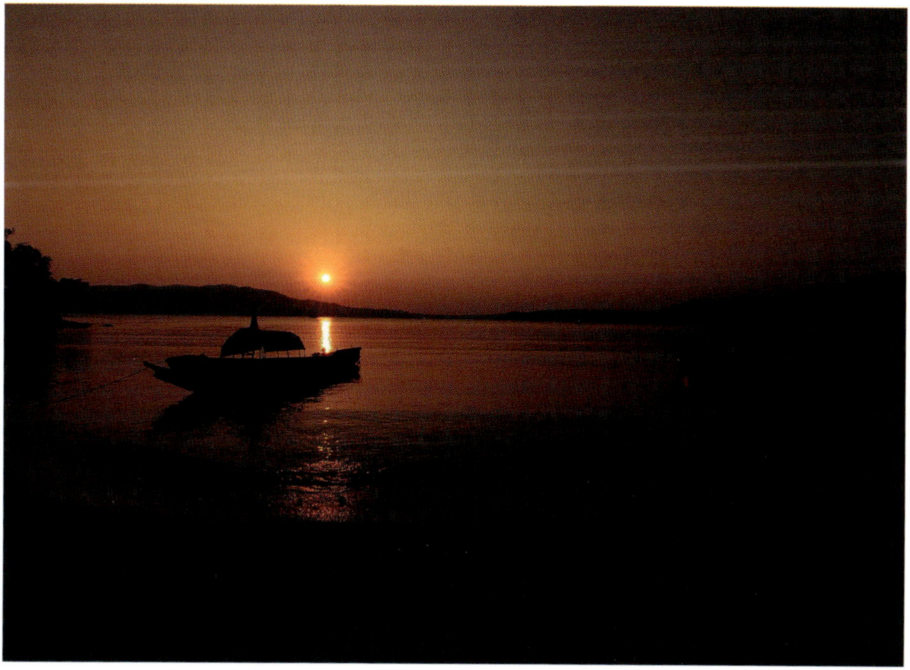

Bhanu Singh

Boat in Wandoor

Saddle Peak National Park

Saddle Bay

The Saddle Peak National Park, with its summit at 737m above mean sea level, is the highest point in the Andaman and Nicobar Islands. The peak is shaped like a double-humped saddle, hence the name. The park runs north to south along the eastern coast of the North Andaman Island. Most of its eastern boundary borders the sea, with a long and rocky beach. The park was established in 1987.

The park is famous for housing some of the most endangered and rarest species of flora and fauna. Its vegetation is unique and is rarely found in other parts of mainland India. The forest land of the islands is covered by luxuriant, lush green, thick tropical rainforests. The vegetation is characterized by a humid, warm and wet tropical climate. The park is home to the Andaman Wild Pig, Water Monitor Lizard and Saltwater Crocodile, and the important birds are the White-headed Starling and Pied Imperial Pigeon.

KEY FACTS

Nearest Major Towns
Port Blair

Habitats
Tropical wet evergreen forest, tropical semi-evergreen forest, mangrove forest

Key Species
Greater Spotted Eagle, Lesser Kestrel, Pallid Harrier, Speckled Piculet, Rufous Woodpecker, Brown-headed Barbet, Common Hoopoe, Plain Prinia, White-rumped Shama, Oriental Scops Owl, Collared Scops Owl, Brown Fish Owl, Painted Francolin, Indian Pitta, Asian Brown Flycatcher, Ultramarine Flycatcher, Verditer Flycatcher, Black-headed Munia, Oriental White-eye

Other Specialities
Andaman Flying Fox, House Mouse, Andaman Island Spiny Shrew, Malay False Vampire

Best Time to Visit
October to March

Andaman Coucal

Andaman Drongo

Andaman Bulbul

White-headed Starling

Birdwatching Sites
The best way to see and photograph birds in Saddle Peak National Park is by foot. There is an 8km-long trail from the entrance of the park to Saddle Peak. The trail offers numerous opportunities for birdwatching. Some of the key bird species found here are the White-headed Starling, Pied Imperial Pigeon, Andaman Crake, Andaman, Andaman Coucal, Andaman Wood Pigeon, Andaman Cuckoo Dove, Andaman Hawk Owl, Andaman Woodpecker, Andaman Drongo and Andaman Treepie. The park is open throughout the year. The best time of the year to visit is November–March.

Access & Accommodation
Visitors can reach the park via ferries from Port Blair, which is well connected to Delhi, Kolkata and Chennai Airports as well as via ships.

Accommodation options are available at two rest houses, the Yatri Nivas and Pristine Beach Resort in Diglipur. Both offer a luxurious stay in the area. Visitors are advised to get in touch with the authorities of the park about accommodation, tariffs and other facilities (Directorate of Tourism, Incharge (Tourism Information) Andaman & Nicobar Administration Kamaraj Road, Port Blair – 744 101, Tel.: 03192-232 694, 232 642, Fax: 03192-232 747, 230 933, Email: accomodation6@gmail.com).

Conservation
Some of the common conservation challenges faced by this park are encroachment on the forest land, intensification and expansion of agriculture, and poaching. Some patches of forest have been cleared for cultivation. Unsustainable agriculture and tilling on the encroached rainforest land has led to the problem of soil erosion. Livestock grazing, hunting, cutting down of trees for firewood and minor forest produce collection are some of the other conservation issues.

Slaty-breasted Rail

ANDHRA PRADESH

1. Nelapattu Bird Sanctuary
2. Rollapadu Bird Sanctuary
3. Pulicat Bird Sanctuary
4. Sri Lankamalleswara Wildlife Sanctuary
5. Kolleru Bird Sanctuary
6. Araku Valley

A state bordering India's southeastern coast, Andhra Pradesh is the gateway to the southern peninsula country from the north and the east. Situated on the Deccan Plateau, the state is bordered by Tamil Nadu in the south, Karnataka in the west, Maharashtra in the north and north-west, Madhya Pradesh and Odisha in the north-east and the Bay of Bengal in the east. Andhra Pradesh is known for its glorious past.

KEY BIRDS

Top 10 Birds
1. Jerdon's Courser
2. Little Tern
3. Spoon-billed Sandpiper
4. Painted Spurfowl
5. Rain Quail
6. Sykes's Lark
7. Black-tailed Godwit
8. Pallid Harrier
9. Jerdon's Baza
10. Short-toed Snake Eagle

The ancient temples, monuments, forts, palaces and old markets are testimony to its rich history, culture and architecture. Major cultural landmarks of the state include Tirumala Venkateswara Temple, an ornate hilltop Hindu shrine in the southern part of the state. It is visited by tens of millions of pilgrims annually. The major northern port of Visakhapatnam has popular beaches like Ramakrishna and Rishikonda lining the Bay of Bengal. The high ranges include the Eastern Ghats, which are ranges of interrupted hills with patches of moist deciduous forests. They run parallel to the coast, with the distance from the coast varying considerably. Fertile valleys formed by the perennial flow of the Godavari and Krishna Rivers break these ranges. The Deccan Plateau lies to the east of these ranges and has an average elevation of 500m.

Climate

The climate of Andhra Pradesh is generally hot and humid. The summer season usually extends from March to June. During these months the moisture level is quite high. The coastal areas have higher temperatures than other parts of the state. In summer, the temperature generally ranges between 20 °C and 40 °C, but in certain places it can be as high as 45 °C on a summer day. The summer is followed by the monsoon season, which starts in June and continues until September. This is the season for heavy tropical rains in Andhra Pradesh. The major role in determining the climate of the state is played by the south-west monsoons. About one-third of the total rainfall in Andhra Pradesh is brought by the north-east monsoons in around

October. The winters, in October–February, are pleasant, and this is the time when the state attracts most of its tourists. Since the state has quite a long coastline, the winters are comparatively mild. The range of winter temperatures is generally 13–30 °C.

Access, Transportation & Logistics

Andhra Pradesh is very well connected to the rest of India and many international destinations. Hyderabad, the capital city, has one international airport, which receives flights from all major cities of the world. Other important airports are located at other major cities, Tirupati, Vijayawada, Visakhapatnam and Rajahmundry, which connect Andhra Pradesh to all prime cities of India. Andhra Pradesh is well connected to all parts of the country by numerous superfast and express trains. Another way to reach the state is by road, and an extensive network of highways and roads connects the state to other major cities of India. The State Transport Service operates regular bus services within the state, connecting its various cities. Andhra Pradesh Tourism also runs deluxe buses to all major tourist attractions and destinations in the state. Private luxury coaches, taxis and cars are available. App-based cab services like Uber and Ola are available in all major cities.

Health & Safety

Andhra Pradesh is a safe state for visitors to India. There are no major health concerns in the state. It is a low- to no-risk area of India when it comes to malaria, but do take precautions, especially after the monsoon. Visit a GP or travel clinic 6–8 weeks before departure to make sure you are up to date with any vaccinations. Avoid roadside food and drink bottled water. During the summer months keep hydrated and carry a stole, scarf or cap to protect yourself from direct sunlight.

Ashy Prinia

Birdwatching Highlights

Andhra Pradesh is rich in biodiversity and avifauna diversity. More than 550 bird species have been recorded in the state, including some that are Critically Endangered, like Jerdon's Courser, Spoon-billed Sandpiper and possibly the Great Indian Bustard. The bird species here represent more than 58 families. There are many wildlife as well as bird sanctuaries and IBAs in the state, as well as numerous birding groups and NGOs, of which Deccan Birders is an NGO founded in 1980 (as the Birdwatchers Society of Andhra Pradesh, renamed Deccan Birders in 2018) by a few enthusiasts of Telangana/Andhra Pradesh with the primary objective of spreading the message about bird conservation. They also organize birdwatching field trips in the state.

Indian Silverbill

Ashy Woodswallows

Nelapattu Bird Sanctuary

Common Redshank

KEY FACTS

Nearest Major Towns
Nellore

Habitats
Swamp forests and southern dry evergreen scrub

Key Species
Black-headed Ibis, Little Cormorant, Little Egret, Great Egret, Indian Pond Heron, Black-crowned Night Heron, Northern Shoveler, Little Grebe, Common Coot, Common Moorhen, Northern Pintail, Fulvous Whistling Duck, Asian Palm Swift, Red-wattled Lapwing, Green Sandpiper, Asian Openbill, Spot-billed Pelican, Glossy Ibis, Black-headed Ibis, Blue-tailed Bee-eater, Green Bee-eater, Rose-ringed Parakeet, Ashy Prinia, Purple Sunbird, Black Drongo, Barn Swallow, Greater Coucal

Other Specialities
Slender Loris, Indian Jackal, Monitor Lizard

Best Time to Visit
October to March

This sanctuary is located in the Nellore district of Andhra Pradesh. It is known as one of the most significant pelican habitats in Southeast Asia. The sanctuary got its name from its neighbouring village, located in the Nellore district of Andhra Pradesh, and was declared a sanctuary in 1997. Large parts of the sanctuary are covered with southern dry evergreen shrub cover and are home to animals such as the Spotted Deer, Indian Jackal, Monitor Lizard, Slender Loris and tortoises. Snakes of various types are also found here. Every year, Nelapattu celebrates a three-day festival called 'Flamingo' to promote tourism. The festival is a much-awaited one and is visited by a large number of people from all parts of the country.

Birdwatching Sites

This sanctuary is spread over a total area of about 4.6km^2 and acts as a crucial breeding site for Spot-billed Pelicans. Other than these pelicans, it is an important breeding ground for birds such as the Black-headed Ibis, Asian Openbill, Black-crowned Night Heron and Little Cormorant. In all, 189 bird species can be found at the sanctuary, of which 50 are migratory. The latter include the Common Teal, Northern Shoveler, Indian Spot-billed Duck, Grey Heron, Black-winged Stilt, Garganey and Gadwall. The ideal time to plan a visit to the sanctuary is October–March, when the weather is the most pleasant in Andhra Pradesh, and visitors get the chance to witness some of the migratory bird species along with residents.

Access & Accommodation

The Nelapattu Bird Sanctuary is easily accessible from major towns and cities in southern India. The sanctuary is just 50km from Chennai and is located in the Nelapattu Village, on the eastern coast of Andhra Pradesh. The nearest railway station to the sanctuary is at Sullurpet, and the final destination is well connected by state transport and bus facilities. You can combine

Asian Palm Swifts

Barn Swallow

Blyth's Reed Warbler

the trip to Nelapattu with one to Pulicat Lake and cover both places easily. There are many accommodation options available near the sanctuary in Nellore. You can also accomplish a visit as a day trip from Chennai, which is around 50km from the sanctuary.

Conservation

The Nellapattu Bird Sanctuary is the biggest pelicanry in India. As it is very well known that the Spot-billed Pelican is an endangered species, everything possible should be done to protect it from extinction, by intelligent management and control of all the factors connected with its survival. There is an urgent need to develop the Nelapattu tank (lake) area as a very sound swamp ecosystem to be the home of winter breeding migrants. Some of the key conservation and management strategies are the creation of an ambush cover on the tank embankment, the creation of wildlife conservation consciousness among people, and protection and improvement of the existing forest in the sanctuary. Nesting space, nesting material, the proximity of water as an insulating medium, fishing grounds and protection from human activity govern the success of the sanctuary and the survival of the Spot-billed Pelican.

Little Cormorant

Eurasian Collared Dove

Rollapadu Bird Sanctuary

Rollapadu Wildlife Sanctuary

Rollapadu is located about 260km south of the state capital, Hyderabad. This is a small village surrounded by grassland with light scrub and scattered trees. The open grassland around the Rollapadu region was declared a sanctuary in 1988 to protect the dwindling populations of the Critically Endangered Great Indian Bustard. Today, however, the bird species that were very common earlier are rarely spotted in Rollapadu. The sanctuary also provides a good habitat for wintering harriers like Montagu's and Pallid Harriers.

Birdwatching Sites

The road from Nandikotkur to Rollapadu is a good place in which to look for the Great Indian Bustard, and if you are lucky, you may find it by scanning the grassland. Rollapadu is also a good place to find Sykes's Lark, an endemic and patchily distributed species throughout the Deccan Plateau. Take the dusty track north from the Nature Education Centre, and keep an eye on the bunds and road edges for this species. You can also spend some time at a water pool that is on the main track for some morning and evening sessions. Many birds visit the pool to drink water. This is a good viewpoint for harriers. The nearby Talamudipi Tank is worth checking for the Small Pratincole and Great Bittern. Some of the important bird species in this area are the Bar-headed Goose, White-eyed Buzzard, Montagu's and Pallid Harriers, Short-toed Snake Eagle, Painted and Grey Francolins, Rain Quail, Jungle Bush Quail, Demoiselle Crane and Chestnut-bellied Sandgrouse.

KEY FACTS

Nearest Major Towns
Kurnool

Habitats
Mixed forests, thorny bushes, semi-arid short grassland

Key Species
Indian Courser, Red-wattled Lapwing, Rufous-tailed Lark, Indian Bushlark, Greater/Sykes's Short-toed Lark, Plain Prinia, Zitting Cisticola, Black Drongo, Ashy Drongo, Oriental Skylark, Sykes's Lark, Ashy-crowned Sparrow Lark, Yellow-wattled Lapwing, Jungle Babbler, Large Grey Babbler, Indian Grey Hornbill, Lesser Kestrel, Blue-faced Malkoha, Grey Francolin, Eurasian Collared Dove, Red Collared Dove, Lesser Whistling Duck, Northern Shoveler, White-throated Kingfisher, Pied Kingfisher, White-eyed Buzzard, Great Egret, Western Reef Egret, Dusky Crag Martin, Common Iora

Other Specialities
Indian Fox, Blackbuck

Best Time to Visit
October to March

Common Hoopoe

Sarus Cranes

Access & Accommodation

Rollapadu Wildlife Sanctuary is located 43km from Kurnool. From Kurnool, the only way to reach the sanctuary is by road. Renting a cab is the best option, but taking a state transport bus can be an affordable one. The route from Kurnool to the reserve goes via NH 40-SH 50. Travellers need to take the New Bus Stand Road on NH 40-SH 50 and head towards Kurnool - Guntur Road. Rollapadu Wildlife Sanctuary is about 36km ahead from Kurnool - Guntur Road. The nearest airport - the GMR Hyderabad International Airport - is located in Hyderabad, and it is about 240km from the sanctuary. The nearest railway station in Kurnool is 45km from the sanctuary, and is well connected to major stations in the state and major cities of India.

There are no accommodation options inside the sanctuary. There are some basic lodges at Nandikotkur, but visitors can find enough places to stay near the sanctuary or in Kurnool City, including upper-bracket luxury resorts and hotels.

Conservation

The intensification of agricultural practices across the area and the construction of the Telugu Ganga Canal, for which a balancing reservoir in the nearby Alaganur village was constructed, has changed the landscape in many ways. There has also been an unprecedented increase in the use of pesticides to promote commercial crops like cotton. Excessive use of pesticides could be leading to a dwindling number of insects and birds, which are the main prey for harriers. There is an urgent need to make the surrounding areas of the sanctuary ecologically friendly. Such zones have been created in some of the larger sanctuaries, but are also necessary for the survival of smaller grassland sanctuaries such as Rollapadu.

Common Iora

Paddyfield Pipit

Pulicat Bird Sanctuary

Pulicat Lake

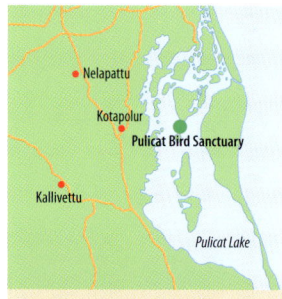

KEY FACTS

Nearest Major Towns
Nellore

Habitats
Wetland

Key Species
Spot-billed Pelican, Grey Heron, Western Reef Egret, Painted Stork, Great Egret, Little Egret, White-bellied Sea Eagle, Spot-billed Pelican, Black-headed Ibis, Garganey, Marsh Sandpiper, Gadwall, Northern Shoveler, Black-tailed Godwit, Little Cormorant, Asian Openbill, Lesser Crested Tern, Great Crested Tern, Caspian Tern, Lesser Noddy, Brown-headed Gull, Common Greenshank, Curlew Sandpiper, Ruddy Turnstone, Black-tailed Godwit, Bar-tailed Godwit, Grey Plover, Lesser Flamingo

Other Specialities
Smooth-coated Otter, Indian Jackal, Water Monitor, Indian Rat Snake

Best Time to Visit
October to March

Pulicat Lake is located on the southeastern coast of India. The greater part of the lake is in Andhra Pradesh, with a smaller southern portion belonging to Tamil Nadu. This is a large, brackish lagoon open to the sea, with its size varying according to the season. The northern part of the lake dries up after the monsoon, making the mudflats a favourite ground for many species of wading bird. The area also attracts a great number of wintering waterfowl.

Birdwatching Sites

Pulicat is on the Eastern Flyway of the Central Asian Flyway, a crucial migratory route for birds across the globe. A good way to explore the lake is by hiring a boat for the day at Tada. The deeper parts at the southern end of the lake have fewer birds. Some of the important birds here are the Spot-billed Pelican, Indian and Little Cormorants, Painted Stork, herons, egrets, Asian Openbill, Bar-headed Goose and Lesser Flamingo. Lesser Flamingos also visit, but in small numbers.

Access & Accommodation

Pulicat Lake is located on the Andhra and Tamil Nadu border, about 66km north of Chennai. The nearest airport is Chennai (two hours) and the nearest railway station is in Sullurpet (13km). Access to Pulicat is from the main road between Chennai and Nellore via a small road at Tada, and via the road from Sullurpet to Sriharikota. There are regular buses on the Chennai – Nellore road. You can also get taxis to the site. For accommodation, there is a forest rest house and a Public Works Department bungalow at Ponneri. Many tourists also cover this site as a day trip from Chennai, where various accommodation options are available.

Red-wattled Lapwing

Little Terns

Conservation

Pulicat is the second largest brackish water ecosystem in India. Many adjoining villages depend on the lake directly or indirectly for their livelihoods. However, this fragile wetland ecosystem is under serious threat because of a plan to build a 'world-class' port and shipbuilding centre, which would lead to massive dredging and also result in many ancillary industries coming up in the area. The Ministry of Environment and Forests has proposed a restricted Eco-Sensitive Zone (ESZ) around Pulicat. This overrides the Andhra Pradesh Government Forest Department's proposal for a 10km ESZ around the lake.

Commmon Snipe

Brahminy Kite

Shikra

Sri Lankamalleswara Wildlife Sanctuary

Sri Lankamalleswara landscape

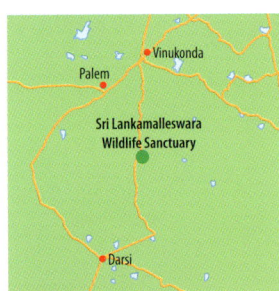

KEY FACTS

Nearest Major Towns
Kadapa, Darsi, Palem

Habitats
Dry deciduous mixed thorn forests with steep slopes

Key Species
Jerdon's Courser, Indian Pond Heron, Cattle Egret, Painted Stork, Indian Black Ibis, Shikra, Tawny Eagle, Black-winged Kite, Common Kestrel, Black Kite, Short-toed Snake Eagle, White-eyed Buzzard, Changeable Hawk Eagle, Jungle Bush Quail

Other Specialities
Indian Fox, Sloth Bear, Indian Leopard, Spotted Deer, Sambar, Chinkara, Nilgai, Wild Pig

Best Time to Visit
October to March

This sanctuary is located in the Kadapa district of Andhra Pradesh. It is the only habitat in the world that provides a home for the most Critically Endangered bird species in India – Jerdon's Courser. The sanctuary is home to a great variety of plant species and unique organisms. The vegetation is dry deciduous mixed thorn forests. Geographically, there are deep gorges and steep slopes. Along with great birdlife, also known for a good population of Indian Leopards, Sloth Bears, Spotted Deer and Sambar, Chowsingha, Chinkara, Nilgai Antelopes, Wild Pigs and Indian Foxes.

Birdwatching Sites

The prime attraction of this site is Jerdon's Courser, a highly endangered bird species. The bird was first discovered in 1848 by the surgeon-naturalist Thomas C. Jerdon, and was thought to be extinct until its rediscovery in 1986. It now inhabits the sparse scrub regions and forests of the Sri Lankamalleswara Sanctuary, where the topography and weather conditions are compatible with its existence. However, there has not been any confirmed record of the bird in recent years. Along with this rare and elusive bird, Sri Lankamalleswara is home to nearly 200 bird species. Some of the key birds are the Painted Stork, Indian Pond Heron, egrets, Jungle Bush Quail, Grey Francolin, Painted Francolin, Grey Junglefowl, and Red and Painted Spurfowls. Among raptor species, important ones are the Short-toed Snake Eagle, Tawny Eagle, Common Kestrel, White-eyed Buzzard and Changeable Hawk Eagle.

Jungle Bush Quail

Greater Coucal

Yellow-legged Green Pigeon

Access & Accommodation

The sanctuary is well connected via road, rail and air. The nearest airport is Tirupati Airport, while Kadapa is the nearest railway station. It can be also reached by bus from Kadapa. The best time to visit is October–March. Kadapa and Siddavatam have forest rest houses providing accommodation facilities to visitors.

Conservation

The main conservation issues of the sanctuary are agricultural expansion and intensification of farming in nearby areas. Hunting and poaching of wild animals and birds in the sanctuary are frequently reported. Other common and serious issues are timber logging and wood harvesting, and mining. Jerdon's Courser, being the most prominent fauna of the sanctuary, is considered to be on the verge of extinction. The sanctuary authorities have taken some measures to preserve it. However, there is a need for more organized efforts to save this very rare bird and revive its population.

Tawny Eagle

Jerdon's Courser

Kolleru Bird Sanctuary

Kolleru landscape

Kolleru, one of the largest freshwater lakes in India, is situated between the Krishna and West Godavari districts of Andhra Pradesh. The lake is fed by the Budameru, Tammileru, Ramileru, Gaderu and Bulusuvagu Rivers as well as several streams. Kolleru Bird Sanctuary was declared a wildlife sanctuary in 1999. Later, it was included under the Ramsar Convention as a Wetland of International Importance.

Birdwatching Sites

Kolleru is one of the most important wetlands in India. The lake supports 200 bird species, including the Critically Endangered Spoon-billed Sandpiper. It serves as a foraging ground for resident as well as migratory birds, and every winter it becomes

KEY FACTS

Nearest Major Towns
Eluru Town

Habitats
Reeds in wetlands

Key Species
Little Egret, Cattle Egret, Pied Kingfisher, Common Kingfisher, Indian Pond Heron, Black-capped Kingfisher, Western Reef Egret, Grey Heron, Black-crowned Night Heron, Little Stint, Common Redshank, Red-wattled Lapwing, Greater Coucal, Purple Heron, Brahminy Kite, Asian Openbill, Little Cormorant, Great Cormorant, Indian Cormorant, Spoon-billed Sandpiper, Painted Stork, Ferruginous Pochard, Red-crested Pochard, Spot-billed Pelican, Common Pigeon, Black-winged Stilt, Brown-headed Gull, Whiskered Tern, Black-headed Ibis, Green Bee-eater, Rose-ringed Parakeet, Indian Golden Oriole, Red-vented Bulbul, Purple Sunbird, Black Drongo

Other Specialities
Water Monitor, Indian Cobra

Best Time to Visit
October to March

Kentish Plover

Lesser Flamingo

Asian Koel

Purple Sunbird

home to thousands of birds from across the globe. Spot-billed Pelicans, Painted Storks and Asian Openbills are found in great numbers here. Some of the other birds found in the area are the Common Redshank, Red-crested Pochard, Glossy Ibis, cormorants and flamingos. The lake also supports about 63 fish species.

Access & Accommodation
Kolleru Lake can be easily accessed via Vijayawada, which is about 60km away. Vijayawada also has an international airport. The nearest railway station is in Eluru. Eluru town is on NH 5 and well connected by bus and railway. From Eluru, you can take a private vehicle to reach Kolleru Bird Sanctuary, Andhra Pradesh Tourism's Haritha Lake Resort at Gudivakalank (http://aptdc.gov.in/aptdc), Eco-Tourism Resort, Hotel Grand Arya (http://hotelgrandarya.in) and Nonsuch Retreat (http://nonesuchretreat.com). There are many other budget hotels around the city if you wish to explore more. Nonsuch Retreat is a popular choice among birdwatchers and wildlife lovers.

Conservation
Kolleru Lake is under threat because of many activities around the area. Most parts of the lake are used for aquaculture, and many small ponds within the lake are created to catch fish. Another half of the area is used for agriculture, which has minimized the lake area. This has, in turn, resulted in pollution of the lake and ecological imbalance in the vicinity. Sewage and industrial pollutants from nearby towns are other threats, as are roads without proper bridges, which affect the free flow of water in the lake. This area has a huge potential to promote sustainable tourism. A sizeable area of Kolleru Lake can be maintained just like Bharatpur in Rajasthan. The local community and landowners can be roped in to promote tourism, which can also be maintained by eco-development committees and locals under the supervision of the forest department.

Asian Openbill

Araku Valley

Araku Valley

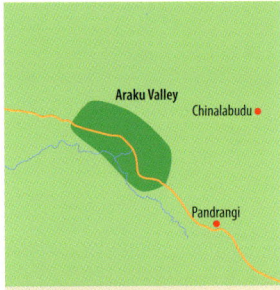

KEY FACTS

Nearest Major Towns
Visakhapatnam (Vizag)

Habitats
Scrub jungle, coffee plantations and well-wooded forests

Key Species
Pale-capped Pigeon, Greater Spotted Eagle, Lesser Kestrel, Pallid Harrier, Red-vented Bulbul, Tickell's Blue Flycatcher, Jungle Babbler, Speckled Piculet, Rufous Woodpecker, Brown-headed Barbet, Common Hoopoe, Plain Prinia, White-rumped Shama, Green Bee-eater

Other Specialities
Indian Leopard, South-western Langur, Macque, Striped Hyena, Water Monitor

Best Time to Visit
Throughout the year

Araku Valley is a small hill station in Andhra Pradesh. It is located about 120km from the coastal city of Visakhapatnam, known popularly as Vizag, and is surrounded by several magnificent hills, among which Galikonda Hills is the highest. Other hills that surround Araku Valley are Sunkarimetta, Raktakonda and Chitamogondi Hills. Araku Valley is known for its coffee plantations, and coffee grown in the region is some of the finest in the country. The whole area is a retreat for nature lovers and birdwatchers. The serene hill station, unspoilt by commercial tourism, is an ideal getaway from the city and can offer wonderful forest species. The places of interest in this area are Ananthagiri, Tyda and Borra Caves. Visitors can also have memorable experiences of the tribal lifestyle and can be a part of tribal folklore.

Birdwatching Sites

More than 150 bird species have been recorded in this area. The birds in Araku Valley include four globally threatened species – the Pale-capped Pigeon, Greater Spotted Eagle, Lesser Kestrel and Pallid Harrier. The frequently sighted birds belong to the families of bulbuls, warblers, flycatchers and babblers. Some of the important bird species of the area are the Speckled Piculet, Rufous Woodpecker, Brown-headed Barbet, Common Hoopoe, Plain Prinia and White-rumped Shama. These species are commonly found nesting in the scrub jungle and wooded forests.

Access & Accommodation

The nearest airport to reach Araku Valley is Visakhapatnam, from where curvy roads to Araku Valley can be relied on. Araku Valley has two railway stations that are on the eastern coastal line. Going by rail is a great way to travel to the valley, as the

Black-headed Cuckooshrike

views are quite enchanting, passing through bridges and tunnels.

For accommodation, the options include Haritha Valley Resort, Punnami Hill Resort, Jungle Bells in Tyda, India Tourism Development Corporation tribal cottages, Mayuri Hill Resort, Ushodaya Resorts, Chandrika Guest House and Holiday Inn Vihar. Pre-booking is always recommended, especially during the peak seasons. A combination of budget, deluxe and premium hotels is available – visitors can make bookings in the type of accommodation that matches their budgets and requirements.

Conservation

The animal diversity of the Araku Valley region is facing serious threats due to habitat loss and habitat degradation. Three mammalian species – the Indian Leopard, Sloth Bear and Indian Giant Squirrel, under Schedule I of the Indian Wildlife Protection Act (1972) – inhabit the area. Podu cultivation, fuelwood collection, customary hunting, forest fires and the cutting of trees are among some of the serious threats. However, mining is a much more serious issue. Large-scale mining for bauxite is proposed almost all along the Eastern Ghats by various organizations. Activities including opencast mining, which adversely affect the ecosystem, are likely to negatively impact the avifauna of the region.

Common Rosefinch

Scarlet Minivet

Golden-fronted Leafbird

White-eyed Buzzard

ARUNACHAL PRADESH

1. Eaglenest Wildlife Sanctuary
2. Mandala, Sela Pass & Sangti Valley
3. Dibang Valley, Mishmi Hills, Roing
4. Pakke Tiger Reserve
5. Namdapha National Park
6. Walong

Arunachal Pradesh – also known as the land of the 'dawn-lit mountains' – is a thinly populated and mountainous tract in the easternmost part of India. It is surrounded on three sides by the international border, with Bhutan to the west, China to the north and Myanmar to the east. The state of Assam lies to the south. Arunachal is one of the most biologically diverse parts of the subcontinent. Its avifauna is a unique blend of Himalayan, Sino-Tibetan and Indo-Burmese forms. Earlier this state was known as NEFA (North-East Frontier Agency). It was renamed as Arunachal Pradesh in 1972, and was administered as a union territory. The present state came into existence in 1987.

Climate
The climate of Arunachal Pradesh (except for the Great Himalayas and higher areas of the Lesser Himalayas and the Mishmi Hills, which may be termed as 'mountain type') is tropical 'monsoon' type with a hot, wet summer and a cool, dry winter. The annual rainfall varies from less than 1,500mm to more than 4,000mm. The temperature generally ranges from 0 °C in winter (minimum) to 35 °C in summer (maximum). The main features of this mountain climate are a sharp contrast between temperatures in sun and shade, high diurnal range of temperatures, inversion of temperature and variability of rainfall, depending upon exposure and elevation. The state has broadly six types of forest, tropical wet evergreen (rainforests), subtropical broadleaved, subtropical conifer, temperate broadleaved, temperate conifer and subalpine forests/alpine scrub. Small savannah grassland patches occur along the major rivers.

Access, Transportation & Logistics
Arunachal Pradesh does not have an airport. The nearest airport to Itanagar, the capital of Arunachal Pradesh, is Lilabari in Assam, which receives flights on four days a week from Guwahati and Kolkata. The distance between Lilabari Airport and Itanagar can be covered in two hours by bus or taxi. Guwahati has an international airport that is well connected with all the major cities of India. Visitors can also get a helicopter ride from Guwahati. Pawan Hans Helicopter services operate from Guwahati and run within Arunachal Pradesh. Recently, railway services have also started in Arunachal, and these will be extended soon. Arunachal Pradesh is easily accessible by road. You can easily get direct buses and a taxi from the various neighbouring cities,

KEY BIRDS
Top 10 Birds
1. Sclater's Monal
2. Hodgson's Frogmouth
3. Chinese Pond Heron
4. Rufous-necked Hornbill
5. Bugun Liocichla
6. Blyth's Kingfisher
7. Pale-headed Woodpecker
8. Temminck's Tragopan
9. Black-tailed Crake
10. Mishmi Wren Babbler

and towns like Guwahati, Tezpur, Jorhat, Dibrugarh, Tinsukia and Nagaon.

Health & Safety
Arunachal is one of the most remote and least travelled mountainous states of India. Some health and safety precautions have to be taken before coming to the region, as the Himalayan hills and dense forests are pervasive throughout. Travellers, especially those from Western countries, might encounter minor health problems like stomach disorders and fever. It is best to carry some medicine as prescribed by medical practitioners, and it is recommended that visitors get a medical check-up before travelling to Arunachal and always carry a personal first-aid box.

As the region has an entirely different culture and cuisine from the rest of India, you need to take basic food-safety precautions while travelling here. Before going for adventurous sports (like rafting and trekking), seek the guidance of an expert guide. Adhere to the photography guidelines, and do not take photographs of places (such as military bases and religious centres) where filming is restricted. All visitors to Arunachal require a permit to enter the state. Foreigners need to apply for a Protected Area Permit (PAP), while Indian travellers have to apply for an Inner Line Permit (ILP). There are several ways of getting a PAP. The easiest and most straightforward way is to visit the Deputy Resident Commissioner Office of Arunachal Pradesh in Guwahati. Indian visitors can apply for an ILP online (https://arunachalilp.com/index.jsp).

Arunachal landscape

Birdwatching Highlights
A considerable geographical part of Arunachal comprises forests. There are two national parks and many wildlife sanctuaries in the state, and it has the second highest number of IBAs in north-east India. Namdapha National Park is an important Tiger Reserve of India. Although some parts of Arunachal are still unexplored, approximately 900 bird species have already been recorded in the region. Some fabulous bird species, like Blyth's Tragopan, Blyth's Kingfisher, Ward's Trogon and the Beautiful Nuthatch, are found in this remote region.

Arunachal landscape

Eaglenest Wildlife Sanctuary

Bugun Liocichla

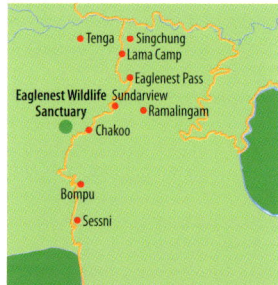

KEY FACTS

Nearest Major Towns
West Kameng

Habitats
Subtropical evergreen and tropical wet evergreen forests

Key Species
Blyth's Tragopan, Rufous-necked Hornbill, Rusty-bellied Shortwing, Beautiful Nuthatch, Satyr Tragopan, Ward's Trogon, Great Hornbill, Hoary-throated Barwing, Streak-throated Barwing, Beautiful Sibia, White-naped Yuhina, Bugun Liocichla, Grey Peacock Pheasant, Hodgson's Frogmouth

Other Specialities
Smooth-coated Otter, Clouded Leopard, Orange-bellied Squirrel, Hoary-bellied Squirrel, Asian Elephant

Best Time to Visit
October to March

Located in the western bend of Arunachal Pradesh, the Eaglenest Sanctuary is covered in tropical, subtropical and temperate forests. Eaglenest (along with Bugun Community Forest) in itself covers 218km² and if we consider its neighbouring forests in Assam (Nameri, Pakke and Doimara Reserve Forest), then it covers an impressive range of 3,500km. Eaglenest has unique biodiversity – to the south are the plains of Assam and its lowland tropical evergreen forests. To the north is the Gori-Chen mountain range (altitude 6,000m), the alpine forests of Dirang and the Sela Pass. In between is a belt of temperate broadleaved and conifer forests. Almost this entire area is accessible through a motorable track, which starts from Tenga, covers Lama, Bompu and Sessni camps, crosses Eaglenest Pass and goes down to the village of Khellong.

Birdwatching Sites

Ramalingam This site is midway between Tenga and Lama Camps and is a good place to look for the White-collared Blackbird, Wallcreeper, Common Buzzard and many species of flycatcher. There are also records of elusive Hodgson's Frogmouths from this area.

Lama Camp This site is situated about 10km beyond Ramalingam. On a clear day, visitors can savour spectacular views of the Gori-Chen mountain range while having their breakfast in the dining room. Lama Camp is one of the top birding locations in the area. The new species Bugun Liocichla (*Liocichla bugunorum*) was discovered in Lama Camp. Other

Red-faced Liocichla

birds that have been seen in the vicinity of the campsite are five species of wren babbler, Fire-tailed Myzornis, Blue-fronted Robin, several ficedula flycatchers, Rusty-bellied Shortwing, many species of laughingthrush, Slender-billed Scimitar Babbler, Yellow-rumped Honeyguide and Brown Bullfinch.

Tragopanda Trail Ascending further up from Lama Camp there is a wonderful trail that passes through an excellent stretch of temperate broadleaved forest, rhododendrons and bamboo. As the name suggests, this trail is a good place in which to look for tragopans and the Red Panda. Ward's Trogon, Green Shrike-babbler, Spotted Laughingthrush, White-browed Shortwing, Rufous-fronted Tit and Yellow-bellied Bush Warbler have also been seen here.

Eaglenest Pass and Sunderview The Eaglenest Pass is the northern boundary of Eaglenest and the highest point. The area has extensive bamboo, along with oaks, birches and rhododendrons. The vegetation is rather scrubby, but the area has a lot of special birds. Sunderview is the equivalent of Lama Camp in terms of altitude, vegetation and excellent birding. The Common Hill Partridge, Temminck's Tragopan and Ward's Trogon can be seen here.

Chakoo-Bompu Road Chakoo lies in a large meadow, and the stretch between Chakoo and Bompu camp has one of the finest areas of Eaglenest for birding. This area is frequented by the Fire-tailed Myzornis, Vivid Niltava, tragopans, Chestnut-fronted Shrike-babbler, Common Hill Partridge, Ward's Trogon, shortwings and the Blue-fronted Robin.

Bompu Bompu was probably named after the bamboo that is found in abundance near the campsite. It has a mixed habitat – open fields on one side and dense forest patches on the other make it a great place to spot birds. Parrotbills, finches, tesias, Blyth's Tragopan, shortwings, and many babblers and warblers are some of the highlights that make this place a hotspot for birding in India.

Sessni The location of Sessni is in the upper tropical zone so you will find an abundance of lush vegetation everywhere. From Bompu the road to Sessni descends the steep hillside in a series of sharp bends, and offers a great birding experience along the roadside itself. The Red-faced Liocichla, Beautiful Nuthatch, Lesser Rufous-headed Parrotbill, Chestnut-headed Tesia, Fire-breasted Flowerpecker, Sultan Tit, Sikkim Wedge-billed Babbler, Rufous-necked Hornbill, Long-tailed Broadbill, Grey-headed Parrotbill, Coral-billed Scimitar Babbler and White-gorgeted Flycatcher are some of the highlights in this area.

Access & Accommodation

Inside Eaglenest only tented accommodation is available. There are three camps, named Lama, Bompu and Sessni. The camps are equipped to provide a comfortable stay. Food options are basic, but basic hygiene is maintained when it comes to both cooking and serving. Contact for bookings: Mr Indi Glow, President, Bugun Welfare Society (Tel.: 94364 26781, 87299 15566; Email: phuarung@gmail.com). Advance booking is essential. Outside Eaglenest Tenga has some basic accommodation available. There is a small market with basic medical facilities. There is also an army medical hospital that can be used in case of emergency.

Conservation

Poaching is still a major conservation problem in general across Arunachal, except in inaccessible areas in the interior. Hunting with guns is not very significant, except in the case of hornbills. However, tragopans, Common Hill Partridges and Kalij Pheasants are regularly snared along the Tenga-Doimara and Bomdila-Bhalukpong roads. Protection measures need to be considerably strengthened. An increased presence of wildlife staff is necessary at Khellong to protect the western part of Eaglenest. With the possible opening of the Tenga-Doimara route for vehicular traffic, protection measures along the road must be strengthened. Camps should be set up at Sessni, Bompu, Chaku and Sundarview, with a larger one at Lama. The wildlife staff should be provided with better infrastructure facilities such as jeeps, motorcycles and modern firearms. Road construction, especially the Tenga-Doimara road, which bisects Eaglenest, has severe impacts on the area. Labourers fell trees for fuel and building materials, and hunt galliformes with snares. The General Road Engineer's Force, which is responsible for the development and maintenance of the road network, should prohibit poaching and substitute coal or other materials for fuel and heating.

Mandala, Sela Pass & Sangti Valley

Sela Pass landscape

KEY FACTS

Nearest Major Towns
West Kameng

Habitats
Subtropical pine forest, subtropical broadleaved hill forest, temperate conifer forest, temperate broadleaved forest

Key Species
Blyth's Tragopan, Ward's Trogon, Rusty-bellied Shortwing, Rufous-throated Wren Babbler, Hoary-throated Barwing, Streak-throated Barwing, Beautiful Sibia, White-naped Yuhina, Black-browed Reed Warbler, Broad-billed Warbler, Rufous-necked Hornbill

Other Specialities
Red Panda, Mithun, Red Fox, Takin, Mouse Deer, Himalayan Black Bear, Brown Bear

Best Time to Visit
Throughout the year

The Mandala-Phudung area, in the West Kameng district of Arunachal Pradesh, forms part of the IBA with the contiguous forest tracts of Shergaon and Kalaktang. This is a region with high biodiversity and is a global biodiversity hotspot as well as an Eastern Himalayas Endemic Bird Area. The Sela Pass is a high-altitude mountain pass located on the border between the Tawang and West Kameng districts of Arunachal. Tawang, being one of the most famous destinations in the state, attracts many visitors from across the globe. This is a good place for high-altitude bird species.

Birdwatching Sites

Mandala Located about 30km above the town of Dirang in the West Kameng district, Mandala shares its borders with the Shergaon and Kalaktang high-altitude regions. The Mandala forest region has been identified as one of the IBAs in Arunachal Pradesh, with more than 300 bird species recorded in the area. It is famous for tragopans and Gould's Shortwing, and there are many more birds here, including, to name a few, Ward's Trogon, Rufous-necked Hornbill, Yellow-vented Leaf Warbler, Broad-billed Warbler, Rufous-throated Wren Babbler, Hoary-throated and Streak-throated Barwings, Beautiful Sibia, Beautiful Nuthatch, Rusty-bellied and Gould's Shortwings, Brownish-flanked Bush Warbler, Red Crossbill, Blanford's Rosefinch, Blyth's and Temminck's Tragopans, Kessler's Thrush and White-bellied Redstart.

Hodgson's Frogmouth

Crimson-browed Finch

Sela Pass Sela Pass separates Tawang from the West Kameng district and due to its high altitude (4,170m), this is a great birding destination, especially for high-altitude species. Some of the target species of the Sela region include the Blood Pheasant, Himalayan Monal, Himalayan Buzzard, Speckled Wood Pigeon, Himalayan Swiftlet, Yellow-billed Blue Magpie, White-browed Tit Warbler, Rufous-fronted and Rufous-vented Tits, Brown-flanked, Hume's and Russet Bush Warblers, Large-billed and Blyth's Leaf Warblers, Fire-tailed Myzornis, Brown Parrotbill, Streak-breasted Scimitar Babbler, Spotted, Grey-sided and Black-faced Laughingthrushes, Beautiful Sibia, Rusty-fronted Barwing, Indian Blue Robin, White-browed Bush Robin, Pygmy Blue Flycatcher, Hodgson's and White-throated Redstarts, Plain-backed Thrush, White-collared and Tibetan Blackbirds, Plain Flowerpecker, Mrs Gould's Sunbird, Brandt's Mountain Finch, Dark-rumped, Spot-winged and Himalayan White-browed Rosefinches, Ruddy Shelduck, Alpine Accentor, Grandala, Common Kestrel, Common Buzzard, Golden Bush Robin, Snow Partridge, Solitary Snipe, Snow Pigeon and White-winged Grosbeak. Some unusual birds have also been recorded near Sela Pass, like the Eurasian Wryneck.

Sangti Valley This quiet valley is located near Dirang, at 1,500m. It contains broadleaved forests, paddy fields and patches of marshy land – a great destination for the Black-necked Crane. Other bird species that can be seen here are the Long-billed Plover, Black-tailed Crake, Yellow-rumped Honeyguide, Slender-billed Oriole, Striated Bulbul, Hume's Bush Warbler, Himalayan Buzzard, Hodgson's Redstart, Golden-spectacled Warbler and White-browed Bush Robin.

Access & Accommodation
For Mandala, Sela Pass and Sangti Valley, Dirang is a good place to stay. It is a small town in the West Kameng district that is located on the way from Tezpur to Tawang highway. Tawang is well known among visitors, and offers varied accommodation options, depending on your budget and requirements. You can easily hire a taxi from Dirang for Mandala, Sela and Sangti.

Conservation
Common conservation issues in this area are poaching of wild animals and birds. Illegal felling of trees on the slopes is still a problem, despite the ban on tree felling by the Supreme Court. Accidental and deliberate forest fires also occasionally damage habitats, especially in dry winters in the higher areas with temperate conifers. Road construction and extension have short-term as well as long-term impacts. The proposed tunnel in Sela Pass will significantly impact the wildlife of this area.

Bar-winged Wren Babbler

Temminck's Tragopan

Dibang Valley, Mishmi Hills, Roing

Tree ferns

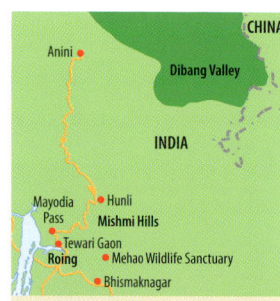

KEY FACTS

Nearest Major Towns
Tinsukia, Dibrugarh

Habitats
Subtropical broadleaved hill forest, Himalayan moist temperate forest, alpine moist scrub

Key Species
Rufous-throated Hill Partridge, Chestnut-breasted Partridge, Snow Partridge, Blyth's Tragopan, Temminck's Tragopan, Himalayan Monal, Sclater's Monal, Beautiful Nuthatch, Ward's Trogon, Beautiful Sibia, Goosander, Ruddy Shelduck, Baer's Pochard, Ferruginous Duck, Gadwall, Mallard, Tibetan Snowcock, Red Junglefowl, Kalij Pheasant, Blood Pheasant, Rock Dove, Asian Emerald Dove, Speckled Wood Pigeon, Pale-capped Pigeon, Pin-tailed Green Pigeon, Wedge-tailed Green Pigeon, Oriental Turtle Dove, Barred Cuckoo Dove, Great Eared Nightjar, Grey Nightjar, Crested Treeswift

Other Specialities
Black-capped Langur, Slow Loris, Assamese Macaque, Hoolock Gibbon, Mishmi Takin, Mithun

Best Time to Visit
October to March

The Mishmi Hills, one of the most famous birdwatching spots in India, is a southward extension of the Great Himalayan Mountain Range, with its northern and eastern reaches touching China. This is breathtaking terrain that is thrown up by the sharp twisting of the Himalayan ranges as they turn from a southeasterly to a southerly direction and descend rapidly to the tropical forests of China and Myanmar. Geographically, the Mishmi Hills area can be divided into two broad sections: the flood plains of the tributaries of the Brahmaputra River, and the Arunachal Himalayas consisting of snow-capped mountains, the lower Himalayan ranges and the Shivalik hills. The Mishmi area is one of the last strongholds for many species dependent on this forest type.

Much of the Mishmi Hills area falls under the Dibang Valley district of Arunachal Pradesh. The district is named after the Dibang River – one of the key tributaries of the Brahmaputra River. The Dibang originates in China and flows through the length of the valley in a north–south direction. Dibang Valley district is divided into the Lower and Upper Dibang Valleys. The headquarters of the two districts are Roing and Anini respectively. Roing is in the plains close to the Lohit River. Anini is at higher elevation in the mountains and is close to the Chinese border.

Birdwatching Sites

Mishmi Hills, situated in the Eastern Himalayan the richest biogeographical province of the Himalayan zone and is one of the mega-biodiversity hotspots of the world. The complex hill system of varying elevations receives heavy rainfall, which can be as much as 4,500–5,000mm annually in the foothill areas. The pre-monsoon showers start in March and the monsoon is active until October. The humidity in

Long-tailed Sibia

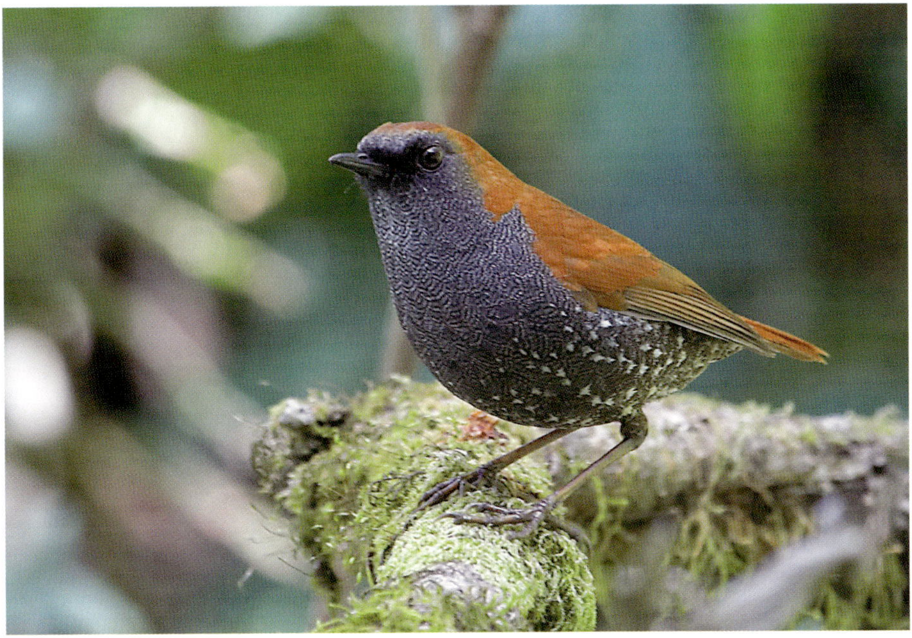

Sachin Rai

Gould's Shortwing

the rainy season is often over 90 per cent. This diversity of topographical and climatic conditions has favoured the growth of luxuriant forests that are home to myriad plant and animal forms.

The area supports more than 700 bird species and it is perhaps one of the least explored birding areas in all of Asia, with many new species awaiting discovery. Specialities include Sclater's Monal, Blyth's and Temminck's Tragopans, Chestnut-breasted Hill Partridge, Rufous-necked Hornbill, Pale-capped Pigeon, Ward's Trogon, Dark-sided Thrush, Green and Purple Cochoas, Rusty-bellied and Gould's Shortwings, Beautiful Nuthatch, Mishmi and Himalayan-billed Babblers, Fire-tailed Myzornis, at least four parrotbill species, Black-headed Greenfinch, Scarlet Finch and Grey-headed Bullfinch.

Prassana Parab

Mishmi Wren Babbler

Biswapriya Rahut

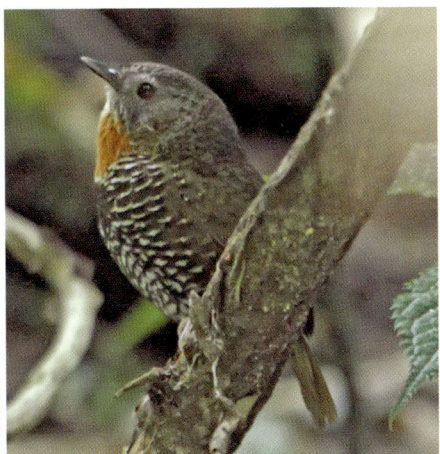

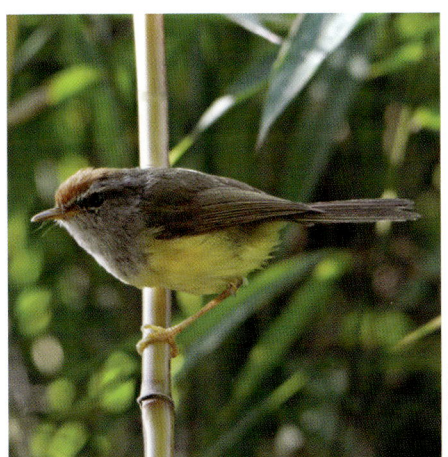

Broad-billed Warbler

Roing area Roing is located in the foothills and its altitude stretches from 200 to 550m. The Mishmi Hills rise around the Dibang River Basin, and the Roing area is a mix of extensive floodplains and outer Himalayan foothills. Birding here is good from the 12th mile (on the road to Mayodia) all the way downhill to the Itapani River. Many Himalayan foothill species can be found in this area. Key birds include the Black Eagle, White-browed Piculet, Pin-tailed Green Pigeon, Red-headed Trogon, Long-tailed Broadbill, White-throated Bulbul, Daurian Redstart, Slaty-backed Forktail, Spot-throated Babbler, Beautiful and Long-tailed Sibias, Hill Blue Flycatcher, Sultan Tit, and numerous Scimitar Babblers, laughingthrushes, shrike-babblers, fulvettas and yuhinas.

Biswapriya Rahut

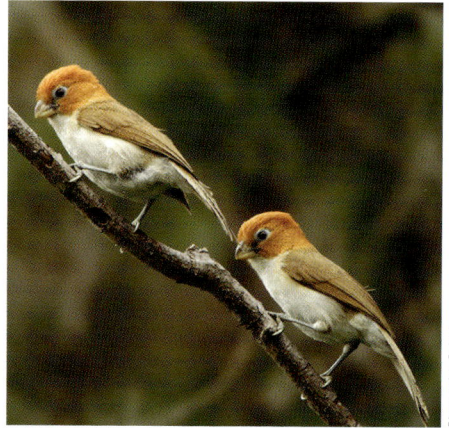

Rufous-headed Parrotbills

Mehao Wildlife Sanctuary This sanctuary stretches over three ecozones, and undisturbed tropical evergreen forests, subtropical and temperate forests, temperate broadleaved forests and temperate conifer forests can be found in the area. The sanctuary is home to key species like Temminck's Tragopan, Rufous-necked and Great Hornbills, and Rusty-throated Wren Babbler . Mammals include the Tiger, Snow Leopard, Clouded Leopard, Red Panda, Mishmi Takin and Hoolock Gibbon.

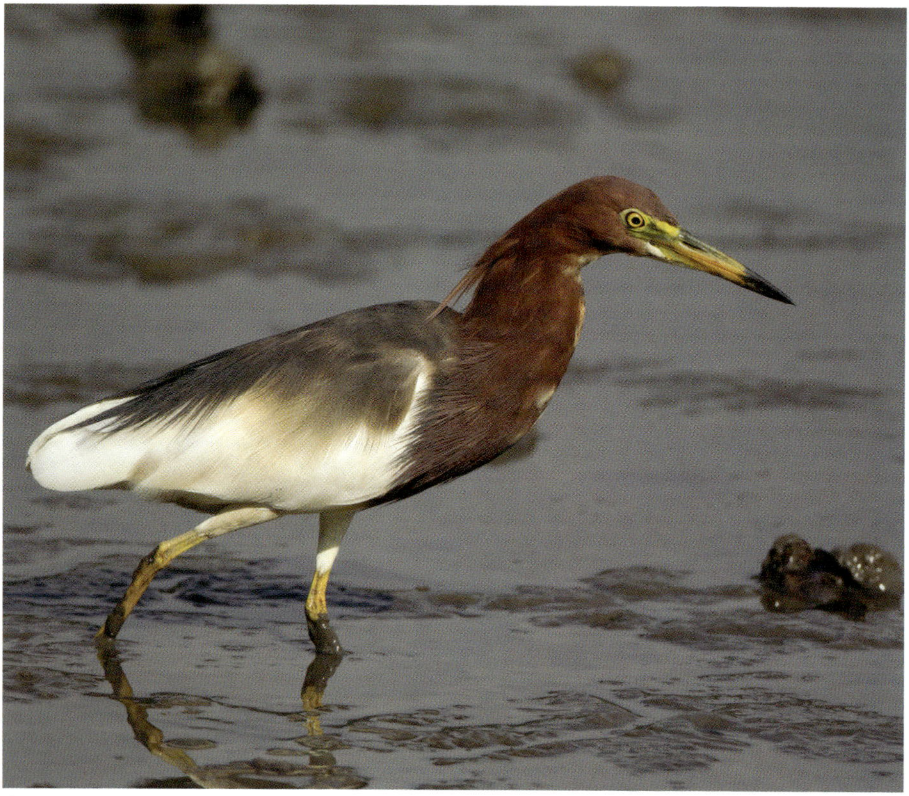

Niranjan Sant

Chinese Pond Heron

Mayodia Pass At 2,666m, this is the highest point on the road from Roing to Anini and is 56km uphill from Roing town. The habitat comprises temperate forests. Strangely, there is a lack of conifer growth even at this height, and extensive bamboo brakes dominate the roadside vegetation. Mayodia is a must-visit place for those seeking the high-altitude Sclater's Monal. It is also one of the few places where Blyth's Tragopan is regularly encountered. Many other rarely seen species like the Darjeeling Pied Woodpecker, Slender-billed Scimitar Babbler, Ward's Trogon, Grey-sided, Spotted and Black-faced Laughingthrushes, Gould's and Rusty-bellied Shortwings and Fire-tailed Myzornis can be found here, and the area is possibly the best place to encounter the now elusive Mishmi Takin.

Access & Accommodation

The Mishmi Hills area can be easily approached by crossing the Bhupen Hazarika Setu (also known as Dhola-Sadiya Bridge), India's longest bridge, connecting Sadiya Ghat with Saikhoa Ghat over the Lohit River. Saikhoa Ghat is about 70km from Tinsukia town in Assam. Tinsukia has a railhead, and the nearest airport is Mohanbari at Dibrugarh 40km away. The recommended access is Dibrugarh – Tinsukia – Sadiya Ghat via NH 37. Roing town is 65km from Sadiya Ghat and is usually the base for a Mishmi Hills trip. Mayodia Pass is 56km from Roing.

Ward's Trogon

Dibang Valley Jungle Camp (www.helptourism.net/dibang-valley-jungle-camp.php), an eco resort run by Help Tourism, offers a basic but comfortable stay option. There is a government inspection bungalow and a coffee house/guest house a couple of kilometres below Mayodia Pass. Both are basic but offer stunning views. Visitors can contact Help Tourism for booking in these places. The best season for birding is November–April. The area is generally inaccessible in May–September/October.

Conservation

The major issue is encroachment and de-reservation for human settlement. As more and more people are looking for flat land, there is tremendous pressure on the area. Poaching, grazing of cattle and buffalo, collection of thatching and felling of trees are other major issues. Construction of roads and sprawling development in nearby areas is an ongoing long-term threat for the conservation of this biodiversity hotspot.

Sclater's Monal

Pakke Tiger Reserve

Tropical forest, Pakke

Also known as Pakhui Tiger Reserve, this is a Project Tiger reserve in the East Kameng district of Arunachal Pradesh. It is bounded to the north and west by the Jia Bhareli River (known as Kameng in Arunachal Pradesh), to the east by the Pakke River and to the south by the Nameri National Park of Assam (an IBA). The Doimara Reserve Forest lies to the west of the sanctuary, while the Papum Reserve Forest lies to the east. The area lies in the foothills of the Himalayas and the terrain is undulating and hilly. A large portion of the northern and central parts of the sanctuary is inaccessible due to dense vegetation, hilly terrain and lack of trails or paths.

Birdwatching Sites

Pakke is undoubtedly one of the hotspots for elusive wildlife, including avian life. More than 300 bird species have been recorded in this area. The pride of Pakke is the four hornbill species found here. The most abundant is the Wreathed

KEY FACTS

Nearest Major Towns
Tezpur

Habitats
Tropical wet evergreen forest, subtropical broadleaved hill forest, tropical semi-evergreen forest

Key Species
White-winged Wood Duck, Rufous-necked Hornbill, Oriental Darter, Black-headed Ibis, White-cheeked Hill Partridge, Great Hornbill, Rufous-throated Wren Babbler, Hoary-throated Barwing, White-naped Yuhina, Black-browed Reed Warbler, Broad-billed Warbler, Blyth's Kingfisher, Ruddy Kingfisher

Other Specialities
Himalayan Black Bear, Arunachal Macaque, Common Palm Civet, Leopard Cat, Indian Leopard, Tiger

Best Time to Visit
October to March

Fire-tailed Myzornis

Mountain Bulbul

Blyth's Kingfisher

Red-headed Trogon

Hornbill, although the Oriental Pied Hornbill is also present in good numbers. Great and Rufous-necked Hornbills are rarer. Pakke also hosts the globally threatened White-winged Wood Duck, among other important birds like Jerdon's Baza, Pied Falconet, White-cheeked Hill Partridge, Ibisbill, Blue-naped Pitta, Mountain Imperial Pigeon, Asian Fairy-bluebird, Green Cochoa, Crow-billed Drongo, Ruddy Kingfisher, Long-billed Plover, shortwings, cuckoos, forktails and other specialties.

There are walking trails inside the park; armed forest guards are recommended for using them. Birding in Pakke is a bit demanding – but it would be very rare for any birder to return from Pakke without seeing some rare birds.

Access & Accommodation

Access to Pakke is from Seijosa, which is about 250km from Guwahati and 50km from Tezpur. The main access is through Soibari, which is on the NH 52. The village of Seijosa is a two-hour drive from Soibari over poor roads. The main airport is Guwahati International Airport, which is approximately 250km away.

Pakke Jungle Camp (www.helptourism.net/pakke-jungle-camp.php) is a good place to stay for birdwatchers. The camp is a community-based tourism initiative jointly implemented and run by Help Tourism, and the local tribal self-help conservation group named Ghora-Aabhe. The camp is very basic, yet gives you a true wildlife experience. You can also stay at the Langka, Khari and Upper Dikori Forest Rest House, but visitors are required to carry basic food items with them. Hotels and tourist lodges are available in Bhalukpong. Accommodation is also available, with advance booking, at the Tipi Forest Rest House, 4km away from Bhalukpong.

Conservation

The biggest threat in recent years has been the increasing encroachment and almost total tree felling of the adjoining reserve forests in Assam, near the state boundary adjoining Pakke Tiger Reserve. Hunting by the local tribal community is mainly for subsistence and local consumption, although sometimes animals are hunted to supplement the community's cash income. Large-scale hunting for commercial purposes is not such a big threat, although incidental hunting of otters, bears and other lucrative wildlife species occurs. The interest and conservation commitment shown by the Nishi tribe needs to be sustained through more incentives, income-generation opportunities tied to wildlife conservation in the area such as employment as nature guides, ecotourism and conservation education programmes. This would increase awareness among villagers about wildlife conservation, and generate pride and enthusiasm among them.

Namdapha National Park

Noa Dihing River

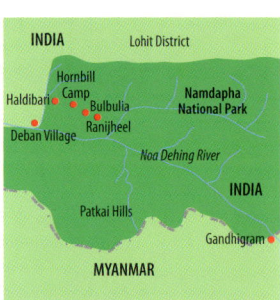

This park in Changlang district is situated at the southeastern tip of Arunachal Pradesh, and is bounded by the international boundary with Myanmar to the south and east. It was declared a Tiger Reserve in 1983. Namdapha comprises the catchment of the Noa-Dihing River, a tributary of the Brahmaputra River. The Noa-Dihing originates in the mountains on the India-Myanmar border, and flows westwards through the park before joining the Lohit and Brahmaputra Rivers in the Assam Valley. The avifauna of the park is a unique mix of Himalayan, Sino-Tibetan and Indo-Burmese forms. The mountainous terrain is crisscrossed by innumerable streams, which support a subtropical humid climate sustaining some of the best mammalian fauna in the subcontinent.

KEY FACTS

Nearest Major Towns
Dibrugarh and Digboi

Habitats
Tropical wet evergreen, subtropical broadleaved, temperate and conifer forests, subalpine scrub, alpine scrub

Key Species
Slender-billed Vulture, White-bellied Heron, White-winged Wood Duck, Greater Spotted Eagle, Blyth's Tragopan, Mrs Hume's Pheasant, Rufous-necked Hornbill, Rusty-bellied Shortwing, Snowy-throated Babbler, Beautiful Nuthatch, White-cheeked Hill Partridge, Ward's Trogon, Great Hornbill, Austen's Brown Hornbill, Sikkim Wedge-billed Babbler, Streak-throated Barwing, Grey Sibia, Beautiful Sibia, White-naped Yuhina, Black-browed Reed Warbler

Other Specialities
Indian Flying Fox, Tiger, Indian Leopard, Assamese Macaque, Golden Cat, Jungle Cat, Fishing Cat, Asian Elephant, Brown Palm Civet, Binturong

Best Time to Visit
November to March

Red-headed Bullfinch *Collared Falconet* *Golden-crested Myna*

Eyebrowed Wren Babbler

White-bellied Heron

Birdwatching Sites

Namdapha is truly a paradise for birdwatchers. More than 511 bird species have been recorded in the area, with some being found only in this area of India. Two important species are the Ibisbill and White-bellied Heron, seen near the Noa-Dihing area. The White-winged Wood Duck has been reported in the lower reaches of the Namdapha River.

Some other important species found in the area are Blyth's Tragopan, Blue-naped Pitta, Snowy-throated Babbler, White-cheeked and Chestnut-breasted Hill Partridges, Beautiful Nuthatch, Ward's Trogon, Rusty-bellied Shortwing and Rufous-necked Hornbill. Many northeastern birds that are uncommon or rarely seen elsewhere are fairly common in Namdapha. For example, the Grey Peacock Pheasant, Wreathed Hornbill, Collared Treepie, Large Scimitar Babbler, Streaked and Eyebrowed Wren Babblers, Rufous-vented Laughingthrush, White-hooded Babbler, Rufous-throated Fulvetta, Green Cochoa and Black-breasted Thrush are fairly widespread in the park, though not necessarily easy to locate.

Access & Accommodation

Dibrugarh Airport and Tinsukia railway station are the nearest points. Assam State Transport Corporation and Arunachal Pradesh State Transport buses ply daily from Dibrugarh to Miao, the entry point for Namdapha. The distance between Dibrugarh and Miao is 160km and is best covered by hired cars. A further drive of 26km is required to reach Deban, a tiny hamlet at the foot of the Patkai range of hills on the banks of the Noa-Dihing River.

The Deban Forest Guest House is one of the popular bases for birding in Namdapha. There is a forest rest house in Namdapha – for bookings contact The Field Director, Project Tiger, Namdapha Tiger Reserve, Miao – 792 122, Changlang District, Arunachal Pradesh, India, Tel./fax: +91-3807-222249.

Conservation

This large IBA is quite remote, largely inaccessible and uninhabited. Most of the forest is largely untouched. The abundance of forest resources around the area fulfills the needs of the people. However, Namdapha is now coming under increasing pressure due to encroachment on its eastern side. Insurgency is another big problem for the administration of the park.

Collared Treepie

Walong

Walong War Memorial

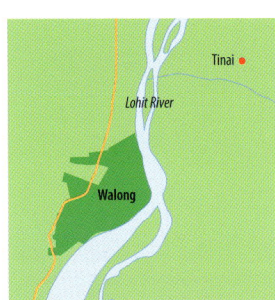

KEY FACTS

Nearest Major Towns
Dibrugarh

Habitats
Montane wet temperate forest, subalpine dry scrub, alpine dry pasture

Key Species
Wood Snipe, Rufous-necked Hornbill, Rusty-bellied Shortwing, Snowy-throated Babbler, Beautiful Nuthatch, White-cheeked Hill Partridge, Yellow-rumped Honeyguide, Spot-breasted Parrotbill, Yunnan Nuthatch, Black-browed Tit, Black-headed Greenfinch

Other Specialities
Assamese Macaque, Capped Langur, Indian Leopard

Best Time to Visit
October to March

Walong is a large wilderness in the Lohit district of Arunachal Pradesh, along the border with China. The Lohit River forms the eastern and southern boundaries of this IBA. The northern mountain ranges along the Chinese border remain snow covered for the greater part of the year. The vegetation of the region is subtropical broadleaved and coniferous forests up to the middle elevation, above which occur temperate broadleaved forests, including pines and other mixed conifers, especially in the north. Alpine and subalpine scrub occur further up. Walong is the perfect base for exploring the coniferous forests of the Lohit Valley.

Yunnan Nuthatch

Derbyan Parakeet

Black-breasted Parrotbill

Black-browed Tit

Birdwatching Sites

Walong is one of the most remote yet excellent places for some rare and elusive birds. More than 300 bird species have been recorded in the area, nearly half of which are restricted to this area alone. The site is in the Eastern Himalayas Endemic Bird Areas. Most of its specialities can be seen on the road to Helmet Top, but the Derbyan Parakeet is less reliable there. Up to Helmet Top is open pine forest with undergrowth of grass and scrub. Above 2,300m lies spectacular dense mixed coniferous forest.

Some of the important species found in the area are the Black-browed Tit, Godlewski's Bunting, Yunnan Nuthatch, Wood Snipe, Black-headed Greenfinch, Ward's Trogon, Green Shrike-babbler, Rufous-necked Hornbill, Rusty-bellied Shortwing, Red-faced Liocichla, Snowy-throated Babbler, Beautiful Nuthatch and Yellow-rumped Honeyguide.

Access & Accommodation

The Walong area is best accessed via Dibrugarh Airport in Assam. From Dibrugarh Airport, it is a long but scenic road journey. Hiring a taxi is recommended as public transport is not reliable in the region. You need to take the Dibrugarh Kakopathar road and enter Arunachal via Dirak Gate.

There are limited accommodation options in the area. Only a few homestays and eco resort accommodation options are available. There is one inspection bungalow in Walong. Hiring a local guide or joining a guided tour is recommended for getting the best birding experience in this area.

Conservation

Hunting by local tribes in this IBA is a serious threat to the birdlife. Hunting occurs mostly with single-barrel shotguns and muzzle-loaders. Organized hunting occurs during winter, when parties of hunters move over different parts of the forest in search of large birds and mammals. In the lower reaches, slash-and-burn (*jhum*) cultivation is a serious problem. There is an opportunity to conduct environmental education programmes with the locals to sensitize them about the importance of this IBA and India's wildlife protection laws.

Godlewski's Bunting

ASSAM

1. Kaziranga National Park
2. Manas National Park
3. Nameri National Park
4. Orang National Park
5. Dibru-Saikhowa National Park
6. Dehing Patkai Wildlife Sanctuary
7. Guwahati

Assam, the land of mysterious blue hills, valleys and the plains of the mighty Brahmaputra River, is located at the northeastern corner of India. This is a place where you can experience temperate, subtropical and tropical moist forests along with the Indo-Gangetic Plain, all at a short distance. This biological diversity had made Assam home to more than 900 varieties of avifauna, including many endangered and endemic species. Assam was largely overlooked for many years due to poor infrastructure. However, significant improvements in tourism infrastructure, and increased foreign as well as local interest in conservation, has helped position Assam as one of the birdwatcher's favourite destinations in the Indian subcontinent.

KEY BIRDS

Top 10 Birds
1. White-winged Wood Duck
2. Goliath Heron
3. Falcated Duck
4. Jerdon's Bushchat
5. Chinese Spot-billed Duck
6. Greater Adjutant
7. Baer's Pochard
8. Finn's Weaver
9. Black Baza
10. Swamp Prinia

Climate

Assam experiences a moist tropical climate, with distinct wet and dry seasons when temperatures vary at 6–40 °C. The warmest and driest time of the year occurs in November–April, which is the best time for birding. During the rainy season in May–October, torrential rain floods a large expanse of Assam and renders many of the unpaved roads impassable, severely limiting access to birdwatching sites.

Access, Transportation & Logistics

Assam's Lokpriya Gopinath Bordoloi International Airport is well connected to all major cities of India, as well as with a few international cities like Bangkok and Paro. Guwahati is also well connected to all the major cities of India via railways. Guwahati Junction is the major railway station of Guwahati, and Kamakhya Junction is the second largest station in the city. NH 37 connects Guwahati with almost all the major cities of Assam. You can hire a cab or self-driven vehicle from Guwahati to visit the birdwatching sites.

Health & Safety

There are numerous natural annoyances in rainforests, particularly in the lowlands of Assam. In addition to terrestrial leeches and ticks, mosquitoes and various species of biting fly occur. The use of long sleeves and insect repellent is recommended for rainforest birdwatching. Vaccinations are also recommended for diseases such as hepatitis A and B, and typhoid. Other mosquito-borne viruses like dengue fever and malaria are present, and appropriate precautions should be taken.

Birdwatching Highlights

Assam is home to more than 900 bird species and you can cover 50 per cent of them in a two-week trip. Its rich birdlife includes the endangered White-winged Wood Duck, Bengal Florican, Red-headed and Long-billed Vultures, Greater and Lesser Adjutants, Long-toed Stint, Marsh Babbler and Black-breasted Parrotbill. Other avian highlights include Jerdon's Babbler, Pale-capped Pigeon, Crested Serpent Eagle, Griffon Vulture, Grey-headed Fish Eagle, Spot-billed Pelican, Baer's Pochard, Great Cormorant, Asian Openbill, Black-necked Stork, Fulvous Whistling Duck, Greylag Goose and Great Crested Grebe.

Gururaj Moorching

Ruby-cheeked Sunbird

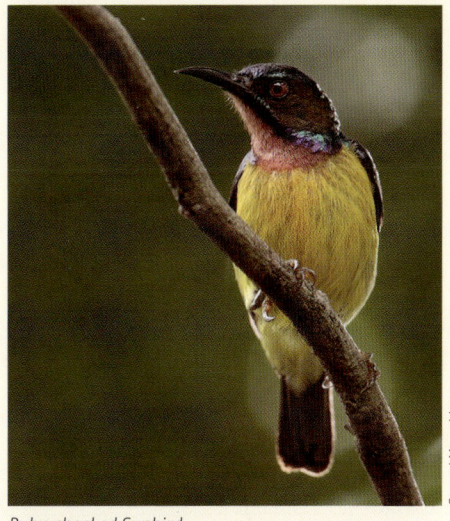

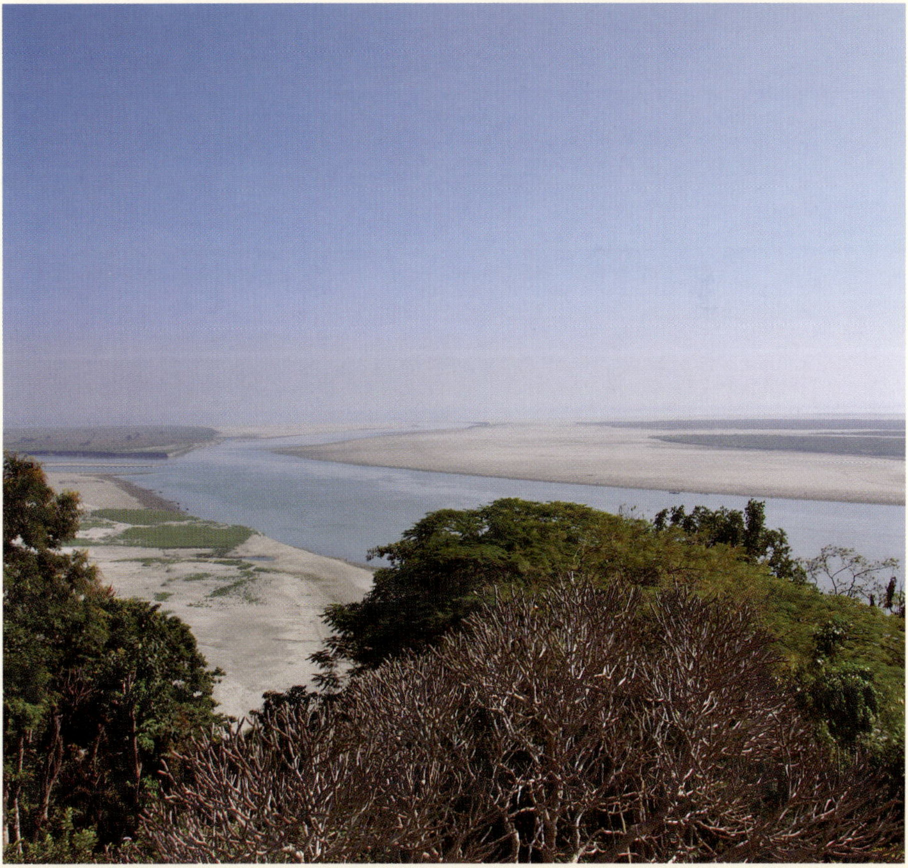

Bhanu Singh

Brahmaputra River

Kaziranga National Park

Kaziranga landscape

This UNESCO World Heritage Site is situated in the Golaghat district of Assam. It has also been recognized as an IBA by Birdlife International, and is home to more than 500 bird species, including resident and migratory birds.

The Critically Endangered Baer's Pochard and Bengal Florican, the Endangered Great Knot, and the Near Threatened Falcated Duck and Black-necked Stork are just a few birds that attract birdwatchers from across the world. Other avian highlights include the Spot-billed Pelican, Jerdon's and Black Bazas, Hen and Pallid Harriers, Pallas's and Grey-headed Fish Eagles, White-tailed Sea Eagle, Great and Wreathed Hornbills, Speckled and White-browed Piculets, Grey-capped Pygmy Woodpecker, Grey-headed, Blossom-headed and Red-breasted Parakeets, Blue-naped Pitta and Silver-breasted Broadbill.

KEY FACTS

Nearest Major Towns
Guwahati and Jorhat

Habitats
Riverine grassland, deciduous forests

Key Species
White-rumped Vulture, Slender-billed Vulture, White-bellied Heron, Greater Adjutant, Bengal Florican, Spotted Greenshank, Spot-billed Pelican, Lesser Adjutant, Lesser White-fronted Goose, Marbled Duck, Baer's Pochard, Pallas's Fish Eagle, Greater Spotted Eagle, Eastern Imperial Eagle, Lesser Kestrel, Swamp Francolin, Indian Skimmer, Pale-capped Pigeon, Hodgson's Bushchat, Marsh Babbler, Jerdon's Babbler, Slender-billed Babbler, Black-breasted Parrotbill, Finn's Weaver, Oriental Darter, Black-necked Stork, Black-headed Ibis, Ferruginous Pochard, White-tailed Sea Eagle, Grey-headed Fish Eagle, Cinereous Vulture, Red-headed Vulture, White-cheeked Hill Partridge, Black-bellied Tern, Blyth's Kingfisher, Great Hornbill, Indian Grassbird

Other Specialities
Asian Elephant, Tiger, Wild Buffalo, Indian Rhinoceros, Hog Deer

Best Time to Visit
October to March

Kaziranga jheels

Jungle Mynas

Grey-headed Fish Eagle

Red-breasted Parakeet

Birdwatching Sites

The park is divided into four tourism zones: the Central Zone or Kohra Zone, Western Zone or Bagori Zone, Eastern Zone or Agaratoli Zone, and Burapahar Zone. The Jeep Safari is organized in every zone, while the Asian Elephant Safari is organized only in the Central and Western Zones. Birdwatchers prefer the Central and Eastern Zones as these are good for spotting birds. Among many endemic and migratory birds, the elusive Swamp Francolin can be seen in the Central Zone. The Eastern Zone is well known for waterfowl like the Lesser White-fronted Goose and Baer's Pochard, and also many forest birds like the Blue-naped Pitta, Green-billed Malkoha and Black-breasted Parrotbill.

Access & Accommodation

Lokpriya Gopinath Bordoloi International Airport, located in Guwahati, is well connected from all major cities of India. The distance from the airport to Kaziranga is about 219km, which can be covered in four hours as the road is good. Taxis or cabs can be hired directly from the airport. For public transport you need to go to the Paltan Bazar bus stand, which is about 23km from the airport.

Kaziranga National Park has a good number of jungle resorts, camps, luxury hotels and government-run lodges. Advance booking is recommended, since during the peak season getting good accommodation sometimes becomes challenging.

Conservation

Kaziranga National Park receives maximum protection under Indian wildlife protection and conservation laws. It has a good conservation history due to efficient and effective government policies with rules and regulations. Some illegal poaching of animals occurs on the remote and less well-patrolled boundaries of this vast park. Along with poaching, regular floods in the monsoon season are the major concerns that result in the loss of precious wildlife.

Violet Cuckoo

Asian Emerald Cuckoo

Manas National Park

Assam

Manas River

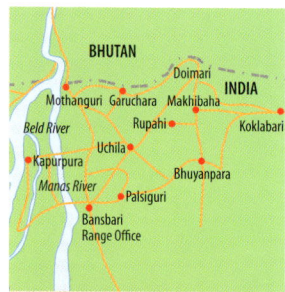

KEY FACTS

Nearest Major Towns
Barpeta and Bongai Gaon

Habitats
Tropical moist deciduous forest, tropical semi-evergreen forest, tropical grassland, riverine vegetation

Key Species
White-rumped Vulture, Slender-billed Vulture, Greater Adjutant, Bengal Florican, Spot-billed Pelican, Lesser Adjutant, Greater Spotted Eagle, Lesser Kestrel, Swamp Francolin, Rufous-necked Hornbill, Hodgson's Bushchat, Jerdon's Babbler, Slender-billed Babbler, Black-breasted Parrotbill, Grey-crowned Prinia, Bristled Grassbird, Finn's Weaver, Bengal Florican, Black-necked Stork, Cinereous Vulture, Red-headed Vulture, Great Hornbill, Pallid Harrier, Marsh Babbler

Other Specialities
Golden Langur, Clouded Leopard, Tiger, Asian Elephant, King Cobra, Hispid Hare, Pygmy Hog, Wild Buffalo

Best Time to Visit
October to March

This park, located in western Assam, is another World Heritage Site. It is a well-known wildlife reserve of north-east India, which was previously called the North Kamrup Wildlife Sanctuary. Manas carries a great reputation among birdwatchers and wildlife enthusiasts due to its unparalleled scenic beauty and endangered wildlife. An added advantage is the presence of the much bigger Royal Manas National Park across the border in Bhutan. For many species, it is a large, contiguous wilderness area.

Manas is home to several globally threatened birds and mammals like the Critically Endangered Pygmy Hog (*Sus salvanius*), and the White-rumped and Slender-billed Vultures. The terrain is mostly flat, gently sloping plain typical of bhabar and terai. There are some hilly patches of Bhutan Himalayas towards the north. Approximately half of Manas is savannah grassland, while the rest is moist deciduous and semi-evergreen forest. There are mainly three types of vegetation: tropical semi-evergreen forests in the northern part of the park; tropical moist deciduous forests (the most common type); and extensive alluvial grassland in the western area of the park.

Birdwatching Sites

The park is divided into three ranges – the western range at Panbari, the central range at Bansbari, and the eastern range at Bhuiyanpara. More than 400 bird species have been reported in these ranges, and this impressive number helped Manas to be recognized as an IBA.

Manas is known for most of the grassland bird species, such as the Swamp Francolin, Marsh, Slender-billed and Jerdon's Babblers, and Bristled Grassbird. Hodgson's Bushchat and the Vulnerable Finn's Weaver can also be seen during winter. One

Pygmy Hog

Bengal Florican

Sultan Tit

of the key attractions of Manas, the Bengal Florican, can be found at two locations – Kuribeel inside Manas National Park, and Koklabari Agricultural Farm, situated on the eastern boundary of Bhuiyanpara Range.

Access & Accommodation

Lokpriya Gopinath Bordoloi International Airport, located in Guwahati, is well connected to all major cities of India. The distance from the airport to Manas is about 138km, which can be covered in three hours. Taxis or cabs can be hired directly at the airport.

The park has many jungle resorts, camps and luxury hotels. There is also a forest lodge inside the park at Mathanguri, which is very popular among birdwatchers and wildlife enthusiasts. The number of visitors to Manas is fewer compared to those going to Kaziranga, but advance booking is recommended to avoid any last-minute disappointments.

Conservation

Manas was once very popular among visitors, but has suffered significantly due to insurgency-related problems since 1989. Between 1989 and 1992, most of the interior camps and bridges were burnt down, and Indian Rhinoceros and Swamp Deer population were almost wiped out. With the breakdown of the administration, professional poachers and timber smugglers took full advantage of the situation. Since 1995 the situation has improved, although there is still a long way to go to revive the original state.

Finn's Weaver

Black-tailed Crake

Nameri National Park

White-winged Wood Duck

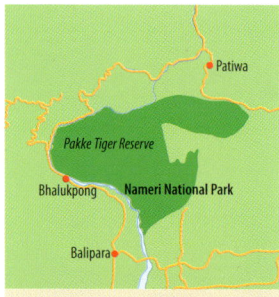

KEY FACTS

Nearest Major Towns
Sonitpur

Habitats
Tropical semi-evergreen forest, tropical moist deciduous forest, tropical grassland, riverine vegetation

Key Species
Slender-billed Vulture, White-bellied Heron, White-winged Wood Duck, Bengal Florican, Lesser Adjutant, Pallas's Fish Eagle, Greater Spotted Eagle, Eastern Imperial Eagle, Lesser Kestrel, Swamp Francolin, Masked Finfoot, Rufous-necked Hornbill, Hodgson's Bushchat, Marsh Babbler, Slender-billed Babbler, Grey-crowned Prinia, White-tailed Sea Eagle, Grey-headed Fish Eagle, Red headed Vulture, Rufous-throated Wren Babbler

Other Specialities
Asian Elephant, Tiger, Wild Buffalo, Pygmy Hog, Hispid Hare

Best Time to Visit
October to March

Also known as Nameri Tiger Reserve, this park is located in the Sonitpur district of northern Assam. The park is on the interstate border with Arunachal Pradesh, which brings an added advantage of a contiguous large wilderness for many endangered mammals and birds. Nameri is a very beautiful park in which you can see the snow-capped Himalayan peaks on clear winter days. The terrain in Nameri is gently sloping plain, typical of bhabar and terai. Towards the north, small, hilly promontories of the Arunachal Himalayas can be seen. There are many bheels (water pools) within the park.

Birdwatching Sites

Nameri is home to approximately 400 bird species, including some that are globally threatened like the White-winged Wood Duck. It is one of the very few places in the world where a noteworthy population of this species has been reported. Some other endangered bird species that have been reported here are the Crow-billed Drongo, Black-backed Forktail, Sultan Tit, Lesser Necklaced, Greater Necklaced and Rufous-necked Laughingthrushes, Grey Peacock Pheasant and Himalayan Golden-backed Woodpecker.

Most of the area inside the park is accessible on foot and recommended for birding. There are many watchtowers and bheels inside the park, where birdwatchers spend a lot of time viewing numerous resident and migratory birds. Potasali watchtower is a good place in which to see grassland birds, and also hill birds during winter. This is also a good place for seeing some of the migratory birds. Kurua Bheel and Borghulli Bheel are good places in which to see the White-winged Wood Duck and other endangered species. Nameri also offers a more leisurely option for birding, on a raft over the Jia Bhoreli. Here

Oriental Hobby

Ibisbill

you can see the elusive Ibisbill and Goosander (Merganser), along with many other migratory birds during winter.

Access & Accommodation
The nearest airport to reach Nameri is Tezpur, which is about 30km from the park. Tezpur is a small airport, so the frequency and connectivity of flights is not very good. Most travellers prefer to fly to the much bigger and better connected Guwahati Airport, then take a taxi to reach Nameri. The road journey from Guwahati Airport to Nameri takes about five hours.

Nameri has a range of accommodation options available to suit different budgets and preferences. There is also an eco camp managed by the Assam Angling and Conservation Association.

Conservation
Nameri is going through a series of conservation issues like encroachment, overgrazing of livestock, illegal fishing and poaching. Considerable commercial timber exploitation and intensive extraction of canebrakes have been taking place in Nameri. Due to poor administration, many cases involving human-animal conflicts have also been reported.

Ruddy KingFisher

Orang National Park

Indian Rhinoceros

KEY FACTS

Nearest Major Towns
Darrang and Sonitpur

Habitats
Tropical moist deciduous forest, wet savannah grassland, wetland

Key Species
Oriental Darter, White-rumped Vulture, Slender-billed Vulture, Greater Adjutant, Bengal Florican, Spot-billed Pelican, Lesser Adjutant, Baer's Pochard, Pallas's Fish Eagle, Swamp Francolin, Bristled Grassbird, Finn's Weaver, Black-necked Stork, Ferruginous Pochard, Grey-headed Fish Eagle, Red-headed Vulture, Blyth's Kingfisher

Other Specialities
Asian Elephant, Indian Rhinoceros, Tiger, Wild Buffalo, Gangetic Dolphin, Capped Langur

Best Time to Visit
October to March

Also known as Rajiv Gandhi National Park, Orang National Park is located in the Darrang and Sonitpur districts of Assam. It was declared a national park in 1999, and is well known as an important habitat for the Indian Rhinoceros. The park is situated on the northern bank of the Brahmaputra River. The Pachnoi and Dhansiri Rivers flow along its eastern and western boundaries respectively. Both these rivers are tributaries of the Brahmaputra. The terrain of the park is flat, being the floodplain of these rivers. There are two distinct alluvial terraces: the lower Orang of relatively recent origin along the Brahmaputra River, and the older upper Orang to its north, separated by a high bank traversing the park from east to west.

Birdwatching Sites

The grassland of the park supports healthy populations of the Swamp Francolin and Bengal Florican. Lesser Adjutants and Pallas's Fish Eagles nest in the park, and there are records of sightings of the Bristled Grassbird. Orang is one of the most important sites for birds of wet, tall grassland of the Indo-Gangetic Plain. Almost all species of conservation concern are found in this small park. An IBA site, it contains more than 225 bird species, including rarities such as Baer's Pochard, Blyth's Kingfisher and Finn's Weaver. Other important species found here are Spot-billed and White Pelicans, Greater and Lesser Adjutants, Black-necked and Woolly-necked Storks, Northern Pintail, Bengal Florican, Ruddy Shelduck, Gadwall, Mallard and Pallas's Fish Eagle.

The Forest Department conducts morning and evening jeep safaris in the park. There are watchtowers and checkposts inside the park where visitors are allowed to get down and experience some birding on foot.

Silver-breasted Broadbill

Orang landscape

Access & Accommodation

The park lies 18km off the national highway from Orang town and 15km off the highway from Dhansirimukh town. The distance from Guwahati to Orang is 140km. The nearest airport is in Tezpur, about 60km from the park. However, for better connectivity, most travellers choose Guwahati Airport to reach Assam. The nearest railway stations are Saloni and Rangapara railway stations. They are well connected by road to the park. The park is also well connected to major cities and other places by the road network. Government and privately operated vehicles go to Orang at frequent intervals. Hiring a private taxi from Tezpur or Orang is recommended for hassle-free travel.

For accommodation, there is one forest guest house near the gate. There is also a tourist lodge that is maintained by the forest department. Advance reservations for stays at these properties are required. Private hotels and guest houses are available in Orang town; for better accommodation visitors can arrange to stay in Tezpur.

Conservation

Erosion by the Dhansiri River, occasional attempts by the surrounding villagers to encroach, and an increasing cattle population in the fringe areas are some of the key conservation issues. Rhino poaching is a constant threat. It is necessary to post highly motivated officers to bring back the old glory of Orang. An environmental awareness campaign should be started in the surrounding villages. A more detailed study on the birdlife, especially threatened species, should be conducted.

Pallas's Fish Eagle

Little Pied Flycatcher

Dibru-Saikhowa National Park

Dibru-Saikhowa wetlands

KEY FACTS

Nearest Major Towns
Tinsukia, Dibrugarh and Dhemaji

Habitats
Tropical moist deciduous forest, wetlands

Key Species
Oriental Darter, White-rumped Vulture, Slender-billed Vulture, White-bellied Heron, Greater Adjutant, White-winged Duck, Bengal Florican, Spotted Greenshank, Spot-billed Pelican, Lesser Adjutant, Baer's Pochard, Pallas's Fish Eagle, Greater Spotted Eagle, Swamp Francolin, Sarus Crane, Masked Finfoot, Pale-capped Pigeon, Indian Skimmer, Jerdon's Babbler, Black-breasted Parrotbill, Finn's Weaver, Black-necked Stork, Ferruginous Pochard, White-tailed Sea Eagle, Grey-headed Fish Eagle, Cinereous Vulture, Black-bellied Tern, Blyth's Kingfisher, Great Hornbill, Swamp Prinia, Black-breasted Parrotbill, Marsh Babbler

Other Specialities
Feral horses, Wild Buffalo

Best Time to Visit
October to March

This beautiful park is located in the Tinsukia district of Assam. It is one of the 19 biodiversity hotspots in the world and also a biosphere reserve. Situated in the flood plains of the Brahmaputra River, the park is bounded by the Brahmaputra River and Arunachal Hills in the north, and by the Dibru and Patkai Hills in the south. It mainly consists of semi-wet evergreen forests, tropical moist deciduous forests, bamboo and grassland.

Dibru-Saikhowa is a safe haven for many very rare and endangered species of wildlife, including avifauna that is both endangered and migratory, as well as various shrub, herb and rare medicinal plant species. Though the park was primarily established for the conservation of the White-winged Wood Duck in its natural habitat, it is also famous for its brightly coloured feral horses.

Birdwatching Sites

Dibru-Saikhowa is very rich in birdlife, with more than 350 species already identified. It is one of the sites in the north-east where the highly endangered and elusive White-bellied Heron is seen. The two Critically Endangered vultures have also been recorded from this region, but both are now very rare. Two more endangered birds that could have significant populations in this IBA are the White-winged Wood Duck and Bengal Florican, the latter a rare resident of the grassland. Another endangered bird of which there are few confirmed records from India is Nordmann's Greenshank. Earlier there were also records of the Masked Finfoot, but there have been no recent sightings. The Black-breasted Parrotbill and Marsh Babbler are also seen in this

Marsh Babbler

area. This is perhaps the best place to see three Indian specialties: Marsh and Jerdon's Babblers, and the Swamp Prinia.

Some other important birds of this region are the Spot-billed Pelican, Baer's Pochard, Pallas's Fish Eagle, Greater Spotted Eagle, Swamp Francolin, Pale-capped Pigeon and Black-breasted Parrotbill.

Maguri Bheel This is a large wetland located 3.8km from Guijan Ghat, the gateway of the Dibru-Saikhowa National Park and Biosphere Reserve. It is home to some of the rarest of birds and attracts varied species from around the globe, as a result of which it has been declared an IBA. Some of the migratory bird species visiting the bheel include the Ruddy Shelduck, Baikal Teal, Chinese Spot-billed Duck, Bar-headed Goose, Falcated and Ferruginous Ducks, Northern Pintail, Eurasian Wigeon, Common Teal, Black-headed and Glossy Ibises, and Eurasian Curlew.

Access & Accommodation
The nearest airport is Mohanbari (Dibrugarh) Airport, which is located about 40km from the park. The airport is well connected by road to Dibru-Saikhowa National Park. The nearest railway station is Dibrugarh station. This place is well connected to major cities by a network. Buses can be taken to the town of Tinsukia, which is well connected to Dibrugarh town by NH 37, and the distance is roughly 55km.

There are many good accommodation options available here. Banashree Eco Camp (http://banashree.com), Dibru-Saikhowa Eco Camp (www.dibrusaikhowaecocamp.com) and Padmini Resort (www.padminiresort.com) are some of the well-known options among birdwatchers and wildlife lovers.

Conservation
The main conservation issue is erosion, which is turning the park into an island. Large areas have been eroded by the Lohit River. The next major issue is the presence of large forest villages inside the park. Most of the forests have been heavily exploited for timber, and very little dense forest remains. Fishing activities cause a considerable amount of disturbance, and there is heavy grazing pressure from domestic livestock. The peripheral areas are being reclaimed for agriculture, and there has been considerable encroachment by the forest villagers. A large portion of the area is under threat because of a natural shift in the course of the Brahmaputra River.

Rufous-gorgeted Flycatcher

Grey-headed Lapwing

Jerdon's Bushchat

Dehing Patkai Wildlife Sanctuary

Dehing Patkai forests

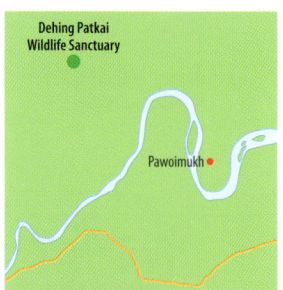

This sanctuary is located in the Dibrugarh and Tinsukia districts of Assam. It forms the largest stretch of tropical lowland rainforests in India. The area has several historic attractions, including Second World War cemeteries, the Stillwell Road and the Digboi Refinery, the oldest in Asia. The sanctuary consists of three parts: Jeypore, Upper Dihing River and Dirok rainforest. It was declared a sanctuary on 13 June 2004, and is also a part of Dehing-Patkai Elephant Reserve. The vegetation of the site is mainly of the tropical wet evergreen rainforest type.

Birdwatching Sites
Dehing Patkai is a paradise for any birdwatcher who is on a quest to locate some infrequently seen bird species. More than 300 bird species have been reported here, including threatened birds such as the White-bellied Heron and Pale-capped Pigeon. The Buffy Fish Owl has also been recorded from this site. Some interesting species of various biomes are the White-cheeked Hill Partridge, Grey Peacock Pheasant, Masked Finfoot, Mountain Bamboo Partridge, Pied Falconet, Crested Goshawk, Pale-headed Woodpecker, Crow-billed Drongo, Black-faced Laughingthrush, White-crowned Forktail, Green Cochoa, Violet Cuckoo and Blue-eared Kingfisher.

Access & Accommodation
You can reach the sanctuary from either Tinsukia (45km) or Dibrugarh. The latter has an airport that connects with all major cities in India. The nearest railway station for the park is Naharkatia, which is about 8km away. Hiring a vehicle that will be at your disposal is recommended, as most locations here are remote, and public commute options are limited and time consuming.

KEY FACTS
Nearest Major Towns
Dibrugarh and Tinsukia

Habitats
Tropical wet evergreen forest

Key Species
Red Jungleflow, Kalij Pheasant, Spotted Dove, Lesser Adjutant, Cattle Egret, Indian Pond Heron, Crested Serpent Eagle, Asian Barred Owlet, Red-headed Trogon, Oriental Pied Hornbill, Greater Golden-backed Woodpecker, Lesser Yellow-naped Woodpecker, Rose-ringed Parakeet, Large Woodshrike, Common Iora, Long-tailed Minivet, Scarlet Minivet, Grey-backed Shrike, Black-hooded Oriole, Maroon Oriole, Black Drongo, Ashy Drongo, Bronzed Drongo, Striated Swallow, Sultan Tit, Cinereous Tit, Red-vented Bulbul, Red-whiskered Bulbul, White-throated Bulbul, Common Tailorbird, Large Niltava, Small Niltava, Scaly Thrush, Chestnut Thrush, Common Hill Myna

Other Specialities
Capped Langur, Leopard Cat, Jungle Cat, Asian Elephant, Spotted Deer

Best Time to Visit
October to March

Sapphire Flycatcher

Oriental Bay Owl

Hodgson's Hawk Cuckoo

There is a Forest Inspection Bungalow at Digboi, contact Divisional Forest Officer, Digboi Division, Digboi, Tel.: (03751)264433. The private lodges are Digboi Tourist Lodge/Restaurant, Golai No.1, Digboi, P.O. Digboi, Tel.: 9435169247, Central Transit Accommodation, I.O.C. Digboi, Tel.: (03751)264025. There are also a few homestay and eco lodges near the sanctuary. It is recommended that visitors hire a local bird guide for the best birding experience.

Conservation

Many of the waterbodies near the oil rigs are heavily polluted, posing a serious threat to the environment. Illegal felling of trees, encroachment by forest villagers and surrounding villagers, and poaching, including the collection of eggs and ducklings of the White-winged Wood Duck, are other major issues. There is an urgent need to control poaching, for better management of the reserve's forests, and for environmental awareness among villagers and the people from the oil refineries. The White-winged Wood Duck could become a focal point for the conservation of these superb tropical moist forests.

Grey Peacock Pheasant

Bay Woodpecker

Guwahati

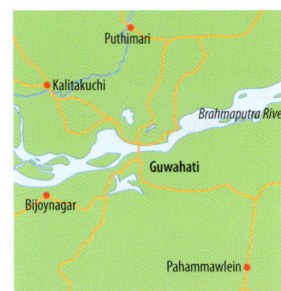

Greater Adjutants at Guwahati rubbish dump

The capital city of the northeastern state of Assam, Guwahati was formerly known as Pragjyotishpur (the City of Eastern Light). Its origin is derived from two Assamese words, *guwa*, meaning areca nut, and *haat*, meaning a market. Wedged between the picturesque hills of the eastern flanks of the Himalayan mountain range, Guwahati houses the political capital of the state, Dispur. Guwahati is an attraction in itself and also acts as the gateway to the seven other northeastern states. It is very rich in biodiversity, with a considerable green belt still intact, but information regarding the diversity and distribution of avifauna in the reserve forests located inside Guwahati is not well documented. Some of the available reports have confirmed that there is a good number of endangered and endemic animal species in and around the city – a number that is comparatively higher than for other Indian cities.

Birdwatching Sites

Guwahati is one of those places where birdwatchers and photographers may experience great birding experiences while staying in comfortable hotels or resorts. Some of the important birding areas near Guwahati are Deepor Bheel, Garbhanga Reserve Forest and Pobitora Wildlife Sanctuary. Deepor Bheel is very near to Guwahati Airport and can offer some great birding sessions during the winter season, when there are migrants from all over the globe in the area. The wetland of Deepor Bheel was once also the haunt of the rare Goliath Heron, but there have been no confirmed sightings of it in recent years.

KEY FACTS

Nearest Major Towns
Guwahati

Habitats
Urban, mixed forests, wetland

Key Species
Common Pigeon, Spotted Dove, Eurasian Collared Dove, Asian Palm Swift, Asian Openbill, Little Cormorant, Black Kite, Indian Roller, Ashy Woodswallow, Long-tailed Shrike, Black Drongo, House Crow, Large-billed Crow, Red-vented Bulbul, Asian Pied Starling, Common Myna, House Sparrow, Oriental Skylark, Greater Adjutant, Great Slaty Woodpecker, Hooded Pitta, Siberian Rubythroat, Green Cochoa, White-browed Piculet

Other Specialities
Capped Langur, Asian Palm Civet, Jungle Cat, Spotted Deer, Asian Elephant, Rhesus Macaque

Best Time to Visit
Throughout the year

Greater Adjutant

Garbhanga Reserve Forest This represents part of the greenbelt region around Guwahati. For the residents of the city of Guwahati, the forest acts as a buffer against the rising pollution and CO_2 levels. Garbhanga Reserve Forest is a hilly area adjoining the hill range of Meghalaya. It covers an area of 14.6km². The forest is bordered on the eastern and northern sides by the Meghalaya hilly ranges, on the western side by the Rani Reserve Forest, and on the southern side by Guwahati city and Deepor Bheel, which are connected with the forest. The water resources are mainly the hill streams coming down from Meghalaya.

Some notable species from this area are the Pied Harrier, Oriental Honey Buzzard, Red-headed Trogon, Rufous-necked Hornbill, Great Slaty Woodpecker, Hooded Pitta, Siberian Rubythroat, Green Cochoa, White-bellied Erpornis, Common Hill Myna, Asian Emerald Cuckoo, White-browed Piculet and Greater Racket-tailed Drongo.

Deepor Bheel Bird Sanctuary Located in the western boundary of Guwahati city, Deepor Bheel Bird Sanctuary is the only Ramsar Site in the state. It is a picturesque wetland comprising 4.14km². Considered one of the largest beels in the Brahmaputra Valley of Lower Assam, it is categorized as a representative of the wetland type under the Burma monsoon forest biogeographic region.

More than 200 bird species have been recorded from the Deepor Bheel area, including some that are Critically Endangered. The major avian species are the Greater Adjutant, Asian Openbill, Lesser Whistling-duck, Northern Shoveler, Northern Pintail and Garganey, and there is a high concentration of Pheasant-tailed Jacanas. The Goliath Heron, too, has been seen here.

Pobitora Wildlife Sanctuary Pobitora or Pobitora Wildlife Sanctuary is a wildlife reserve in the Morigaon district, about 30km east of Guwahati. It is high on the agenda of visitors to north-east India. The sanctuary is known for its dense population of Indian Rhinoceroses. Pobitora can be divided into three distinct categories: forest, grassland, and waterbodies or beels. Pobitora grassland and wetlands harbour most of the representative birds of the Brahmaputra floodplains, such as Striated Babbler and Finn's Weaver. The wetlands of Pobitora attract thousands of waterfowl during winter. Some of the important birds recorded here are Greater and Lesser Adjutants, Spot-billed Pelican, Pallas's Fish Eagle, Greater Spotted Eagle, Swamp Francolin and Marsh Babbler.

Access & Accommodation

Lokpriya Gopinath Bordoloi International Airport in Guwahati is most convenient for all travellers to this state. It is very well connected to the major cities of India, like Delhi, Mumbai, Kolkata and Bangalore. It also operates flights from international destinations like Bangkok and Paro. For visitors travelling to Guwahati by train, Guwahati junction operates regular trains from different parts of India. There is also good connectivity via the state as well as the national highways.

There are various accommodation options available in Guwahati, from budget hotels and lodges, to the most luxurious brand hotels. You can hire a vehicle to travel to nearby birding destinations. App-based cabs like Ola and Uber are also available in the city.

Conservation

Guwahati has witnessed many changes, such as a rapid increase in population, depletion of forest cover, and the spread of diseases, which resulted in many environmental problems linked to land, air, water and the society. Most of the changes have taken place due to the effects of changing the natural environment and the tremendous population growth. Guwahati is bounded on three sides by hills and on the remaining side by the mighty Brahmaputra River, hence the horizontal expansion is restricted; as a consequence, many multi-storied buildings have been built in recent years just to accommodate the ever-increasing population. Deforestation, soil erosion, hill-slope destabilization, siltation in low-land areas and encroachment on natural resources are some of the long-term and ongoing issues.

Rufous-bellied Niltava

GOA

Goa is a tiny coastal state located in the Konkan region of the central west coast of India. It is wedged between the Arabian Sea on the west and the Western Ghats (Sahyadri range) on the east. The Western Ghats, which runs north–south, forms the entire eastern boundary, where evergreen and semi-evergreen forests provide a habitat to many threatened and restricted range bird species. Moist deciduous forests, mostly secondary and degraded, occur along the foothills of the Sahyadris. Goa is India's richest state, and tourism is its leading industry, attracting 12 per cent of all international travellers to India. Tourists visit the beaches and archaeological sites, and many come for the rewarding birdwatching experience.

Climate

The climatic conditions of Goa are influenced by two global biomes – the marine biome of the Arabian Sea and the terrestrial forest biome of the Western Ghats. Physiographically, Goa can be divided into three ecological zones:

Coastal plains Coastal belt with sandy beaches broken by the wide mangroves of the Mandovi and Zuari Rivers.

Middle plateau The midland region, distinguished by large lateritic tablelands with stony outcrops, and marked by thorny scrubland and plantations.

Western Ghats region The Sahyadris of the Western Ghats, covered by mixed moist deciduous forests, and semi-evergreen and evergreen forests interspersed with bamboo and canebrakes.

Goa has a calm tropical climate. Temperatures range from 16 °C to 34 °C in winter, and from 27 °C to 37 °C in summer. The average annual rainfall is 3,100mm, received mainly in June–September during the south-west monsoon period.

KEY BIRDS
Top 10 Birds
1. Blue-eared Kingfisher
2. Sri Lanka Frogmouth
3. Nilgiri Wood Pigeon
4. Malabar Parakeet
5. Vigors's Sunbird
6. Malabar Barbet
7. Malabar Whistling Thrush
8. Black-throated Munia
9. Oriental Dwarf Kingfisher
10. Malabar Trogon

1. Bhagwan Mahavir Wildlife Sanctuary
2. Cotigao Wildlife Sanctuary
3. Bondla Wildlife Sanctuary
4. Carambolim Lake
5. Zuari River

Access, Transportation & Logistics

Goa is a well-established tourist destination and is well connected with the rest of India via airways, railways and roadways. Dabolim Airport near Panjim is the easiest entry point into Goa for both foreign and Indian visitors. The main railway stations in Goa are located in Madgaon and Vasco da Gama. These are well linked to Mumbai and other major parts of India. A Mumbai to Goa train journey via the Konkan Railway is much more comfortable than going by bus, and the scenery is spectacular. For getting around in Goa, hired cabs are the preferred means of transport. Many taxi drivers are familiar with key birding sites.

Health & Safety

There are no major health concerns in Goa. It is a low- to no-risk area of India when it comes to malaria, but do take precautions especially after the monsoon. Visit a GP or travel clinic 6–8 weeks before departure to make sure you are up to date with any vaccinations. Consult a doctor and carry your medicines with you. Avoid too much spicy food if you are unfamiliar with it. Always keep some medicine handy for indigestion. Goans are friendly people so befriend them and enjoy their culture and traditions.

Goan coastal scene

Birdwatching Highlights

Goa's diverse ecosystems, comprising coastal areas, mangroves, estuarine grassland, wetland and Western Ghats habitats, favour a great diversity of bird species. The bird list consists of more than 400 species, including 17 that are endemic or near endemic to the region. Key birds include the Red Spurfowl, Malabar Trogon, Nilgiri Wood Pigeon, Malabar Parakeet, Malabar Grey and Malabar Pied Hornbills, Sri Lanka Bay Owl, Mountain Imperial Pigeon, Sri Lanka Frogmouth, Grey-headed Bulbul, White-bellied Woodpecker, White-bellied Blue Flycatcher, eight kingfisher species including Collared and Blue-eared Kingfishers, and a host of shorebirds and surprise rarities.

Vernal Hanging Parrot

Sri Lanka Frogmouths

Jungle Nightjar

Bhagwan Mahavir Wildlife Sanctuary

Stork-billed Kingfisher

This sanctuary is located in the Sanguem taluka (administrative division) on the eastern border of Goa, along the Western Ghats. It was earlier called the Mollem Game Sanctuary but was renamed in 1976 after the great saint, Bhagwan Mahavir. Due to excessive tree growth in this region, the canopy is almost closed and the availability of grass is very limited. Evergreen vegetation is mainly seen at higher altitudes and along riverbanks. The main vegetation types are west coast tropical evergreen forests, west coast semi-evergreen forests and moist deciduous forests.

KEY FACTS

Nearest Major Towns
Panjim and Mollem

Habitats
Tropical wet evergreen forest, tropical moist deciduous forest, tropical semi-evergreen forest

Key Species
Nilgiri Wood Pigeon, Malabar Parakeet, Malabar Grey Hornbill, Grey-headed Bulbul, White-bellied Treepie, Malabar Trogon, Malabar Pied Hornbill, White-cheeked Barbet, Malabar Barbet, Indian Scimitar Babbler, Loten's Sunbird, Blue-eared Kingfisher, Black-capped Kingfisher, Stork-billed Kingfisher, Cinnamon Bittern, Jerdon's Nightjar, Savanna Nightjar, Sri Lanka Bay Owl

Other Specialities
Tiger, Indian Leopard, Rhesus Macaque, Spotted Deer, Gaur, Sloth Bear, Slender Loris, Indian Jackal

Best Time to Visit
Throughout the year

Blue-eared Kingfisher

Lesser Golden-backed Woodpecker

Bhagwan Mahavir Wildlife Sanctuary

Birdwatching Sites

Nearly 200 bird species have been reported in the sanctuary, including Western Ghats endemics. The Malabar Pied Hornbill is found all over the sanctuary. The White-bellied Woodpecker can be seen in most parts of the sanctuary, and was seen nesting in the Dudhsagar area. Other key species to be seen here are the Sri Lanka Frogmouth, Grey, Jerdon's and Savanna Nightjars, Nilgiri Wood Pigeon, Malabar Parakeet, Malabar Grey Hornbill, Grey-headed Bulbul and Crimson-backed Sunbird. The White-bellied Treepie has been reported from the Dudhsagar area, a popular tourist and birdwatching destination.

Tambdi Surla This is located quite near the Bhagwan Mahavir Wildlife Sanctuary. It includes a twelfth-century Shaivite temple of the Lord Mahadeva, which is an active place of Hindu worship – it is the only remaining temple of the Kadamba period. The area is also famous for birding. Some of the key species found here are the Blue-eared Kingfisher, Grey-fronted Green Pigeon, Asian Fairy-bluebird, Malabar Barbet, Vernal Hanging Parrot, Malabar Parakeet, Mountain Imperial Pigeon and Greater Racket-tailed Drongo.

Access & Accommodation

Bhagwan Mahavir Wildlife Sanctuary is well connected by road to other parts of Goa, and you will find plenty of private and government-operated buses and vehicles going to the park at frequent intervals. Panaji is about 70km from the park, and you can book a private car directly from here. Kulem or Collem is the town nearest to the place, about 6km away. The nearest railway station to the sanctuary is Collem or Kulem railway station, from where you can head to the park by car. Some of the main trains, like the Amravati Express, Goa Express and Poorna Express, stop here.

Some accommodation options specialize in handling birdwatchers or are located in suitable areas for birdwatching. These include Backwood Camps (email: backwoodsgoa@hotmail.com) and Avocet & Peregrine (www.avocet-peregrine.com) +91-7588927753.

Conservation

At present, there is no major threat to the sanctuary, and there are very few settlements inside it. There is a need for a more focused attempt to bring this birdwatcher's paradise onto the main map. Some plans have been announced by the government to build an all-weather tar road for visitors.

Cotigao Wildlife Sanctuary

Cotigao landscape

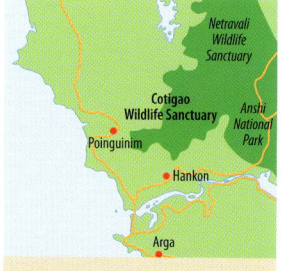

KEY FACTS

Nearest Major Towns
Panjim

Habitats
Tropical wet evergreen forest

Key Species
Nilgiri Wood Pigeon, Malabar Parakeet, Malabar Grey Hornbill, Grey-headed Bulbul, White-bellied Blue Flycatcher, Jerdon's Nightjar, Malabar Trogon, Malabar Pied Hornbill, White-cheeked Barbet, Yellow-browed Bulbul, Malabar Whistling Thrush, Indian Scimitar Babbler, Loten's Sunbird, Black-throated Munia, Malabar Barbet, Malayan Night Heron, Brown Fish Owl, Crimson-backed Sunbird, Dark-fronted Babbler, Flame-throated Bulbul, Mountain Imperial Pigeon, Green Imperial Pigeon

Other Specialities
Indian Civet, Indian Porcupine, Indian Pangolin, Wild Pig, Gaur, Slender Loris

Best Time to Visit
Throughout the year

This sanctuary in the Canacona taluka is 10km south-east of Chaudi. The sanctuary lies partly in the Western Ghats. It was established in 1969 to protect a remote and vulnerable area of forest lining the Goa-Karnataka interstate border. The terrain is hilly and includes undulating uplands. Many small rivulets flow through the sanctuary in the monsoon but dry up in summer. The main vegetation types are west coast tropical evergreen forests, west coast semi-evergreen forests and moist deciduous forests. The evergreen forests are mainly present on higher altitudes and on riversides.

Birdwatching Sites
A special feature of the sanctuary is a treetop watchtower positioned 25m above a watering hole where animals go to drink. The best times to visit the watchtower are dawn and dusk, when many birds and other animals are most likely to be visiting. Some of the key bird species reported from this area are the Nilgiri Wood Pigeon, Malabar Parakeet, Grey-headed Bulbul, White-bellied Blue Flycatcher, White-bellied Treepie, Green-billed Malkoha, Jerdon's Nightjar, Malabar Trogon, Malabar Pied Hornbill, White-cheeked and Malabar Barbets, Yellow-browed Bulbul, Malabar Whistling Thrush and Indian Scimitar Babbler.

Access & Accommodation
Cotigao Wildlife Sanctuary is situated in the Canacona taluka, about 2km from Poinguinim, which is 10km away from Chaudi, the main town of Canacona on NH 17. You can hire local transport from Poinguinim to reach the sanctuary. The nearest railway station is Canacona, which is located about 10km away. There are various accommodation options available around the sanctuary.

Malabar Parakeet

Conservation

There are no major conservation issues in Cotigao Wildlife Sanctuary. It is a fully notified wildlife sanctuary. There are some minor issues like the grazing of livestock inside the sanctuary, and there have also been some incidents of encroachment in the fringes.

Niranjan Sant

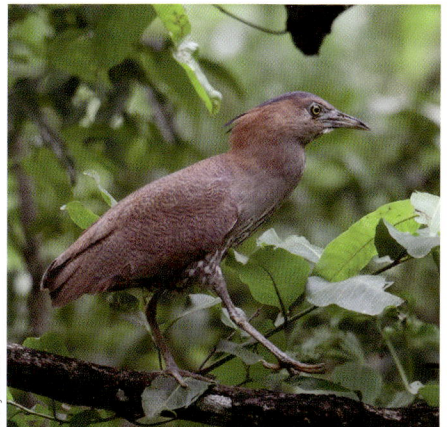

Malayan Night Heron

Nikhil Devasar

Black-crowned Night Heron

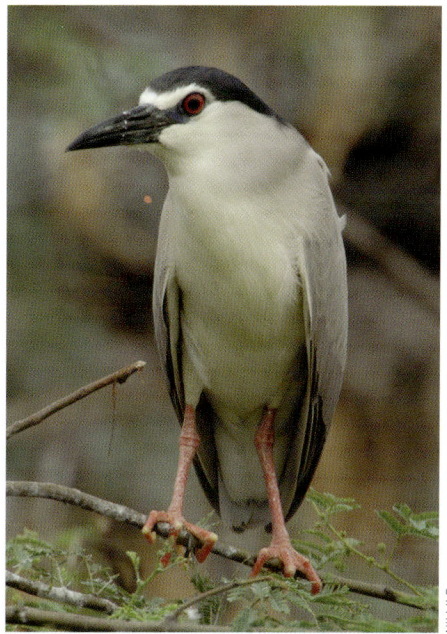

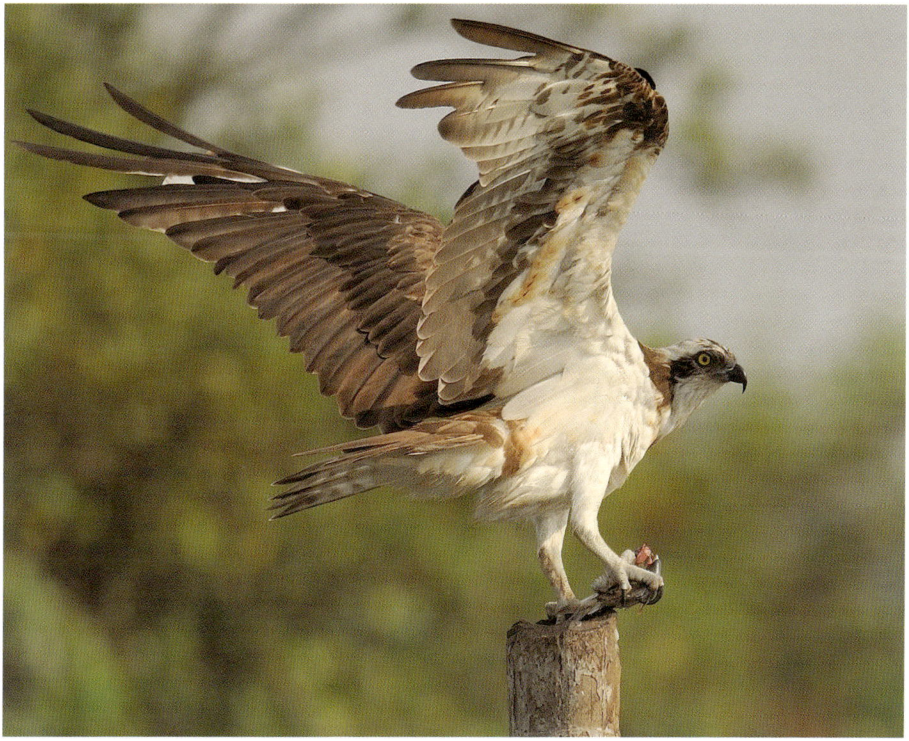

Bhanu Singh

Osprey

Bondla Wildlife Sanctuary

Bondla River

KEY FACTS

Nearest Major Towns
Panjim and Ponda

Habitats
Moist deciduous forests, evergreen forests

Key Species
Grey Junglefowl, Spotted Dove, Asian Koel, Little Egret, Crested Treeswift, Black Drongo, Jungle Babbler, Jungle Myna, Orange-headed Thrush, Asian Emerald Dove, Crested Serpent Eagle, Greater Flameback, Bar-winged Flycatcher-shrike, Yellow-browed Bulbul, Dark-fronted Babbler, Ultramarine Flycatcher, White-rumped Shama, Purple Sunbird, Nilgiri Flowerpecker, Oriental Dwarf Kingfisher, Malabar Trogon, Crested Hawk Eagle, Red Spurfowl, Blyth's Starling

Other Specialities
Gaur, Indian Leopard, South-western Langur, Indian Chevrotain, Rhesus Macaque, Wild Pig, Malabar Giant Squirrel

Best Time to Visit
October to March

This sanctuary lies 50km east of Panjim and is a short drive from Bhagwan Mahavir Wildlife Sanctuary and Mollem National Park. The sanctuary is only 8km² in area but is covered with good mixed forests at the foothills of the Western Ghats. More than 150 bird species have been recorded from this small area. Bondla is a paradise for the botanist. At the start of the monsoon, rare and endangered species of wild orchid flourish here, and many enthusiasts and study groups visit to see them and collect samples.

Birdwatching Sites

Bondla does not offer jeep safaris, and most visitors arrive in private vehicles. Covering this area on foot as much as possible is recommended for the best birding experience. Walking on the uphill road is slightly strenuous, but it snakes through a forest that has a varied palette of colours and an extraordinary birdlife. Some of the forest birds here include the Black Eagle, Nilgiri Wood Pigeon, Malabar Trogon, Green-billed Malkoha, Grey Junglefowl, Orange-headed Thrush, Indian Pitta, Asian Fairy-bluebird, Forest Wagtail, Brown-cheeked Fulvetta, woodpeckers, barbets, orioles, wood shrikes, parakeets, flycatchers and many more.

Access & Accommodation

Bondla Wildlife Sanctuary is located just 35km north-east of Madgaon and 50km east of Panjim. The best way to reach the park is by taxi, which can be hired from either of these places. You can also cover Bondla along with Bhagwan Mahavir Wildlife Sanctuary, which is just an hour's drive from this place.

Vigors's Sunbird

Rajan Hatiskar

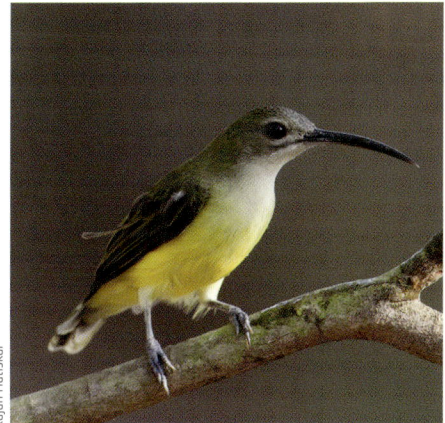

Little Spiderhunter

Clement M Francis

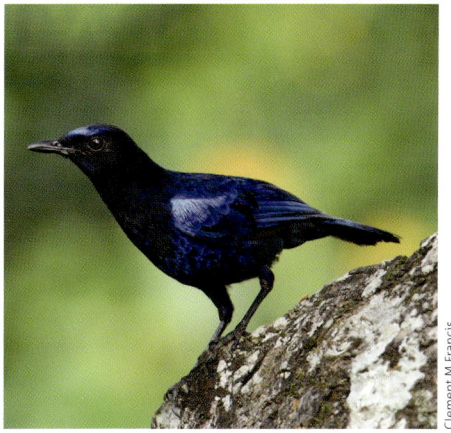

Malabar Whistling Thrush

The best place to stay inside Bondla Wildlife Sanctuary is its Eco Complex, where two dormitories can accommodate 48 people and seven ecotourism cottages can accommodate 14 guests. There is also one canteen that serves vegetarian and non-vegetarian meals. It is essential to make advance bookings (contact 0832-2935800; website: www.goatourism.gov.in) before your arrival. Nature's Nest Camp (naturesnestgoa.com) also arranges a guided birdwatching day tour.

Conservation

There are no major conservation issues in the Bondla area as this is a small yet notified sanctuary. However, large-scale quarrying on the fringe of Bondla is one conservation issue that is affecting the birdlife of Bondla. Birds and other wild animals are naturally affected, as there is a lot of blasting in the basalt quarries. The blasts produce heavy noise and vibration, which drives the wild animals from their natural habitat.

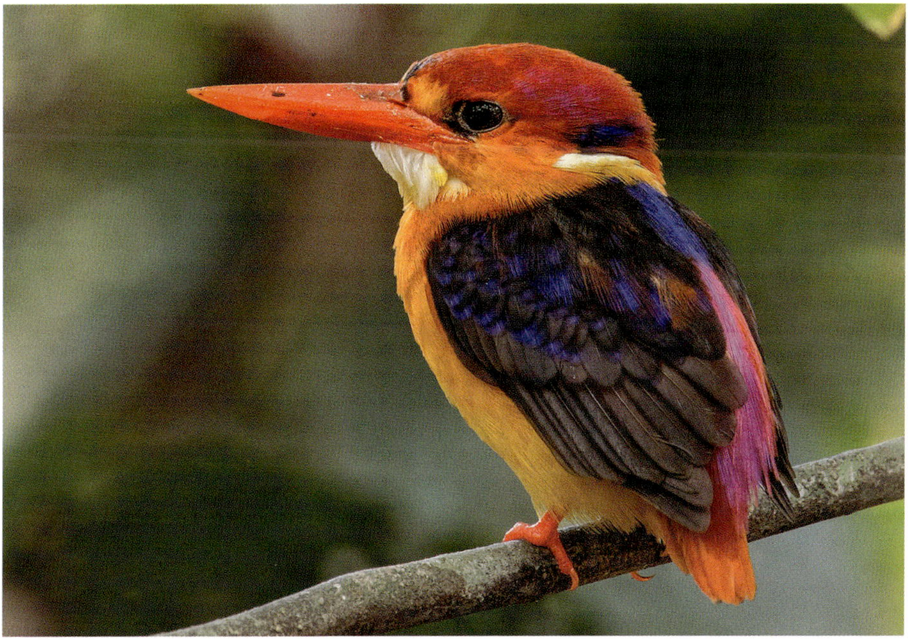

Rahul & Khushboo Sharma

Oriental Dwarf Kingfisher

Carambolim Lake

Carambolim Lake

This lake lies about 12km from Panjim, the capital of Goa. It is spread over an area of 0.07km², of which 0.4km² is under water for most of the year. The western embankment of the lake serves as the base for the Konkan railway track, while the eastern side has been partially cleared of its scrub forests, and mango and cashew groves, to pave the way for the construction of residential buildings. Coconut and mango trees fringe the remaining sides of the lake. The water depth is 1.5–3m, depending on the season. The lake lies between the estuarine zone of the Mandovi and Zuari Rivers. It is large, marshy and lotus covered, and is a good place to catch up with Indian waterbirds and wintering wildfowl.

KEY FACTS

Nearest Major Towns
Panjim

Habitats
Freshwater swamp and wetland

Key Species
Long-billed Vulture, Lesser Adjutant, Greater Spotted Eagle, Indian Peafowl, Spotted Dove, Greater Coucal, Grey-headed Swamphen, White-breasted Waterhen, Intermediate Egret, Cattle Egret, Indian Pond Heron, Black Kite, Small Minivet, Red-vented Bulbul, Jungle Myna, Purple Sunbird, Brahminy Kite, Spotted Owlet, Oriental Darter, Bronze-winged Jacana, Red-wattled Lapwing, Asian Koel, Great Egret, Pheasant-tailed Jacana, White-bellied Sea Eagle, Osprey, Brown Hawk Owl

Other Specialities
Rhesus Macaque, Jungle Cat

Best Time to Visit
Throughout the year

Black-headed Munia

Indian Roller

Brown-headed Gull

Birdwatching Sites

Carambolim Lake attracts thousands of birds, especially waterfowl. Some of the common birds seen in the area are kingfishers, terns, cormorants, jacanas, lapwings, storks, pratincoles and other common waders. Regular wildfowl include the Lesser Whistling Duck, Cotton Pygmy-goose, Garganey, Northern Pintail, Northern Shoveler, and Comb and Indian Spot-billed Ducks. It is also a good place for raptors like the Western Marsh Harrier, Greater Spotted, Indian Spotted, Tawny and Booted Eagles, and Osprey. The woodland holds various species of wintering warblers, woodpeckers, barbets, orioles and cuckoos. The scrub holds wagtails, pipits, munias, larks, starlings and similar birds.

Access & Accommodation

The best way to reach the lake is by hired taxi or motorbike from Panjim or Madgaon. The lake is located about 15km from Panjim and about 30km from Madgaon.

There are various accommodation options in Panjim and Madgaon. Visitors may choose one depending upon their needs and budgets. Visiting is recommended during the winter months, when many migratory birds arrive from all over the place.

Conservation

There are no major conservation threats in the region, but there are some common issues like deforestation, illegal cutting of trees and intense grazing pressure. The Konkan Railway Corporation Ltd planted a large number of trees on the track embankments to muffle the sound of the moving trains and declared the area a 'no hooting zone'. These steps have helped in restoring the birdlife of this important Goan IBA. The long-term and irreversible disturbance is from private encroachment, night soil generated by the migrant human population and silt deposition, leading to the development of marshy conditions.

Jerdon's Nightjar

Zuari River

White-bellied Sea Eagle

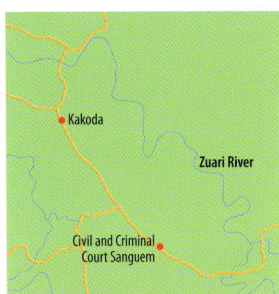

KEY FACTS

Nearest Major Towns
Panjim

Habitats
Wetland

Key Species
Common Pigeon, Spotted Dove, Red-whiskered Bulbul, Red-rumped Swallow, White-spotted Fantail, Indian Golden Oriole, Common Kingfisher, Black-headed Ibis, Black-crowned Night Heron, Rufous Treepie, Collared Kingfisher, White-bellied Sea Eagle, Greater Crested Tern, Lesser Crested Tern, Peregrine Falcon, Collared Kingfisher, Black-capped Kingfisher, Stork-billed Kingfisher, Woolly-necked Stork, Eurasian Curlew, Terek Sandpiper, Malabar Grey Hornbill, Wire-tailed Swallow, Western Reef Egret, Striated Heron

Other Specialities
Marsh Crocodile, Dolphin

Best Time to Visit
Throughout the year

The Zuari River is the largest river in Goa. Along with the Mandovi River, it forms the backbone of the river system of Goa. The two rivers are connected by the Cumbarjuem Canal, which enables navigation using a ferry that connects different islands. The canal is tidally controlled so the water is saline in nature. The mangrove habitat along the canal supports the varied and rich birdlife in this region.

Birdwatching Sites

The best way to do birding in Zuari River is by boat. The boat ride here is well managed and provides a great way to appreciate Goa's mangrove habitat. The main attraction of the ride is that it enables you to get closer to some of the birds, and can allow you to take good photographs with a clean green background. Some of the birds found here are different varieties of kingfisher (like Collared, Black-capped, Stork-billed and Common Kingfishers), Osprey, White-bellied Sea Eagle, Brahminy Kite, Lesser and Greater Crested Terns, Common Greenshank, Lesser Adjutant and Peregrine Falcon.

Access & Accommodation

The boat safari in Zuari River starts from Cortalim Fishing Jetty, Cortalim, which is located about a 50km (one-hour) drive from Bhagwan Mahavir Wildlife Sanctuary and around 15km from Panjim. The boat ride needs to be pre-booked by calling Mrs Kamat at +91 98 22 127 936. There are two boat rides a day, one in the morning and one in the afternoon. First timers could try both, as they may encounter different species.

There are various accommodation options in Panjim, depending on the need and budget. Nature's Nest Camp, Goa

Blue-bearded Bee-eater

(www.naturesnestgoa.com) also coordinates and arranges boat rides for its guests.

Conservation

The Zuari River, along with many other rivers of Goa, is the lifeline of the state and supports a rich marine life as well as birdlife. Huge piles of corroded metal and other waste in the shipyards are polluting the waters and adversely impacting the fish life and the birdlife. There is a need for some strict measures to check and regulate this to protect Zuari and its inhabitants.

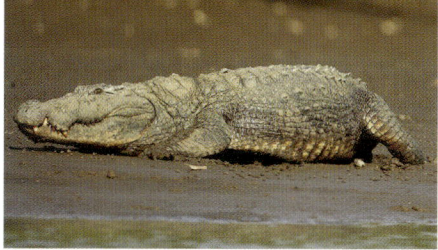

Savio Fonseca

Marsh Crocodile

Bhanu Singh

Collared Kingfisher

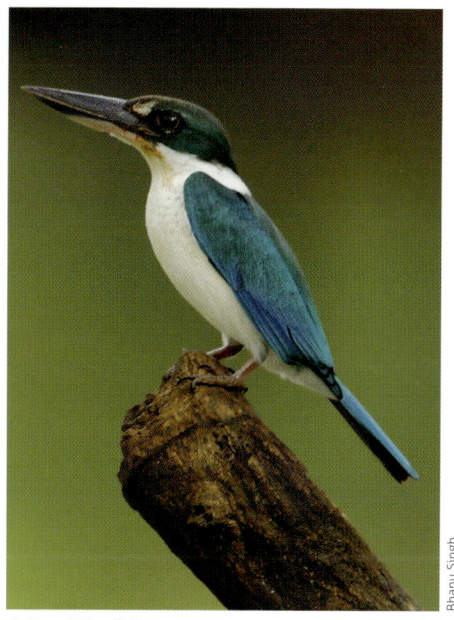

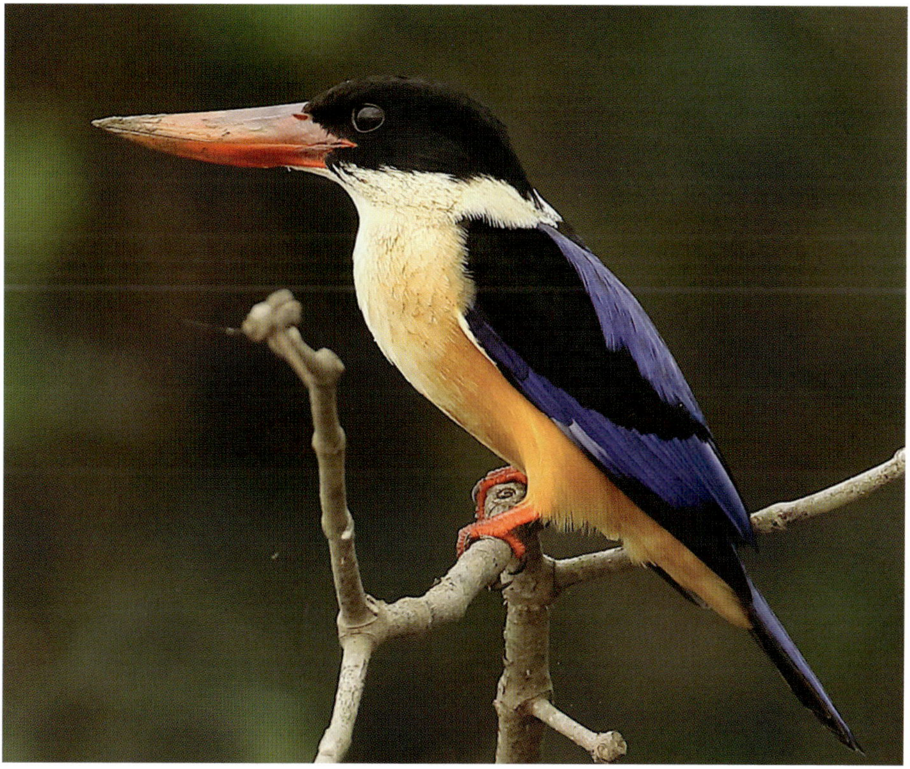
Savio Fonseca

Black-capped Kingfisher

GUJARAT

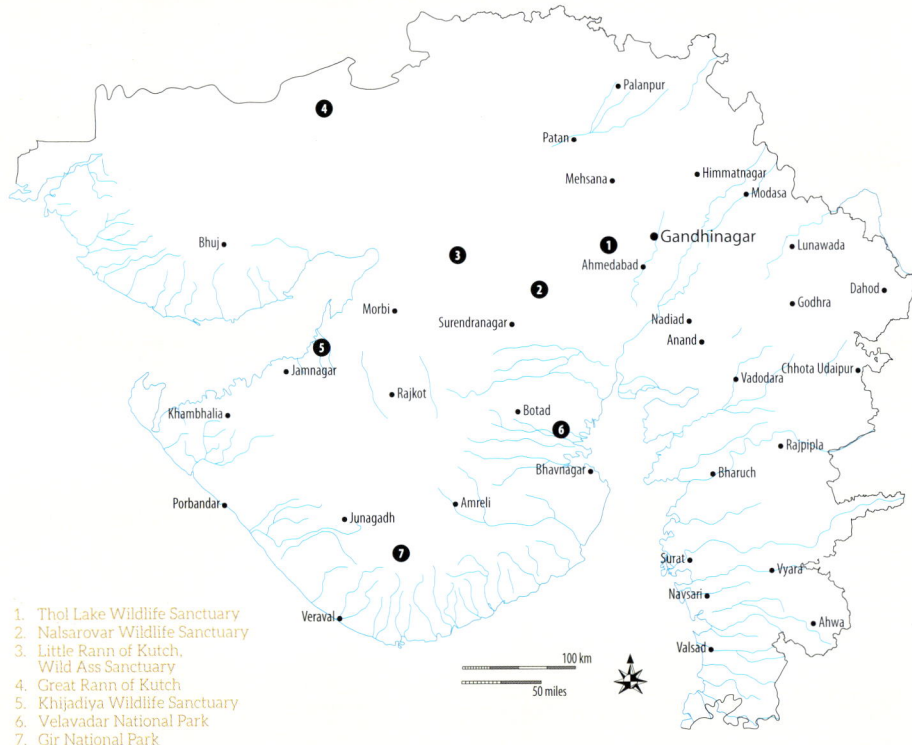

1. Thol Lake Wildlife Sanctuary
2. Nalsarovar Wildlife Sanctuary
3. Little Rann of Kutch, Wild Ass Sanctuary
4. Great Rann of Kutch
5. Khijadiya Wildlife Sanctuary
6. Velavadar National Park
7. Gir National Park

Gujarat occupies the northern extremity of the western seaboard of India, with the international border with Pakistan on the north-west. It comprises three geographical regions:

1. The peninsula, traditionally known as Saurashtra.
2. Kutch on the north-east, which is dry and rocky and contains the famous Rann (desert) of Kutch, the Greater Rann in the north and the Little Rann in the east.
3. The mainland extending from the Rann of Kutch and the Aravalli Hills to the River Damanganga.

KEY BIRDS
Top 10 Birds
1. Eurasian Oystercatcher
2. Crab-plover
3. Grey Hypocolius
4. Laggar Falcon
5. Lesser Flamingo
6. White-naped Tit
7. Bimaculated Lark
8. Stoliczka's Bushchat
9. Spotted Flycatcher
10. MacQueen's Bustard

Gujarat is bounded by the Arabian Sea in the west, the Aravallis in the north-east, the Thar Desert in the north, the Vindhyas and the Satpura Range in the east, and parts of the Western Ghats in the south. Its natural ecosystems range from marine areas and wetlands, to deserts, grassland and moist deciduous forests. The coastline of Gujarat is the longest among the Indian states and shelters diverse coastal ecosystems such as mangroves, coral reefs, estuaries and mudflats. The plains of Gujarat are watered by four major rivers, the Sabarmati, Mahi, Narmada and Tapti. There are no major rivers in the Saurashtra and Kutch regions.

Climate

The climate of Gujarat is generally hot. The average annual rainfall ranges between 400 and 1,000mm, and the mean temperature from 25º C to 27.5 ºC. As the Tropic of Cancer passes through the northern border of Gujarat, the state has intensely hot and cold climates. However, the Arabian Sea and Gulf of Cambay in the west and the forest-covered hills in the east soften the rigours of climatic extremes.

There are four forest types, tropical moist deciduous, tropical dry deciduous, tropical thorn and littoral and swamp forests. The forests are mostly distributed in the southern part of the state, whereas the middle and eastern parts bear bamboo forests of inferior quality. The main forest formations are of teak, bamboo and mangroves.

Access, Transportation & Logistics
Gujarat is well connected with the rest of the country via airways, railways and roadways. The Sardar Vallabhbhai Patel International Airport in Ahmedabad is the biggest airport. It is served by regular non-stop flights from all major cities of India as well as other countries. The state is well known for its good rail connectivity with different parts of India. The main railheads are Surat, Ahmedabad, Rajkot and Vadodara. Vadodara Railway Station is one of the busiest railway stations in India because it is part of the Delhi-Mumbai railway line. There is also an extensive road network covering the state, connected to different parts of India via national and state highways.

Health & Safety
There are no major health concerns in Gujarat. The state is a low- to no-risk area of India when it comes to malaria. Visit a GP or travel clinic 6–8 weeks before departure to make sure you are up to date with any vaccinations. Avoid roadside food and drink bottled water. During the summer months the temperature of Gujarat is very high; keep hydrated and carry a stole, scarf or cap for protection from direct sunlight.

Short-eared Owl

Birdwatching Highlights
The birdlife of Gujarat is very rich. More than 600 species have been identified in this region. Some of the important birds are Greater and Lesser Flamingos, Dalmatian and Great White Pelicans, Common, Demoiselle and Sarus Cranes, Great Indian Bustard, Lesser Florican, Sociable Lapwing, Grey Hypocolius, Stoliczka's Bushchat, and harriers and many other raptors. Gujarat is one of the states in India where both species of flamingo breed. The Great White Pelican, a winter visitor in north India, occasionally breeds in the Great Rann of Kutch. The Dalmatian Pelican is frequently seen in the Little Rann of Kutch. The grassland of the Velavadar National Park (IBA) and the surrounding areas are known for large harrier roosts. There are many sanctuaries in the state that fulfill the IBA criteria and support bird protection. Nalsarovar, Thol and Khijadiya, along with the Little and Greater Rann of Kutch, are some of the famous birding destinations.

Lesser Flamingos

Thol Lake Wildlife Sanctuary

Thol Lake

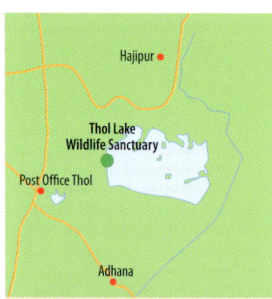

KEY FACTS

Nearest Major Towns
Mehsana

Habitats
Reservoir, wetland

Key Species
White-rumped Vulture, Long-billed Vulture, Greater Spotted Eagle, Sarus Crane, Indian Skimmer, Dalmatian Pelican, Oriental Darter, Painted Stork, Black-necked Stork, Black-headed Ibis, Lesser Flamingo, Baillon's Crake, Pheasant-tailed Jacana, Ruddy Shelduck, European White Stork, White-tailed Lapwing, Demoiselle Crane

Other Specialities
Water Monitor, Indian Cobra, Grey Mongoose

Best Time to Visit
Throughout the year

The Thol Lake is an artificial lake located in the Mehsana district of Gujarat. This freshwater lake was constructed as an irrigation tank in 1912 and was declared the Thol Bird Sanctuary in 1988. The lake is predominantly an open sheet of shallow water surrounded by cropland. Thol is an important inland wetland in north Gujarat and provides excellent habitat for waterfowl during post-monsoon to the winter season. A great number of waterfowl can be seen at the site in winter.

Birdwatching Sites

Thol Lake is home to many varieties of waterbird, such as cranes, geese, ducks, flamingos and pelicans. More than 200 bird species are reported from the sanctuary, of which about 90 are waterbirds. The site is important for pre-breeding congregation and nesting of the Sarus Crane. Thol also supports one of the biggest congregations of Ruff. Thousands of flamingos congregate in the Thol Lake. Some of the key bird species of the area are the Greater Spotted Eagle, Sarus Crane, Indian Skimmer, Dalmatian Pelican, Oriental Darter, Painted Stork and Ferruginous Pochard.

Access & Accommodation

Thol Lake is located just 30km from one of the most popular cities of Gujarat, Ahmedabad. This is very well connected with all the major cities of India via air, railways and highways. Sardar Vallabhai Patel International Airport in Ahmedabad is also connected to major cities of the world.

Sarus Crane

There are many hotels and resorts in Ahmedabad to serve the needs of different types of customer. Caspia Hotel (www.caspiahotels.com) and Aarya Grand (www.aaryagrand.com) are two of the known hotels in the area. Visitors can make online bookings for most of the hotels in Ahmedabad.

Garima Bhatia

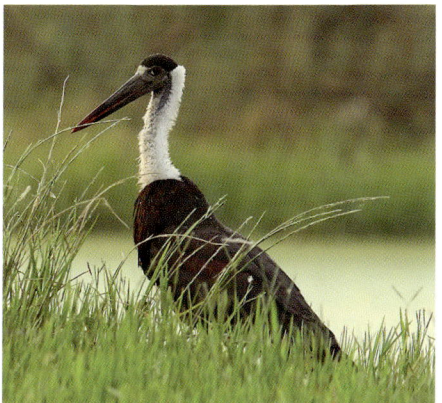

Woolly-necked Stork

Conservation

Thol Lake was constructed to provide irrigation facilities to farmers, so this area is surrounded by farmland. Excessive use of pesticides in the surrounding paddy croplands may be toxic to the birds feeding on them. Illegal cultivation inside the protected area and the withdrawal of excessive water for cultivation is an ongoing problem.

Kunan Naik

Ferruginous Duck

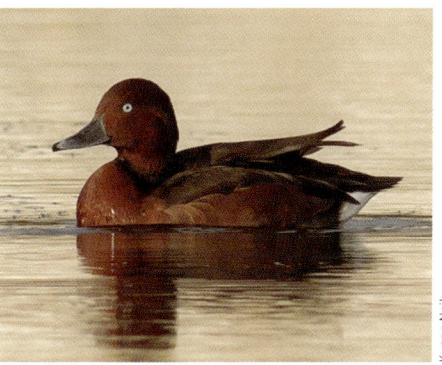

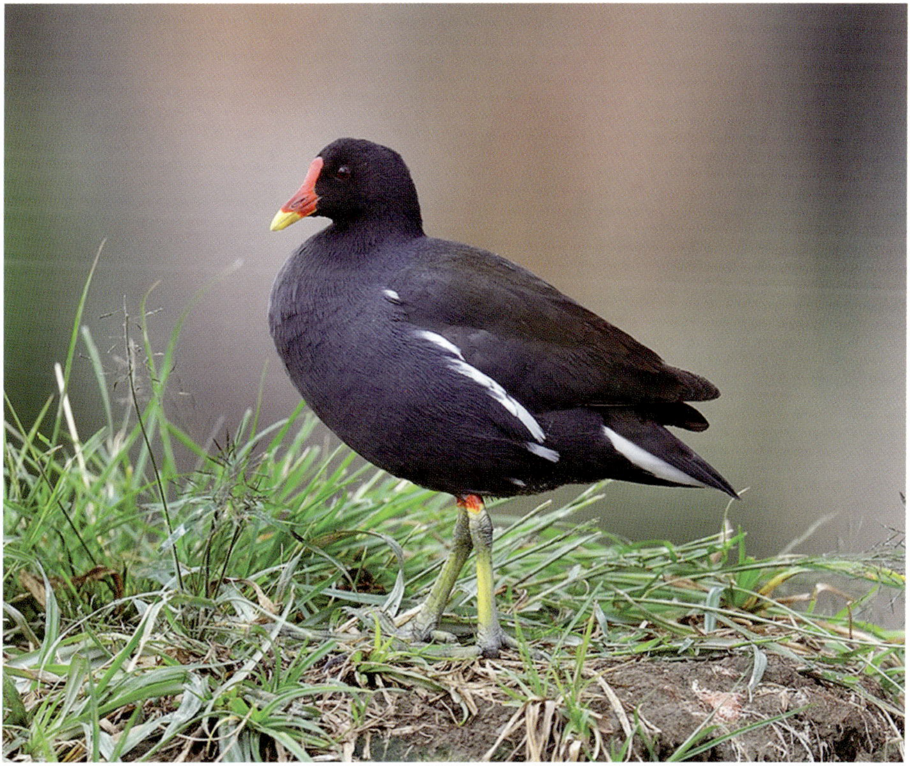

Clement M Francis

Common Moorhen

Nalsarovar Wildlife Sanctuary

Nalsarovar Lake

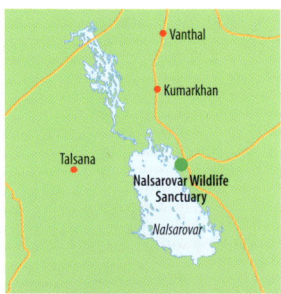

KEY FACTS

Nearest Major Towns
Ahmedabad and Surendranagar

Habitats
Wetland, seasonal marsh

Key Species
Pallas's Fish Eagle, Greater Spotted Eagle, Eastern Imperial Eagle, Lesser Kestrel, Sarus Crane, Indian Skimmer, Dalmatian Pelican, Oriental Darter, Painted Stork, Black-necked Stork, Black-headed Ibis, Lesser Flamingo, Ferruginous Pochard, Cinereous Vulture, Red-headed Vulture

Other Specialities
Water Monitor, Indian Cobra, Grey Mongoose

Best Time to Visit
Throughout the year

This sanctuary is spread over an area of 120km² and is one of the largest shallow freshwater lakes in India. The lake is in the lowest lying area between Central Gujarat and East Saurashtra. It represents a sea link that once existed between the Little Rann and Gulf of Khambhat. Nalsarovar was declared a wildlife sanctuary in 1969 and proposed as a Ramsar Site in 2012.

Birdwatching Sites

Nalsarovar is a birdwatchers' paradise. About 250 bird species have been recorded in this area, of which 158 are waterbirds. You can find Spot-billed Pelicans, Lesser and Lesser Flamingos, crakes, Ruddy Shelducks, Grey-headed Swamphens, herons, European White Storks, various species of bittern, and grebes in the lake. The best time to visit is in winter, in November–February. Migratory birds start arriving in October and stay until April, but their population reaches its peak in mid-winter. The best time to see the birds is early in the morning and in the evening.

Access & Accommodation

Nalsarovar Wildlife Sanctuary is located about 60km from Ahmedabad. You can visit the sanctuary by personal vehicle or taxi. Nalsarovar sanctuary timings are 6 a.m. to 6 p.m., but the ticket window closes at 5.30 p.m. Private vehicles are not allowed past the ticket window area, but vehicles are available to the lake site, which is about 1km away. Boating is permitted on Nalsarovar Lake and boat rides offer good photography opportunities.

Indian Paradise-flycatcher

Spot-billed Pelican

Gujarat Tourism offers luxury tent accommodations near the lake. There are few resorts near the sanctuary. Ahmedabad city offers better accommodation options.

Conservation

Some of the common conservation issues faced by the sanctuary are excessive illegal fishing, poaching of birds and other wild animals, illegal grazing of cattle inside the protected area, and pumping of water for irrigation. The increasing population of tourists is putting pressure on birds and other wildlife in the sanctuary.

European White Stork

Ruddy Shelduck

Little Rann of Kutch, Wild Ass Sanctuary

Asiatic Wild Asses

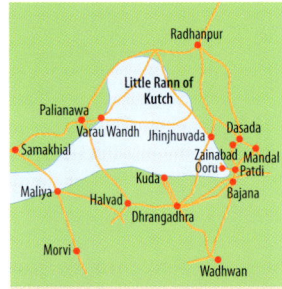

KEY FACTS

Nearest Major Towns
Ahmedabad

Habitats
Saline desert plains, rocky and thorn scrub, arid grassland, plateaus to lakes and marshes

Key Species
Lesser Flamingo, Greylag Goose, Northern Shoveler, Eurasian Wigeon, Common Crane, Pied Avocet, Painted Stork, Grey Heron, Eurasian Spoonbill, Crested Lark, Little Grebe

Other Specialities
Asiatic Wild Ass, Nilgai, Desert Fox, Indian Jackal, Striped Hyena, Desert Cat, Jungle Cat, Collared Hedgehog, Spiny-tailed Lizard, Caracal, Indian Leopard, Small Indain Civet, House Shrew, Indian Desert Jird, Five-striped Palm Squirrel, Indian Pangolin, Small Indian Mongoose

Best Time to Visit
November to February

The Wild Ass Sanctuary encompasses the Little Rann of Kutch and its peripheral areas in Surendranagar, Rajkot, Mehsana, Banaskantha and Kutch districts of Gujarat. The sanctuary holds the last population of the Endangered Asiatic Wild Ass in India. The sanctuary area is characterized by vast, salt-impregnated, sun-baked mudflats, which are dotted with small patches of uplands, locally called *bets* (meaning islands). The mudflats remain submerged for about 4–5 months of the year, under fresh water received from a few rivers and saline water from the Arabian Sea. The vegetation here consists mainly of grasses and dry thorny scrub.

Birdwatching Sites

The Little Rann of Kutch is a place to see birds in numbers. Due to its unique geographical location, Kutch is considered to be at the crossroads of Palaearctic migration streams and witnesses great waves of migratory birds in winter. Its location is near the sea and its vast, low-lying expanses, which get periodically flooded by marine water, create a unique habitat that attracts a wide variety of birds.

More than 200 bird species have been recorded in the area. Breeding birds include India's only known nesting colony of the Lesser Flamingo. Other key birds are the Eurasian Roller, MacQueen's Bustard, Lesser Florican, Demoiselle and Common Cranes, Spotted Sandgrouse, Red-necked Phalarope, Cream-coloured Courser, Sociable Lapwing, Collared Pratincole, White-rumped Vulture, Indian Spotted Eagle, various harriers, Lesser Flamingo, Dalmatian Pelican, European White Stork, and Greater Hoopoe and Sykes's Larks.

Common Crane

Cream-coloured Courser

Indian Spotted Eagle

Access & Accommodation
The nearest airport is located in Ahmedabad, which is about 130km from the sanctuary. The nearest railway station is Dhrangadhra, around 16km from the sanctuary. Road conditions are good and there are regular buses as well as taxis that will take you to the sanctuary. The sanctuary has three main access points, at Dhrangadhra, Range Bajana and Range Adesar (Aadeshwar). Range Bajana is the best place to enter in the winter as the wetlands that house migratory birds are closest to it.

Dhrangadhra is the easiest place in which to find transportation and accommodation. Rann Riders (www.rannriders.com/index.html) at the Little Rann of Kutch is quite a famous eco resort among birdwatchers. Bird and wildlife viewing is easy at the site due to the presence of excellent birding infrastructure. Most of the top-rated tourist establishments have knowledgeable guides who can show you all the regular birds.

Conservation
The Little Rann of Kutch has been identified by the Indian government as an important site for a Biosphere Reserve. Illegal pumping out of water by neighbouring farmers has been an ongoing issue. This adversely impacts a large number of birds like egrets and cormorants that regularly nest here. Illegal encroachment of the sanctuary area by salt traders, farmers and fishermen is another common issue. With drastic deterioration in the habitat of the Asiatic Wild Ass, its future is severely threatened.

MacQueen's Bustard

Great Rann of Kutch

Great Rann of Kutch

KEY FACTS

Nearest Major Towns
Ahmedabad

Habitats
Saline desert plains, rocky and thorn scrub, arid grassland, plateaus to lakes and marshes

Key Species
Common Pigeon, Crested Lark, Eurasian Collared Dove, Grey Wagtail, Northern Shoveler, Common Teal, Common Crane, Black-winged Stilt, Laggar Falcon, Pallid Harrier, Short-toed Snake Eagle, Common Crane, European White Stork, Western Reef Egret, Grey Hypocolius, Great Indian Bustard, Bimaculated Lark

Other Specialities
Asiatic Wild Ass, Nilgai, Desert Fox, Indian Jackal, Striped Hyena, Desert Cat, Jungle Cat, Collared Hedgehog, Spiny-tailed Lizard, Caracal, Indian Leopard, Indian Civet, House Shrew, Indian Desert Jird, Five-striped Palm Squirrel, Indian Pangolin, Small Indian Mongoose

Best Time to Visit
October to March

The Great Rann of Kutch is located in the Thar Desert in the Kutch District of Gujarat. Kutch is the largest district in India, covering an impressive area of 45,674km^2, and is part of the Kathiawar Peninsula occupying the northwestern part of Gujarat. It is a land of deserts, dry, salty alluvial mudflats, extensive grassland and great stretches of water in the *dhand*s (shallow wetlands) left by the monsoons. Dry thorn forests and mild hillocks punctuate the flat, limitless stretches of land, and a great variety of birds finds refuge in these seemingly hostile surroundings. Kutch can be divided into four distinct regions:

1. The deserts of the Great Rann, to the north.
2. The grassland of Banni.
3. The mainland, consisting of plains, hills and dry river beds.
4. The coastline along the Arabian Sea in the south with mangrove creeks to the west.

Due to its unique geographical location and habitat, Kutch is considered to be at the crossroads of Palaearctic migration streams and witnesses great waves of migratory birds in winter.

Birdwatching Sites

With more than 200 resident bird species, the Great Rann of Kutch is an important stopover for migratory birds, which makes it one of the top birding sites in India. Some of the key birds seen here are Marshall's Iora, Grey Hypocolius, Spotted Flycatcher, Rufous-tailed Scrub Robin, shrikes, White-naped Tit, Stoliczka's Bushchat, Sykes's Nightjar, Tawny Eagle, Greater Spotted Eagle, Long-legged Buzzard, Laggar Falcon, White-bellied Minivet, Short-eared Owl, Indian Eagle Owl, Greater Hoopoe Lark,

Greater Hoopoe Lark

Merlin, MacQueen's and Great Indian Bustards, Common Crane, Chestnut-bellied and Painted Sandgrouses, pelicans, and of course the majestic flamingos. The tropical thorn forest area is good for forest birds. Some of the birds found here are the Spotted Flycatcher, Rufous-tailed Scrub Robin, Red-backed Shrike and Common Whitethroat. The Banni Grassland area is another important site for birding. This is a great place for many types of raptor and water-dependent birds. Other key birds include the Grey Hypocolius, White-naped Tit, MacQueen's Bustard, Dalmatian Pelican, Spotted Flycatcher, European Nightjar, Rufous-tailed Scrub Robin, Red-backed and Red-tailed Shrikes, Common Whitethroat, and Eurasian Cuckoo.

Access & Accommodation

The Great Rann of Kutch is best approached via Bhuj. Dhordo, about an hour and a half north of Bhuj, is being developed by the Gujarat government as the Gateway to the Rann.

Dhordo is on the edge of the salt desert. It is most convenient to stay there, or in nearby Hodka. There are many resorts and hotels near Dhordo and Bhuj. Some of the popular choices among travellers are the Gateway to Rann Resort (www.kutchrannresort.com) and Shaam-e-Sarhad Village Resort (www.hodka.in). The Gujarat government has also set up tourist accommodations at the Toran Rann Resort (booking.toranrannresort.com), opposite the army checkpoint near the entrance to the salt desert.

Conservation

One of the primary conservation issues of the Great Rann of Kutch is desertification of the land. This is mainly the result of population increase of both humans and livestock. In general, the problems of the Rann environment are usually due to developmental initiatives like the construction of roads and bunds that could block the natural flow of the sea and fresh water into the Rann. The fringe areas of the Rann are facing threats from overgrazing, increasing vehicular traffic, intensive salt farming and industrialization, which cumulatively cause some extent of change to the fragile ecosystem. The above activities result in disturbances to the unique wildlife, especially to the Asiatic Wild Ass, and floricans, bustards, flamingos, pelicans and other birds.

Stoliczka's Bushchat

Merlin

Grey Hypocolius

Khijadiya Wildlife Sanctuary

Black-headed Gulls

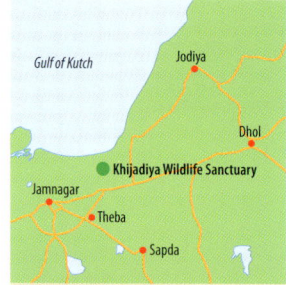

KEY FACTS

Nearest Major Towns
Jamnagar

Habitats
Freshwater swamp, wetland

Key Species
Baer's Pochard, Greater Spotted Eagle, Sarus Crane, Indian Skimmer, Dalmatian Pelican, Oriental Darter, Painted Stork, Black-necked Stork, Black-headed Ibis, Lesser Flamingo, Ferruginous Pochard, Red-headed Vulture

Other Specialities
Nilgai, Indian Jackal, Jungle Cat, Grey Mongoose

Best Time to Visit
Throughout the year

The Khijadiya Wildlife Sanctuary is located in the Jamnagar district of Gujarat. It lies on the south shore of the Gulf of Kutch. This IBA site consists of a group of three shallow freshwater lakes and extensive marshes adjacent to a large area of salt pans and salt marshes. Khijadiya has a dry tropical monsoon climate, with rainfall mainly concentrated in July and August. Habitats in these seasonal inland wetlands include freshwater shallow marshes, intertidal mudflats, creeks, salt pans, saline land and mangroves.

Birdwatching Sites

The sanctuary is located in the extreme western part of the country and is an important stopover site and wintering ground for migratory birds. More than 200 bird species have been recorded in the area, which is a unique combination of seasonal freshwater wetland and coastal wetland ecosystems. It supports over 90 species of waterfowl and waders, some of which, like the Great Crested Grebe, breed here. Some of the important bird species in this area are the Dalmatian Pelican, Baer's Pochard, Oriental Darter, Sarus Crane, Painted and Black-necked Storks, Black-headed Ibis and Indian Skimmer.

Access & Accommodation

The nearest airport and railway station for the sanctuary is Jamnagar, which is about 20km away. Jamnagar is easily accessible by road from all parts of Gujarat. There are state transport buses, private buses and private vehicles that you can use to reach Jamnagar from Rajkot, Ahmedabad or Dwarka. Travel by train from Ahmedabad is the most affordable way to reach Jamnagar.

Ashy Drongo

There is no accommodation inside the sanctuary and camping is not permitted. You can find accommodation in Jamnagar, where hotels range from budget to luxury, as well as dormitories and guest houses. December is the peak season and finding good accommodation is a challenge during this time.

Conservation

The biggest danger to this small sanctuary is from the unauthorized drawing of water for irrigation, which reduces water levels during the peak winter season. Livestock overgrazing is a curse as in most of the sanctuaries of Gujarat, but in the case of Khijadiya limited grazing is beneficial to the freshwater wetland as it removes the biomass, which otherwise would accumulate excessively. Due to its easy accessibility, Khijadiya attracts a lot of visitors, and this is putting pressure on its flora and fauna. There is an urgent need to sensitize the visitors to the importance of wildlife.

Bhanu Singh

Painted Storks

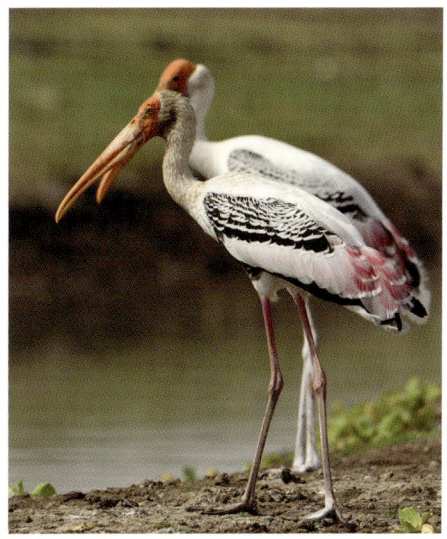

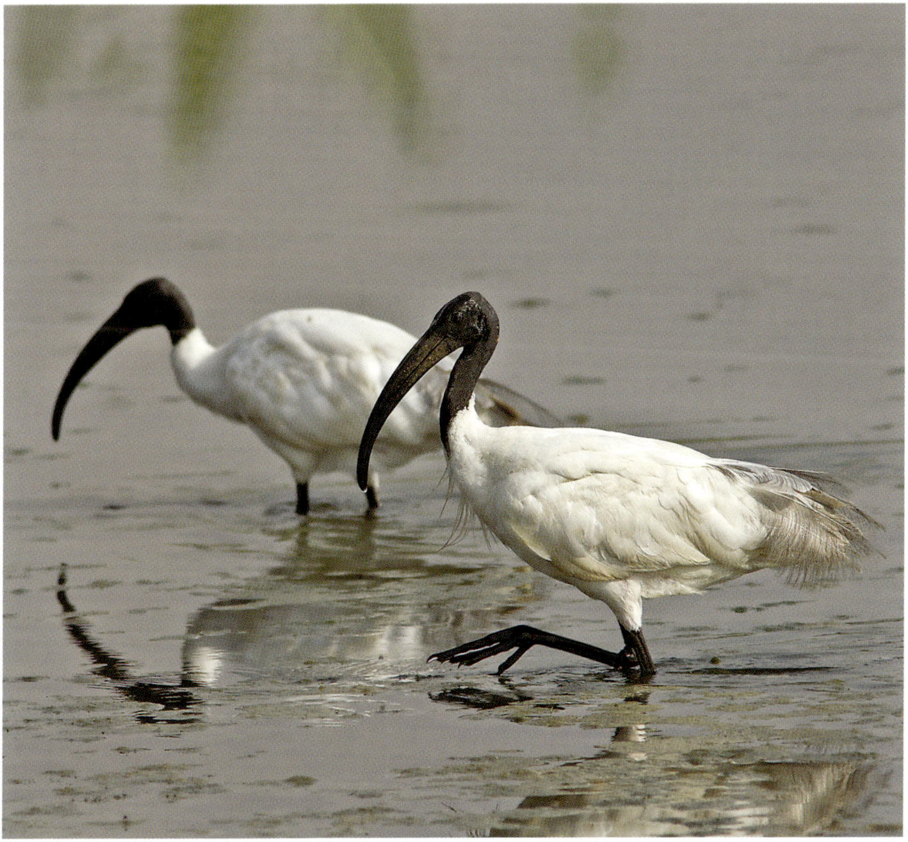

Garima Bhatia

Black-headed Ibises

Velavadar National Park

Blackbuck

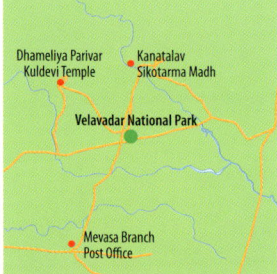

KEY FACTS

Nearest Major Towns
Bhavnagar

Habitats
Tropical grassland

Key Species
Lesser Florican, Lesser Adjutant, Greater Spotted Eagle, Eastern Imperial Eagle, Sarus Crane, Stoliczka's Bushchat, Northern Shoveler, Lesser Flamingo, Red Collared Dove, Jacobin Cuckoo, Demoiselle Crane, Lesser Florican, Painted Stork

Other Specialities
Indian Fox, Indian Jackal, Jungle Cat, Wild Pig, Nilgai, Black-naped Hare, Blackbuck

Best Time to Visit
October to March

This park is situated in the Bhavnagar district of Gujarat, 144km from Ahmedabad. It is considered a unique grassland ecosystem and probably the only tropical grassland in India reckoned as a national park. Velavadar is famous not only for the largest concentration of Blackbuck, but also for the largest population of harriers in the world. During winter, Montagu's, Pallid and Western Marsh Harriers can be viewed in abundance. The park extends over an area of 35km² and comprises mainly flat grassland. It lies between two rivers some distance away from the Gulf of Cambay. The fertile soils are believed to have arisen from the sea. This is a unique national park with an exclusive grassland habitat. The savannah-type grassland extends uniformly, interspersed with dry thorny scrub. The grasses grow on an average to about 30–45cm tall.

Birdwatching Sites

With more than 200 birds recorded in this area, Velavadar is a perfect site for birdwatchers to see many tropical Indian grassland birds like Stoliczka's Bushchat (winter) and the Rufous-tailed Lark. Velavadar is also known for the largest breeding populations of the Lesser Florican. Although a rare and shy bird, it is relatively easy to spot at Velavadar, and you may see several in a day without making much of an effort. The park staff even mark its territories with a pole, so you may see many males doing their high jumps near their territory, even from the Tourist Lodge. Velavadar is also a good place in which to see grassland species, including numerous varieties of lark, pipit, sandgrouse, francolin, quail and bushchat. Winter is the best time to visit, with many migratory waterbirds and waders making the park their home during these months (November–March). Prominent among them are flamingos, pelicans, cranes,

Crested Lark

storks and a multitude of duck species. Majestic birds of prey make this a raptors' paradise. They include Steppe, Tawny and Eastern Imperial Eagles. There is also a huge number of roosting harriers, including Pallid, Montagu's, Hen and Western Marsh Harriers, as well as Short-eared and Indian Eagle Owls.

Access & Accommodation
The nearest airport to the park is at Bhavnagar, which is located about 65km away. There is also a railhead at Bhavnagar, and there are direct trains every day from Ahmedabad to Bhavnagar. Buses ply between Bhavnagar and Velavadar every day.

Accommodation can be arranged at the Kaliyar Bhavan Tourist Lodge at the park. However, advance bookings are a must. For details and reservations, contact Assistant Forest Conservator, Velavadar, Blackbuck National Park, PO Vallabhipur, Bhavnagar, Tel.: 0278-428644. Another posibility is Bhavnagar, where several accommodation options are available: Nilambagh Palace Hotel, Tel.: 0278-424241; Hotel Blue Hill, Tel.: 0278-426951; Jubilee Hotel, Tel.: 0278-430045, and Hotel Apollo, Tel.: 0278-245251. Lower budget alternatives are Shital Guest House, Tel.: 0278-428360; Vrindavan Hotel, Tel.: 0278-518928, and Hotel Mini, Tel.: 0278-24415.

Conservation
Some of the common conservation issues faced by this park are illegal livestock grazing inside the protected area, and excessive use of pesticides in the neighbouring farmland. Poaching is not a very big problem, especially of birds, but the use of pesticides in the surrounding agricultural fields is unregulated and could be having negative impacts on the harriers and floricans, as both feed on insects.

Pied Harrier

Tawny Pipit

Greater Spotted Eagle

Montagu's Harrier

Gir National Park

Asiatic Lion

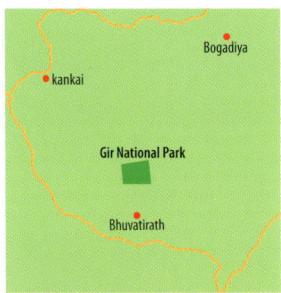

KEY FACTS

Nearest Major Towns
Junagadh and Amreli

Habitats
Semi-arid forests

Key Species
White-rumped Vulture, Long-billed Vulture, Lesser Florican, Spot-billed Pelican, Baer's Pochard, Greater Spotted Eagle, Eastern Imperial Eagle, Sarus Crane, Indian Skimmer, Dalmatian Pelican, Indian Blackbird, Pied Bushchat, Blue Rock Thrush, Black Redstart, Ultramarine Flycatcher, Indian Blue Robin, Rusty-tailed Flycatcher, Oriental Magpie Robin, Jungle Myna, Brahminy Starling

Other Specialities
Asiatic Lion, Striped Hyena, Indian Leopard, Jungle Cat, Desert Cat, Rusty-spotted Cat, Common Palm Civet, Four-horned Antelope, Indian Pangolin, Honey Badger, Kutch Rock Rat, Grey Musk Shrew, Pale Hedgehog, Bandicoot Rat

Best Time to Visit
October to March

This park is situated in the Junagadh district of Gujarat. It is one of the oldest sanctuaries in India and is famed for being the last remaining habitat of the Asiatic Lion in the world. Gir forest has become a very stable ecosystem with tremendous regenerating, self-supporting and self-sustaining capacity, due to its compactness and richness of biodiversity. Gir forest has a topography made up of successive rugged ridges, isolated hills, plateaus and valleys. Besides being the last abode of Asiatic Lions, Gir forest forms a unique habitat for many mammal, reptile, bird and insect species, along with a rich variety of flora.

Birdwatching Sites

Gir is home to more kinds of bird than any other park in Gujarat, yet somehow it is not known for its birdlife. It includes more than 300 bird species, many of which can be seen year round. Some of the highlights of the bird species are Greater Spotted and Eastern Imperial Eagles, Baer's Pochard, Malabar Whistling Thrush, Indian Paradise-flycatcher, Crested Serpent Eagle, Spot-billed Pelican and Painted Stork.

Access & Accommodation

The nearest airports to Gir National Park are Keshod Airport and Rajkot Airport. Keshod Airport is located about 70km from the park, whereas Rajkot Airport is about 160km away. Take a cab or bus service to reach the park from either of these two airports. Junagadh and Veraval railway stations are the nearest railway stations to the park.

Forest guest houses are Maneland Jungle Lodge and Sinh Sadan Guest House. There are also many private hotels and resorts outside the park. The resorts arrange safari and birdwatching trips inside as well as outside the park.

Marshall's Iora

Conservation

Some of the common conservation issues of Gir National Park are illegal grazing inside the protected area and encroachment on forest land. There are 14 forest settlements of Maldharis, livestock-owning communities that live in the sanctuary. Their increasing population, and that of the cattle and additional cattle that migrate from peripheral villages, exerts tremendous grazing pressure on the sanctuary. Rapidly expanding pilgrimage sites attract thousands of visitors throughout the year, causing heavy traffic and immense damage to the area.

Grey Francolins

Indian Paradise-flycatcher

Mottled Wood Owl

HIMACHAL PRADESH

Himachal Pradesh is situated in the north-west of India in the Himalayan ranges. It is bounded by Jammu and Kashmir in the north, Uttarakhand in the south-east, Haryana in the south and Punjab in the west, and in the east it forms India's international boundary with Tibet. The state is mountainous, with altitudes of 460–6,600m. It has a deeply dissected topography, a complex geological structure and a rich temperate flora in subtropical latitudes. It is drained by several snow-fed perennial rivers, of which the most important are the Chenab, Ravi, Beas, Sutlej and Yamuna Rivers. The state is well known for its hill stations and attracts visitors from across the world, especially in the summer months. It is also known for its rich wildlife, especially for rare species such as the Himalayan Musk Deer, Asiatic Ibex, Himalayan Tahr, Asiatic Brown Bear and Snow Leopard. Some of the pheasant species that are very important in the state include the Himalayan Monal, Western Tragopan, Koklass Pheasant and Himalayan Snowcock.

1. Great Himalayan National Park
2. Pong Dam Wildlife Sanctuary
3. Chail Wildlife Sanctuary
4. Kibber Wildlife Sanctuary

Climate

Himachal Pradesh has a varied climate, topography and altitudinal ranges. There are six major forest types: tropical dry deciduous, subtropical pine, subtropical dry evergreen, Himalayan moist temperate, Himalayan dry temperate, subalpine and alpine. The amazing diversity in habitat from the plains up to the numerous mountain peaks results in marvellous avifaunal diversity. The average rainfall in the state is 1,800mm. The mean annual temperature range is 20–22.5 °C. November–May is the season for birding. January–February are the peak winter months and visitors should expect medium to heavy snow at higher elevation ranges. This time is best for winter migrants, whereas April–May are best for observing breeding birds.

Access, Transportation & Logistics

The easiest way to reach Himachal Pradesh is via airways. There are three major airports in Himachal Pradesh – Shimla, Kangra and Kullu. However, Chandigarh Airport in Punjab is bigger and well connected with the rest of India, as well as with many international cities. Since Himachal Pradesh is a hilly region, it is quite tedious to lay railway tracks and operate trains in the state. Apart from the heritage train that runs within the state, there are no ideal railway routes that will take you through the destinations in Himachal Pradesh. Only one major railway station is connected with other Indian cities – Kalka railway station, which is located about 90km

KEY BIRDS

Top 10 Birds
1. Western Tragopan
2. Twite
3. Brambling
4. Whooper Swan
5. Ruddy Turnstone
6. Fire-capped Tit
7. Greater White-fronted Goose
8. Cheer Pheasant
9. Red-necked Grebe
10. Sociable Lapwing

Western Tragopan

from Shimla. You can also reach Himachal by road. Himachal is one of the most serene places for a road trip, and you can either take your car or hire a cab from New Delhi.

Health & Safety
There are no major health concerns in Himachal. The state is a low- to no-risk area of India when it comes to malaria, but do take precautions especially after the monsoon. Visit a GP or travel clinic 6–8 weeks before departure to make sure you are up to date with any vaccinations. If you have altitude sickness, consult a doctor and carry your medicine with you.

Birdwatching Highlights
The state of Himachal is a paradise for birdwatchers, with the Great Himalayan National Park and Pin Valley National Park identified as IBAs. More than 776 bird species have been recorded from these areas. Out of that list, the Western Tragopan is a Vulnerable species. Himachal Pradesh is very important for the protection of many pheasant and forest bird species. It lies in the Western Himalayas Endemic Bird Area (EBA). Eleven species are confined to this EBA, of which 10 are known to occur in this state, with confirmed records. They are the Western Tragopan, Cheer Pheasant, Brook's and Tytler's Leaf Warblers, Kashmir Flycatcher (vagrant), White-cheeked and White-throated Tits, Kashmir Nuthatch, Spectacled Finch and Orange Bullfinch.

Grey-headed Woodpecker

Great Himalayan National Park

Tirthan River

KEY FACTS

Nearest Major Towns
Kullu

Habitats
Subalpine dry scrub, alpine moist pasture, subtropical broadleaved hill forest

Key Species
Cheer Pheasant, White-cheeked Tit, White-throated Tit, Spectacled Finch, Orange Bullfinch, Western Tragopan, Himalayan Monal, Koklass Pheasant, Kalij Pheasant, Variegated Laughingthrush, Scaly-breasted Wren Babbler, Spectacled Finch, Golden Bush Robin, Blue-capped Redstart, Black-throated Thrush, Black-throated Accentor, Brown Dipper, Golden Eagle, Bearded Vulture, Himalayan Vulture

Other Specialities
Asiatic Ibex, Bharal, Himalayan Tahr, Himalayan Black Bear, Goral, Indian Leopard

Best Time to Visit
Throughout the year, though winters are exceedingly cold

The sprawling Great Himalayan National Park in Kullu district has relatively undisturbed areas that support diverse Himalayan wildlife. The park lies in the upper catchment area of the Tirthan, Sainj and Jiwa Rivers, which flow westwards and feed the Beas River. The park includes parts of Tirthan Sanctuary and is bordered by Pin Valley National Park in the north-east, Kanawar Sanctuary in the north-west and Rupi Bhabha Sanctuary in the east (all IBAs). These constitute Himachal Pradesh's largest protected area concerning wildlife. The eastern part of the park lies above the snowline and has both glaciers and permanent ice.

Birdwatching Sites

The area is particularly noted for its prolific pheasant populations. The park is home to more than 300 bird species, and contains an excellent representation of West Himalayan avifauna. The Himalayan Monal, Koklass and Kalij Pheasants, and Common Hill Partridge are common, while the Cheer Pheasant and Western Tragopan have more restricted ranges. Chukar Partridge and Snow Partridges, and Himalayan Snowcocks, occur in suitable habitats all over the park. The Himalayan Vulture and Bearded Vulture are common, and seen daily in all seasons. Golden Eagles and Common Buzzards are seen frequently in all seasons in the subalpine and alpine zones. The Eurasian Sparrowhawk is common below the tree line. Black and Booted Eagles are less frequently seen. There have been rare sightings of the Peregrine Falcon. The Eurasian Woodcock and Solitary Snipe both occur in summer and breed within the park. The Speckled Wood Pigeon and Snow Pigeon are both common, as is the Oriental Turtle Dove in summer.

Chukar Partridge

Snow Partridge

Black-and-yellow Grosbeak

Access & Accommodation
The nearest airport to the park is Bhuntar in Kullu, about 60km away. There are regular flights from Delhi and Chandigarh to this airport, but flights are often cancelled so not very reliable. Train and road options are much more reliable. Joginder Nagar, near Mandi, is the nearest railhead to the park; it is about 143km away. One of the convenient options is to take a bus from Delhi. Regular Volvo buses connect Delhi and Manali in Himachal Pradesh. They all leave Delhi in the evening and reach the Kullu Valley (near the park) in the morning.

There are few accommodation options available near Sainj and Tirthan Valleys, and inside the park only tented accommodation is possible. Advance permission is required to enter the park and pitch tents. There are a few local groups and resorts at Tirthan that organize birding treks in the park.

Conservation
Despite its large size and protected status, the park is not free from human disturbances. There are villages and numerous settlements inside the park, which use the land for livestock grazing. To promote the growth of new grass, graziers start fires, which sometimes get out of control. Some poaching is known to occur, and a few cases have been registered. The locals are known to deliberately start fires in the forest area to catch escaping animals. However, the greatest and irreversible threat to the park comes from denotification of certain areas for development projects.

Black-throated Sunbird

White-throated Dipper

Pong Dam Wildlife Sanctuary

Mallards

The Pong Reservoir, also called the Maharana Pratap Sagar, was created in 1976 on the Beas River in the foothills of the Himalayas. It contains several deforested islands that attract a large number of waterbirds. The northern edge is very flat, with mudflats and wet grassland, and attracts major concentrations of birds. Besides providing a home to a variety of birds, this place is also famous for its tourist attractions and water sports.

Birdwatching Sites

Pong Dam Sanctuary is famous for its waterbirds. More than 220 bird species have made this sanctuary their home. The lake is an important wintering ground for waterfowl, and every

KEY FACTS

Nearest Major Towns
Dharamshala

Habitats
Aquatic

Key Species
White-rumped Vulture, Slender-billed Vulture, Tickell's Thrush, Orange-headed Thrush, Desert Wheatear, Isabeline Wheatear, Black Redstart, White-capped Water Redstart, Red-breasted Flycatcher, Blue Whistling Thrush, Rufous-bellied Niltava

Other Specialities
Wild Pig, Indian Jackal, Indian Leopard, Grey Mongoose, Small-clawed Otter, Northern Plains Langur, Nilgai, Sambar, Indian Chevrotain

Best Time to Visit
Throughout the year, though winters are exceedingly cold

Common Greenshank

Cattle Egret

Western Marsh Harrier

Green Sandpiper

year thousands of ducks visit the place. The Mallard, Northern Pintail, Common Teal and Common Pochard are some of the prominent species. Comparatively uncommon Red-necked Grebes and Great Black-headed Gulls have also been recorded from this area. Waders such as the Greenshank, Green and Common Sandpipers, and Temminck's Stint occur in considerable numbers. A great variety of raptors can also be seen here, including the Osprey, Pallas's Fish Eagle, Western Marsh Harrier and Tawny Eagle.

Access & Accommodation

The sanctuary is located about 8km from Kangra, which is easily accessible from Masrur, Pathankot and Nurpur. There is a regular bus service connecting Dharamshala and Pathankot; private taxis and cabs are also available. The major rail junction to reach the sanctuary is Pathankot, which is about 35km away. Pathankot is well connected to all the major cities of India. The nearest airport is Gaggal Airport in Kangra Valley. The airport is 12km from Dharamshala and about 45km from the sanctuary.

There are many resorts and hotels near Pong Dam Wildlife Sanctuary as this place is quite well known. You can find luxurious as well as budget hotels, depending on your needs and budgets. The Pong Eco Village is quite famous among birdwatchers and photographers. The time to visit is November–March, the best season for birdwatching.

Conservation

Apart from its importance as a source of water for irrigation and domestic use, Pong Dam attracts a large number of migratory waterfowl. The Himachal Pradesh government wants to develop the reservoir as a new tourist paradise, with water sports as the main attraction. There is a need to regulate tourist activities and create a sustainable model to promote wildlife as well as tourism.

Long-legged Buzzard

Wood Sandpiper

Hen Harrier

Chail Wildlife Sanctuary

Chail landscape

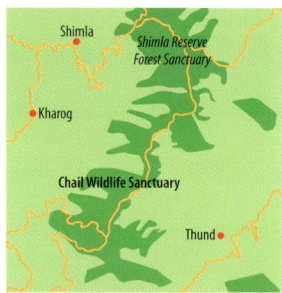

KEY FACTS

Nearest Major Towns
Solan, Shimla and Chail

Habitats
Subalpine forest, montane grassy slopes, subtropical broadleaved hill forests

Key Species
White-rumped Vulture, Cheer Pheasant, Red-headed Vulture, Pallid Harrier, Blue-capped Redstart, Rufous Sibia, Green-backed Tit, Bar-tailed Treecreeper, Himalayan White-browed Rosefinch, Slaty-headed Parakeet, Himalayan Bulbul, Grey-winged Blackbird, Black-headed Jay, Dollarbird, Pied Kingfisher, Lesser Kestrel, Indian Roller, Blue-cheeked Bee-eater, Coppersmith Barbet, Great Barbet, Rufous-bellied Woodpecker, Grey-capped Pygmy Woodpecker, Eurasian Wryneck, Yellow-rumped Honeyguide, Common Hoopoe

Other Specialities
Goral, Rhesus Macaque, Indian Leopard, Himalayan Langur

Best Time to Visit
Throughout the year, though winters are exceedingly cold

This sanctuary at the foothills of the Himalayas is about 3km from Chail town of Himachal Pradesh. Chail is in the Shimla district and hence attracts many tourists. It was publicly announced and accepted as a wildlife sanctuary in 1976. Covered in dense green forests with snow-capped mountain ranges in the backdrop, the sanctuary is home to numerous bird and other wildlife species. It is famous for different species of pheasant, and the Cheer Pheasant Breeding and Rehabilitation Programme was introduced in the sanctuary. Visitors are welcome to visit the centre.

Birdwatching Sites

More than 150 bird species have been reported in this site, including five species of pheasant–Cheer Pheasant, Koklass, Kalij, Peafowl and Red Junglefowl. While the Cheer Pheasant occurs only in grassland, Kalij and Koklass Pheasants are found in the oak forest. This IBA is extremely important for the protection of the globally threatened Cheer Pheasant. Other important bird species found here include the White-rumped and Red-headed Vultures, Slaty-headed Parakeet, Blue-capped Redstart, Rufous Sibia, Green-backed Tit and Black-headed Jay. The uneven, wide stretch of land distribution in the sanctuary makes it very difficult to do a jeep safari. The best option is trekking, which is also promoted by the forest department.

Access & Accommodation

The Toy Train from Kalka takes about five hours to reach Kandaghat, and then it is a 45-minute taxi ride. Visitors can also

Rufous Sibia

opt for the road passage, which is 40km from Shimla and 3km from Chail town. Shimla can be reached via flight or train from the major cities. However, the Toy Train journey is truly recommended – it gives travellers a lifetime experience through the dense forests, sharp bends and several tunnels.

There are several private as well as government resorts around this area, where a comfortable stay is provided. The best times to visit are April–July and September–November.

Conservation

The surrounding area of the sanctuary is densely populated. This small sanctuary therefore is under tremendous human pressure due to fuelwood collection, livestock grazing, quarrying and other unfavourable activities. Residents of Chail demand that the sanctuary be denotified, as their private land lies within it. Two pheasant breeding centres, at Karium and Blossom, are located in the sanctuary, and captive breeding of Cheer Pheasant, Kalij and Red Junglefowl is being attempted here.

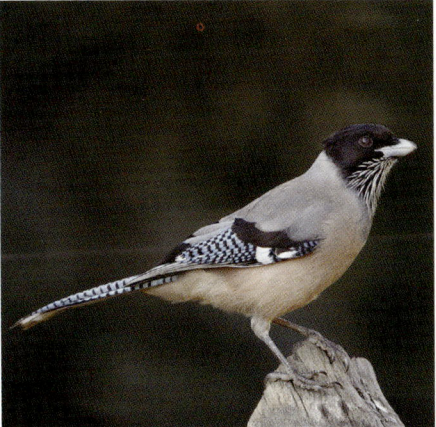

Clement M Francis

Black-headed Jay

Nitin Srinivasamurthy

Red-billed Leiothrix

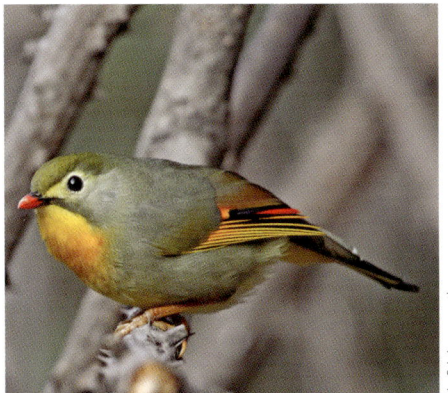

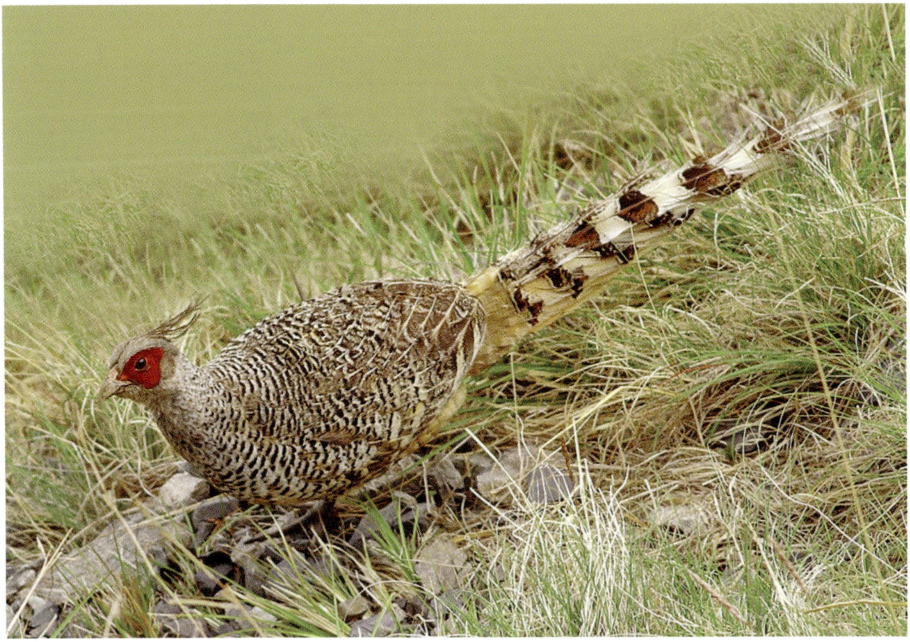

Sonu Anand

Cheer Pheasant

Kibber Wildlife Sanctuary

Snow Leopards

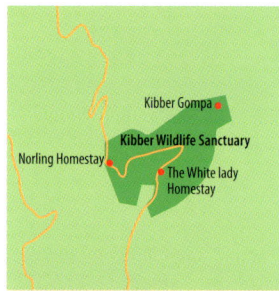

KEY FACTS

Nearest Major Towns
Keylong and Shimla

Habitats
Alpine dry scrub, alpine moist pasture

Key Species
Himalayan Vulture, Himalayan Snowcock, Snow Pigeon, Guldenstadt's Redstart, Great Rosefinch, Alpine Chough, Bearded Vulture, Chukar Partridge, Red-fronted Serin

Other Specialities
Snow Leopard, Asiatic Ibex, Bharal, Red Fox, Tibetan Woolly Hare, Himalayan Wolf, Lynx, Large-eared Pika

Best Time to Visit
Throughout the year, though winters are exceedingly cold

This sanctuary, in the Trans-Himalayan district of Lahaul and Spiti, is situated in the cold desert area of the Himalayas, and has the unique flora and fauna characteristic of this area. At altitudes of 3,600–6,700m, the sanctuary is spread over a vast 2,220km^2 area. It is by far the largest sanctuary in Himachal, and also occupies some of the highest regions of the state. Kibber happens to be the only cold desert wildlife sanctuary in India. The site falls in the rain-shadow area of the Himalayas, so the rainfall is very low. Most of the moisture is provided by snow. Summers are very dry, while winters are very cold, with the mercury dropping to -32 °C.

Birdwatching Sites

Kibber Wildlife Sanctuary is a very understudied forest. Practically no work has been done on the avifauna of the sanctuary, except for stray observations. Recently, due to regular sightings of the Snow Leopard in the area, it has started gaining impetus. Most of the high-altitude birds, such as the Himalayan Snowcock, Golden Eagle, Bearded Vulture, Himalayan Vulture, Snow Pigeon, European Goldfinch, Gold-naped Finch, Red-fronted Serin, Twite, Pied Wheatear, Alpine Swift and Alpine Chough, have been reported from this area. The Chukar Partridge is common at lower elevations.

Access & Accommodation

There are two routes to Kibber, one via Manali and the other via Shimla. The nearest airport is Bhunter and Jabbar Airport near Shimla. Kibber is accessible by road from Manali on the Manali-Leh road, and also linked by NH 22 via Shimla. Buses and taxis are available from Manali and Shimla.

There are only a few homestay options available in Kibber,

Yellow-billed Blue Magpie

Bearded Vulture

Red-fronted Serin

so making advance bookings is recommended before visiting the place. Kibber is very cold during the winter months, so proper care should be taken to protect yourself from the harsh weather conditions.

Conservation
The preservation of vegetation is a major issue in the Kibber Wildlife Sanctuary due to intensive grazing by goats, sheep and domestic yaks. In the prevailing geographical and climatic conditions, these animals are indispensable. The winter being extremely severe, the local people need fuelwood to keep their houses warm, so the scanty vegetation becomes the major victim. Poaching is not a major issue, as most of the people are Buddhists and do not kill animals. As the area lies on the international border, military and paramilitary forces regularly patrol the area and conduct exercises in it.

Alpine Chough

Gold-naped Finch

JAMMU & KASHMIR

1. Hokersar Lake
2. Dachigam National Park
3. Hemis National Park
4. Tso Moriri, Tso Kar & Hanle

Jammu and Kashmir is the northernmost Union Territory of India. It is located in the Indian Himalayas and shares borders with the states of Himachal Pradesh and Punjab to the south. The Line of Control separates it from the Pakistani-administered territories of Azad Kashmir and Gilgit-Baltistan in the west and north respectively, and a Line of Actual Control separates it from the Chinese-administered territory of Aksai Chin in the east. This hilly state is divided into three geographical regions, namely the Kashmir Valley, the Ladakh region and the Jammu region. The Kashmir Valley is well known for its beautiful mountainous landscape, and Jammu's numerous shrines attract tens of thousands of Hindu pilgrims every year, while Ladakh is renowned for its remote mountain beauty and Buddhist culture. The state's higher regions are covered by Pir Panjal, Karakoram and the inner Himalayan ranges. The Chenab, Ravi and Jhelum are the important rivers of this state. The capital city is Srinagar, but in winter the administration offices move to Jammu.

Climate

Jammu and Kashmir have a varied climate, topography and altitudinal ranges. The average annual rainfall and temperature are in the range of 600–800mm and 15–17.5 °C respectively. The climatic conditions vary significantly from subtropical in the Jammu region to cold and arid in Ladakh. Broadly, the state has five types of vegetation, namely subtropical dry evergreen, Himalayan moist temperate, Himalayan dry temperate, subtropical pine, and subalpine and alpine forests. Forests are largely distributed in the Kashmir Valley and the Jammu region. Leh and Kargil

KEY BIRDS

Top 10 Birds
1. Kashmir Nuthatch
2. Orange Bullfinch
3. Ground Tit
4. Black-necked Crane
5. Little Crake
6. Rook
7. Eurasian Jackdaw
8. Snow Partridge
9. Tibetan Partridge
10. Little Owl

are devoid of forest vegetation. This area is a cold desert. January and February are the peak winter months, and visitors should expect medium to heavy snow at the higher elevation ranges. This time is best for winter migrants, whereas April–May are best for observing breeding birds. September–October are good for recording some passage migrants.

Access, Transportation & Logistics

The easiest way to reach Jammu and Kashmir is via airways. There are three major airports in the state, covering three of its regions – Jammu, Srinagar and Leh have one airport each. These airports are well connected with Delhi, Mumbai and other major Indian cities. Jammu-Tawi is the major railway station for all travellers visiting the state via the railways. This station has frequent connectivity with the rest of the big stations of India. The state has an outstanding road network that connects it to neighbouring cities like Delhi, Katra, Amritsar, Chandigarh, Shimla, Leh and Ambala. Many of Jammu's cities are well connected with NH 1A. Visitors can also take Jammu and Kashmir State Road Transport Corporation buses to visit different parts of the state. This state is a hotspot for tourists from across the globe, so transportation, accommodation and logistics are well managed and easily available almost everywhere within it.

Health & Safety

There are no major health concerns in the state. It is a low- to no-risk area of India when it comes to malaria, but do take precautions especially after the monsoon. Visit a GP or travel clinic 6–8 weeks before departure to make sure you are up to date with any vaccinations. If you have altitude sickness, consult a doctor and carry your medicine with you. Despite the highly visible military presence in the state, it is generally a safe place for foreigners and domestic tourists, although local protests may occur in some regions, especially in the Kashmir Valley. Avoid areas where there is a strike, and if a curfew is imposed, remain in your hotel. Driving in India instills fear in even the most seasoned driver, so hiring a car and driver together is recommended.

Birdwatching Highlights

Jammu and Kashmir is known not only for its picturesque landscapes, but also for its rich and diverse birdlife. Bird diversity varies seasonally, and as many as 708 species, including 31 that are globally threatened, have been recorded in the state. The state lies in the Western Himalayas Endemic Bird Area, where many restricted range species have been listed.

In the Kashmir Valley, many protected areas support restricted range species, and some waterbodies support large congregations of migratory waterbirds. The restricted range species occur mainly in temperate coniferous or broadleaved forests, subalpine forests and montane grassland. For example, the Kashmir Flycatcher, which is one of the globally threatened species, is found at an altitude of 1,800–2,700m in temperate mixed broadleaved forests, especially where there is a dense growth of *Parrotia*. Other similar species, such as Tytler's Leaf Warbler, and White-throated and White-cheeked Tits, are found at 1,500–3,600m in pine, oak, mixed and deciduous forests. The White-throated Tit can also be seen easily in rhododendrons and willow scrub, near the tree line, as well as in riverine tamarisk scrub. Other restricted range species that can be seen in or near the valley are the Kashmir Nuthatch, Spectacled Finch and Orange Bullfinch. The finches are found in open coniferous, mixed and deciduous forests, and occasionally birches. The Changthang and Tsomoriri region in Ladakh is an important breeding ground for waterbirds. Apart from hosting the largest breeding congregation of Bar-headed Geese in India, these regions also support the largest population of the endangered Black-necked Crane. Hemis National Park (IBA) in Ladakh is important for all the high-altitude birds of the Western Himalayas.

Himalayan Rubythroat

Hokersar Lake

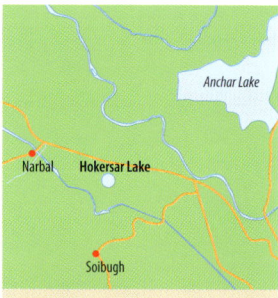

Hokersar wetlands

Hokersar, a renowned waterfowl reserve, lies very near the capital city Srinagar on the Srinagar-Baramulla highway, on the banks of the Jhelum River. It is fed by the perennial Doodhganga River, which makes its way through the village of Hajibabh, situated on its south-east, to meet the Jhelum. There are many floating gardens in the lake. The planting of *Salix* has been taken up along the shoreline, while rice is grown in the surrounding areas. These crop fields also provide foraging areas for birds.

Birdwatching Sites

Hokersar is the most accessible and best known of Kashmir's wetlands, which include Hygam, Shalibug and Mirgund. A great number of migratory birds has visited Hokersar in recent years. Those from Siberia and Central Asia use the wetlands as their transitory camps in September–October and again around spring.

Some of the important birds recorded in this area are the Greylag Goose, Ruddy Shelduck, Northern Pintail, Common Teal, Mallard, Red-crested Pochard, Northern Shoveler, Ruddy Shelduck, Eurasian Wigeon, Common Snipe, Garganey, Little Bittern, lapwings, kingfishers and herons.

Access & Accommodation

Hokersar Lake is located near Srinagar, so access to it is very easy. Srinagar has one airport, which has daily flights from Delhi and Jammu. It is also well connected through the road network.

There are various accommodation options available near Srinagar. You can book hotels online, depending on your budget and needs.

KEY FACTS

Nearest Major Towns
Srinagar

Habitats
Wetlands

Key Species
Garganey, Little Grebe, Black Kite, Eurasian Eurasian Jackdaw, Cinereous Tit, Citrine Wagtail, White Wagtail, Common Coot, Grey-headed Swamphen, Common Cuckoo, Little Bittern, Grey Heron, Little Egret, Black-crowned Night Heron, Common Kingfisher, European Bee-eater, Eurasian Roller, Long-tailed Shrike, Indian Golden Oriole, Clamorous Reed Warbler, Kashmir Flycatcher, Tickell's Thrush, Common Starling, European Goldfinch, Long-tailed Duck

Other Specialities
Snow Leopard

Best Time to Visit
Throughout the year, though winters are exceedingly cold

Little Bittern

Eurasian Wigeon

Pheasant-tailed Jacana

Garganey

Conservation

The main threats to this site are increased siltation, eutrophication and encroachment of agricultural land into the marshes peripheral to the lake. Fertilizer run-off from the nearby agricultural land into the lake accelerates the rate of eutrophication. The lake receives a heavy load of silt from the Doodhganga catchment area, and the expanses of open water are decreasing in size as the lake silts up and the reedbeds expand. It is felt that the lake has shrunk to one-third of its former size. Encroachment is also becoming a major threat to the site. Poaching is a common issue, but is less frequent due to vigilant media and pressure on the Wildlife Department.

Red-crested Pochards

Dachigam National Park

Hangul or Kashmir Stag

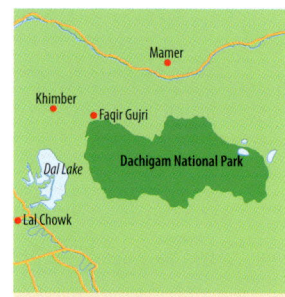

KEY FACTS

Nearest Major Towns
Srinagar and Anantnag

Habitats
Alpine moist scrub, Himalayan moist temperate forests, alpine moist pasture

Key Species
White-rumped Vulture, Eastern Imperial Eagle, Kashmir Flycatcher, Tytler's Leaf Warbler, Orange Bullfinch, Himalayan Vulture, Grandala, Tickell's Leaf Warbler, Red-mantled Rosefinch, Red-fronted Rosefinch, Koklass Pheasant, Himalayan Monal, Himalayan Rubythroat, Streaked Laughingthrush, Variegated Laughingthrush, Long-billed Bush Warbler, Western Crowned Leaf Warbler, Rusty-tailed Flycatcher, Fire-capped Tit, Rufous-naped Tit, Green-backed Tit, Bar-tailed Treecreeper

Other Specialities
Hangul, Himalayan Black Bear, Asiatic Brown Bear, Himalayan Langur, Himalayan Musk Deer

Best Time to Visit
Throughout the year, though winters are exceedingly cold

Located about 20km from Srinagar, this park was established in 1910 as a hunting reserve by the Maharaja of Kashmir. After the merger of Jammu and Kashmir with independent India in 1948, the management of the park was handed over to the Fisheries Department and subsequently to the Forest Department. The park is the catchment area of the Dal Lake, which supplies water to Srinagar. It has an area of some 140km^2 and covers a wide altitudinal range from the edge of the Kashmir Valley to 4,200m in upper Dachigam. The largest extant population of the highly endangered Kashmir Stag or Hangul is found in Dachigam.

Birdwatching Sites

Dachigam National Park is very rich in high-altitude birds. The park is of key importance for the globally vulnerable Kashmir Flycatcher. Along with this rare species, more than 150 bird species have been recorded in the park. Some of the other important bird species found here are the Besra, Northern Goshawk, Himalayan Vulture, Bearded Vulture, Himalayan Monal, Koklass Pheasant, Solitary Snipe, Collared Owlet, Tawny Owl, Scaly-bellied and Himalayan Pied Woodpeckers, and Slaty-blue and Ultramarine Flycatchers. A tarmac road runs in the park for several kilometres, then there are trails following the course of the river. With advance permission from the Forest Department, it is possible to walk inside the park. Birding is good along the main trail.

Eurasian Wren

Access & Accommodation

The park is about 20km from Srinagar, which is well connected with the rest of the state and other cities via a strong road network. The entrance to the park is a five-minute walk past the village of Harwan. You can easily get a bus or taxi to the entrance of Dachigam from Srinagar. To visit Dachigam, permission must be obtained from the office of the Chief Wildlife Warden in Srinagar. It is necessary to take a park ranger as a guide.

There is a forest guest house in Dachigam, but to stay there you need to book well in advance. Since the park is quite near Srinagar, you may find better accommodation options in Srinagar.

Conservation

Dachigam is vital not only as a refuge for the Hangul but also as an undisturbed catchment area for the Harwan Reservoir, which is the main fresh water supply for Srinagar and contributes major water supply to the Dal Lake. The two most common conservation challenges to this park are overgrazing and unregulated tourism. Poaching is not a big issue, although occasional cases are known to have occurred.

Collared Owlet

Orange Bullfinch

Red-billed Blue Magpie

Himalayan Pied Woodpecker

Hemis National Park

Tibetan Wild Asses

KEY FACTS

Nearest Major Towns
Ladakh

Habitats
Alpine dry pasture, riverine vegetation, alpine moist scrub

Key Species
Himalayan Vulture, Himalayan Snowcock, Tibetan Partridge, Ibisbill, Brown-headed Gull, Tibetan Sandgrouse, Snow Pigeon, Grey-backed Shrike, Robin Accentor, Brown Accentor, Guldenstadt's Redstart, Stoliczka's Bushchat, Red-mantled Rosefinch, Streaked Rosefinch, Great Rosefinch, Alpine Chough

Other Specialities
Tibetan Wolf, Eurasian Brown Bear, Red Fox, Mountain Weasel, Himalayan Mouse Hare, Himalayan Marmot

Best Time to Visit
Throughout the year, though winters are exceedingly cold

This park is located in the Trans-Himalayan Ladakh district on the south bank of the Indus River. It extends from the southern side of the Indus Valley, southwards across the Zanskar Range as far as the Tsarap Chu, and eastwards to the Buddhist monastery, Hemis Gompa, after which the park is named. The Markha and Rumbala Valleys and the Zanskar River are located within the park. Hemis is the largest protected area in the Indian Himalayas. Its large size and altitudinal ranges, from valley floors to mountain peaks, ensure that it is fully representative of the Trans-Himalayan ecosystem of central Ladakh. Important features of the park are remnant patches of juniper scrub and riverine woodland, the Snow Leopard and associated prey populations, with an uninhabited and little-disturbed core area. Almost all the high-altitude birds of the Western Himalayas are found in the park.

Birdwatching Sites

More than 100 high-altitude bird species have been recorded in this site, including many Vulnerable and Endangered species. They include the Golden Eagle, Himalayan Vulture, Bearded Vulture, Tibetan Snowfinch, Robin and Brown Accentors and Tickell's Leaf Warblers, Fork-tailed Swift, Red-fronted Serin, Himalayan Snowcock, Chukar Partridge and Red-billed Chough. Along with birds, Hemis is a great place to see some rare and elusive mammals like the Snow Leopard, Great Tibetan Sheep, Bharal, Asiatic Ibex, Tibetan Wolf, Red Fox, Eurasian Brown Bear, Himalayan Marmot, Mountain Weasel, Mountain Mouse Hare and Tibetan Wolf.

Access & Accommodation
The nearest airport to service the park is Leh, about 10km away.

Black-billed Magpie

Wallcreeper

Ground Tit

The nearest railway station is Jammu Tawi railway station. The nearest city, Leh, is well connected to major cities and places by a road network. There is a daily bus service from Leh to Hemis. Private vehicles and taxis can also be taken up to Hemis.

There are no hotels in or near the park. Six villages exist within its confines: Rumbak, Kaya, Sku, Shingo, Urutse and Chilling. The villagers offer basic homestay options. Hemis is mainly visited by trekkers. There are no facilities for concrete accommodation, but dormitories and tents may be used for an overnight halt. The Hemis Monastery provides accommodation for visitors, while Leh has many resorts and hotels to suit any budget.

Conservation

There are no major threats and challenges to this remote national park. However, from a long-term sustainability perspective, there is an immediate need to develop the park infrastructure, eliminate current land use and disturbance in the core area, and develop strategies in consultation with local people for managing resources in the buffer zone for the benefit of residents, but without detriment to the habitat. A reduction in grazing and the establishment of fuelwood plantations are high priorities. Extending both the core and buffer zones, and designating the whole area as a biosphere reserve, is suggested.

Black-necked Cranes

Tso Moriri, Tso Kar and Hanle

Tso Kar Lake

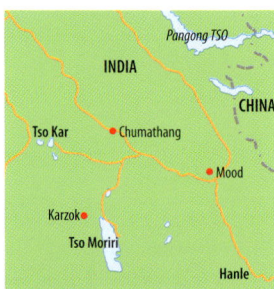

KEY FACTS

Nearest Major Towns
Leh and Ladakh

Habitats
Wetland, alpine moist pasture

Key Species
Black-necked Crane, Ferruginous Pochard, Common Stonechat, Tickell's Thrush, Blyth's Rosefinch, House Sparrow, Chukar Partridge, Bar-headed Goose, Alpine Swift, Golden Eagle, Great Rosefinch, Fire-capped Tit, Himalayan Rubythroat, White-throated Dipper, Rock Bunting, White-winged Redstart, Wallcreeper, Plain-backed Snowfinch

Other Specialities
Tibetan Wild Ass, Tibetan Gazelle, Lynx, Nayan, Bharal, Snow Leopard, Tibetan Wolf

Best Time to Visit
Throughout the year, though winters are exceedingly cold

Ladakh is a place of high-altitude lakes and passes. There are several remote lakes (*tso* means lake in the local language) in eastern Ladakh. Tso Moriri and Tso Kar are quite famous among birdwatchers and photographers, as they serve as breeding grounds for many bird species. Tso Moriri is situated at an impressive altitude of 4,552m and spreads over an area of 19km. This is the largest high-altitude Trans-Himalayan lake, situated in the scenic Rushpu Valley, and is a part of the larger Changthang Cold Desert Sanctuary.

Birdwatching Sites

Tso Kar About five hours from Leh, this is the easiest lake to visit from there, by driving along the Leh-Manali road crossing Tanglang La, which is a good place in which to look for the Tibetan Snowcock. There have also been reports of breeding Blanford's Snowfinches in this area. Tso Kar Lake is surrounded by marshland that hosts amazing birdlife, which includes the Bar-headed Goose, Ruddy Shelduck and Great Crested Grebe. However, the main attraction among the birdlife is the Black-necked Cranes, which come to Tso Kar to lay eggs. These birds make an amazing sight when they take off over the green, picturesque plains against the backdrop of snow-clad mountains. Besides birdwatchers, Tso Kar also attracts a lot of other wildlife lovers. The most easily spotted mammals here are Tibetan Wild Ass, the largest of the wild asses. The hills and mountains around Tso Kar are also home to Asiatic Ibex, Snow Foxes and Snow Leopards.

Chukar Partridge

Golden Eagle

Bar-headed Goose

Tso Moriri This lake is situated in the middle of the elevated district of Rupshu. The 137km drive from Leh to Rupshu can be covered in 7–9 hours. The area is rich in wildlife, including the Tibetan Wild Ass, and Black-necked Cranes and Bar-headed Geese flock to the lakeside for breeding during the summer months. This serene lake and the area around it attract a wide range of wildlife, including migratory birds, marmots, wild asses and the rarely spotted Tibetan Wolves.

Hanle This is one of the easternmost villages of Ladakh. It was closed to tourists for years, but has recently opened for domestic tourists only. Some of the rare wildlife of Ladakh is found in and around this high-altitude, remote region. There have been frequent records of elusive Pallas's Cats in the area.

Access & Accommodation
The only way to visit the above places is via a road journey, and you can hire a taxi from Leh to do so. You may also book a taxi online via www.ladakhtaxiunion.com. Due to the high altitude, it is recommended that visitors stay a couple of nights in Leh to acclimatize before starting their journeys. This is essential to avoid high-altitude sickness.

There are only a few good homestay options available at these remote places. Book in advance before travelling to them in order to avoid any last-minute trouble. Padma Homestay, run by naturalist Sonam Dorjay (http://padmahomestay.in), is a good place to stay around Hanle. He can also arrange accommodation near Tso Kar and Tso Moriri. Tso Kar Eco Resort (www.tsokarecoresort.in) in Tso Kar is well known among birdwatchers, and Hotel Grand Dolphin (http://hoteltsomoriri.com), located near Korzok Gömpa, is a convenient place to stay near Tso Moriri.

Conservation
This area used to be closed to tourists and the general public due to its proximity to the international border, but since 1994 it has been opened up. The main conservation problems that are now appearing are mostly due to increasing tourism pressures. Korzok, the only village situated near the lake, has become an important tourist centre. There is an immediate need to regulate tourism to ensure the sustainability of this fragile ecosystem.

Great Crested Grebe

KARNATAKA

Karnataka is situated in the western part of the Deccan Peninsular Region. The state is edged by a roughly 320km-long coastline. Karnataka shares its border with Maharashtra and Goa in the north, Telangana and Andhra Pradesh in the east, Tamil Nadu in the south-east and Kerala in the south-west. Bengaluru (formerly Bangalore), the capital of Karnataka, has a flourishing information technology industry and is known as India's Silicon Valley. The name Karnataka is derived from Karunadu, which means lofty land. Most of the state is plateau land, which justifies the name. The state can be divided into two regions – the hilly region comprising mainly the Western Ghats, and the plains region comprising the inland plateau of varying heights. The major rivers of the state are the Cauvery, Tungabhadra, Krishna, Sharavati and Kalinadi. All flow eastwards and fall into the Bay of Bengal. Karnataka has some of the most spectacular and best-known protected areas of India. The Bandipur Tiger Reserve is one of them. The hilly Nilgiri area of the state has been included in the Nilgiri Biosphere Reserve. In recent years, Nagarhole National Park has gained the attention of all wildlife enthusiasts due to frequent sightings of the Black Panther (melanistic Leopard) in the area.

Climate

Karnataka experiences mainly three types of climate. The state has dynamic and erratic weather that changes from place to place within its territory. Due to its varying geographic and physiographic conditions, it experiences climatic variations that range from arid to semi-arid in the plateau region, sub-humid to humid tropical in the Western Ghats, and humid tropical monsoon in the coastal plains. The rainfall varies from 450mm in the drier northeastern plateau to 3,200mm in the Western Ghats. The average high temperature during summer is 34°C across the state. The average day temperature is 29 °C in the monsoon season. During winter temperatures range from 32 °C to below 20 °C.

1. Bandipur National Park
2. Nagarahole National Park
3. Dandeli, Ganeshgudi and Anshi National Park
4. Hampi
5. Nandi Hills
6. Ranganathittu Bird Sanctuary

Access, Transportation & Logistics

Karnataka is very well connected with the rest of India and many international destinations through its six airports. Kempegowda Airport (in Bangalore) and Mangalore Airport (in Mangalore) are two

KEY BIRDS
Top 10 Birds
1. Blue-breasted Quail
2. Red-billed Tropicbird
3. Jerdon's Nightjar
4. Brown-backed Needletail
5. Brown Booby
6. Eastern Grass Owl
7. White-bellied Woodpecker
8. Ashy Minivet
9. Painted Francolin
10. Grey Junglefowl

Asiatic Wild Dogs in Kabini Forest

Common Kingfisher

Coppersmith Barbet

international airports. Karnataka's major railway station is in Bangalore. Being connected to Mumbai, Delhi, Chennai, Hyderabad, Sikkim, Chennai, Kolkata and Jammu, the railway head in the city offers the maximum number of trains for commuting every day. Almost every town and city in Karnataka can be reached by rail. The Karnataka State Road Transport Corporation connects the state with all other states and towns of South India. The North Eastern Karnataka Road Transport Corporation and North Western Karnataka Road Transport Corporation are two bus services that can take you to the remote areas of Karnataka, making commuting in the state easier. A large number of national highways links Karnataka with Maharashtra, Kerala, Andhra Pradesh and Tamil Nadu.

Health & Safety

Karnataka is a safe state for visitors in India, and there are no major health concerns here. It is a low- to no-risk area of India when it comes to malaria, but do take precautions, especially after the monsoon. Visit a GP or travel clinic 6–8 weeks before departure to make sure you are up to date with any vaccinations. Avoid roadside food and drink bottled water. During the summer months keep yourself hydrated and carry a stole, scarf or cap for protection against direct sunlight. The Karnataka government has banned night traffic through the road passing through Bandipur National Park from 9 p.m. to 6 a.m., so if you have to visit or pass through this area, plan accordingly.

Birdwatching Highlights

The birdlife of Karnataka is unique and remarkable, and there are many IBAs in the state. A substantial part of the Western Ghats Endemic Bird Area lies in the state, and most of the endemic bird species of the Western Ghats are found in the IBAs of Karnataka. More than 542 bird species have been identified in this region, including Western Ghats endemics like the Nilgiri Wood Pigeon, Malabar Parakeet, Malabar Grey Hornbill, Grey-headed Bulbul, White-bellied Shortwing, Wayanad Laughingthrush, Black-and-orange Flycatcher and Nilgiri Flycatcher. Other important bird species found here include the Spot-billed Pelican, Greater Spotted Eagle, Yellow-throated Bulbul, Broad-tailed Grassbird, Pallid Harrier, Malabar Pied and Great Hornbills, Grey-headed Fish Eagle and Eastern Imperial Eagle.

Brown Fish Owl

Bandipur National Park

Bandipur landscape

KEY FACTS

Nearest Major Towns
Mysore

Habitats
Tropical moist deciduous forests

Key Species
White-rumped Vulture, Lesser Adjutant, Yellow-throated Bulbul, Cinereous Tit, Nilgiri Wood Pigeon, Malabar Parakeet, Malabar Grey Hornbill, Grey-headed Bulbul, White-bellied Blue Flycatcher, Crimson-backed Sunbird, White-bellied Treepie

Other Specialities
Tiger, Indian Leopard, Asian Elephant, Gaur, Spotted Deer, Striped-necked Mongoose, Sloth Bear

Best Time to Visit
October to March

This is one of the best-known national parks not only in Karnataka, but also in the whole of India. Contiguous to Mudumalai Wildlife Sanctuary (IBA) in Tamil Nadu in the south, to Wynaad Wildlife Sanctuary (IBA) in Kerala in the south-west, and to Nagarhole National Park (IBA) in the north-west, it holds most of the representative species of South India biodiversity. The park is a part of the Nilgiri Biosphere Reserve. The terrain of Bandipur is undulating and broken by chains of hills, flat-topped hillocks and watercourses. The park is drained by the perennial Kabini, Nugu and Moyar Rivers. One of its main waterbodies is the Kabini Reservoir, which separates Bandipur from Nagarhole National Park. The underlying rocks are mainly metamorphic, and the soil is a mixture of red laterites and black cotton soil.

Birdwatching Sites

Bandipur is rich in birdlife, and many of the region's endemics can be seen on a trip to the park. It is a good place in which to see the Red Spurfowl, Grey Junglefowl, Malabar Pied and Malabar Grey Hornbills, Malabar Parakeet, Heart-spotted Woodpecker, Changeable Hawk Eagle, Spot-bellied Eagle Owl, Mottled Wood Owl, Grey-fronted Green Pigeon, Indian Swiftlet, Jerdon's Bushlark, White-bellied Drongo, Tawny-bellied and Rufous Babblers, Yellow-browed Bulbul, Indian Scimitar Babbler and Malabar Trogon. Regular jeep safaris are run by the Forest Department. There are also some birdwatching trails outside the national park. The road to Ooty is good for the White-bellied Minivet, Grey-headed Bulbul, Malabar Lark and Malabar Whistling Thrush.

Malabar Grey Hornbill

Access & Accommodation

Bandipur National Park is located on the highway connecting Mysore and Udhagamandalam (Ooty). It takes two and a half hours to reach Bandipur from Mysore, and about three hours from Ooty. The nearest important rail junction is Mysore and the nearest international airport is Bangalore.

One of the most famous and convenient accommodation options for birdwatchers and wildlife photographers is Jungle Lodges (www.junglelodges.com/bandipur-safari-lodge), which also arranges safaris in the park. Additionally, various private hotels and resorts near the park can be booked online.

Ramki Sreenivasan

Asian Koel

Conservation

Bandipur National Park contains the entire area of the Tiger Reserve, and is split into core, tourism and restoration zones. There is no human settlement inside the park, but many villages are situated adjacent to the park. Illegal collection of fuelwood and smuggling are major problems. A national highway, Mysore-Sultan Bathery-Kozhikode, and one state highway, Gundlupet-Ooty, pass through the park, both with very heavy traffic, which causes many accidental deaths of wild animals every year.

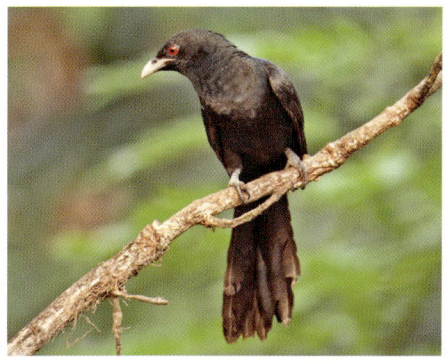

Clement M Francis

Changeable Hawk Eagle

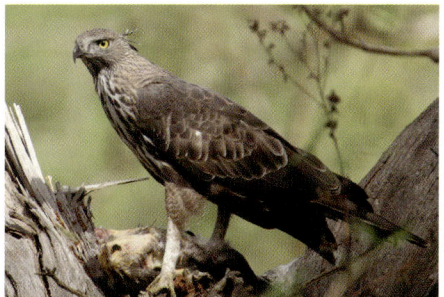

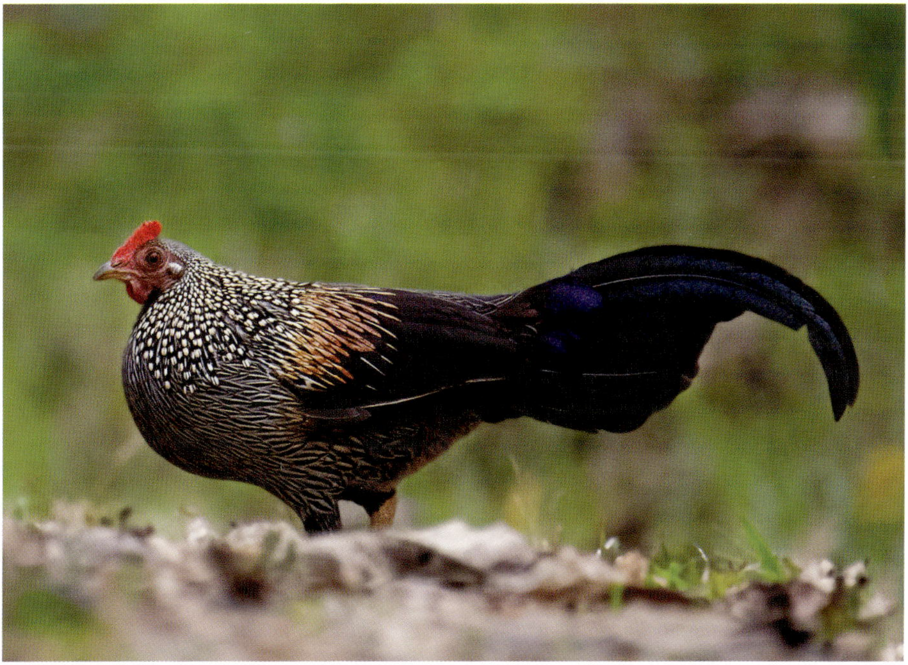

Amit Sharma

Grey Junglefowl

Nagarahole National Park

Karnataka

Asian Elephant

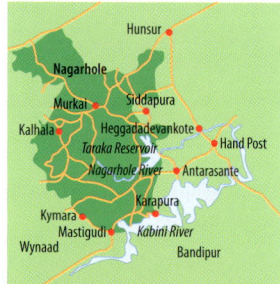

KEY FACTS
Nearest Major Towns
Mysore and Mercera

Habitats
Tropical moist deciduous forest, tropical dry deciduous forest

Key Species
White-rumped Vulture, Lesser Adjutant, Greater Spotted Eagle, Nilgiri Wood Pigeon, Oriental Darter, Black-headed Ibis, Grey-headed Fish Eagle, Red-headed Vulture, Malabar Parakeet, Malabar Grey Hornbill, White-bellied Treepie, Green-billed Malkoha, Jerdon's Nightjar, White-cheeked Barbet, Yellow-browed Bulbul, Malabar Whistling Thrush, Indian Scimitar Babbler, Loten's Sunbird, Painted Bush Quail, Indian Peafowl, Yellow-legged Green Pigeon, Plum-headed Parakeet, Sirkeer Malkoha, Indian Grey Hornbill, Yellow-crowned Woodpecker, Small Minivet

Other Specialities
Tiger, Indian Leopard, Asian Elephant, Gaur, Spotted Deer, Striped-necked Mongoose, Sloth Bear, Marsh Crocodile

Best Time to Visit
October to March

Situated within the southeastern parts of Kodagu (Coorg) and southwestern parts of the Mysore district, Nagarahole National Park borders Kerala state. It is very close to Mysore city, which is about 50km to its north-east. The southern border is contiguous with the reservoir of the Kabini River Dam. Nagarahole forms a part of the Nilgiri Biosphere Reserve and, together with the Bandipur Tiger Reserves and Mudumalai National Park in the south-east, and the Waynaad Wildlife Sanctuary in the south-west, forms the largest protected forest tract in peninsular India.

Birdwatching Sites
The park is famous for its Indian Leopard population, and especially for the Black Panther (melanistic Leopard) sightings in recent years. It is also home to various birdlife, and more than 300 bird species have been recorded here. Some of the important ones are the Lesser Adjutant, Greater Spotted Eagle, Nilgiri Wood Pigeon, Grey-headed Fish Eagle, Green-billed Malkoha, Jerdon's Nightjar, Yellow-browed Bulbul and Loten's Sunbird. Jungle lodges in Nagarahole organize jeep as well as boat safaris in the park in conjunction with the Forest Department. There are also birding trails outside the park.

Access & Accommodation
The nearest airport to the park is Mandakalli Airport, Mysore, which is about 98km away. The airport has regular flights connected to Bengaluru. The nearest international airport is in Bangalore, about 268km away. It is well connected to important Indian and international cities through regular flights. Rent a taxi from the airport to reach the park.

Indian Golden Oriole

Shikra

Loten's Sunbird

Kabini River Lodge (www.junglelodges.com/kabini-river-lodge), run by Jungle Lodges, is the preferred place to stay by many wildlife photographers and birdwatchers because jungle safaris start from it. There are also many private lodges near the park.

Conservation

Large-scale felling and smuggling of sandalwood and teak trees take place in Nagarhole National Park. Poaching of elephant tusks, apart from birds and other animals, also takes place in this IBA. Forest fires set by humans are regular during the summer. Severe drought and water shortages are reported to be forcing wild animals to migrate to the forests of neighbouring areas in Kerala. There have also been reports of villages on the park fringe being raided by wild animals, particularly elephants, and this demands some organized effort to minimize the human-animal conflict.

Indian Leopard

Dandeli, Ganeshgudi & Anshi National Park

Kali River

KEY FACTS

Nearest Major Towns
Hubbali and Dharwad

Habitats
Tropical semi-evergreen forest, tropical wet evergreen forest

Key Species
Nilgiri Wood Pigeon, Malabar Parakeet, Malabar Grey Hornbill, White-bellied Blue Flycatcher, Crimson-backed Sunbird, White-bellied Treepie, Sri Lanka Frogmouth, Malabar Trogon, Malabar Pied Hornbill, Yellow-browed Bulbul, Malabar Whistling Thrush, Black-headed Babbler, Loten's Sunbird, Black-throated Munia

Other Specialities
Tiger, Indian Leopard, Gaur, Marsh Crocodile

Best Time to Visit
Throughout the year

Anshi National Park lies in the Western Ghats, adjoining the state of Goa, to the south of Dandeli Wildlife Sanctuary. Earlier, it was a part of the sanctuary. The step to alter the limits and boundaries of the sanctuary was considered necessary due to the hydroelectric project, naval base, rehabilitation of displaced persons, roads, transmission lines, and mining and other industries that existed within it. The park area is less disturbed than the sanctuary. To the west, the park adjoins the Cotigao Wildlife Sanctuary in Goa. Anshi can be reached from Bangalore by road or rail to Dharwad (470km), then by road to Dandeli (62km), and on to Anshi (38km). Anshi National Park contains deep valleys, steep hills and rich wet evergreen and semi-evergreen forests.

Birdwatching Sites

The Dandeli and Anshi area is a birdwatchers' paradise and one of the prime areas for birding in South India. More than 300 bird species have been recorded here. Some of the key species include the Malabar Pied Hornbill, Malabar Trogon, Black-naped Monarch, Orange Minivet, Dark-fronted Babbler, Sri Lanka Frogmouth, Red Spurfowl and Blue-faced Malkoha. The Ganeshgudi area has some bird hides where birders can enjoy armchair birding and bird photography from a close distance without much effort. There are also birding trails for some of the elusive birds like the Malabar Trogon, Heart-spotted Woodpecker and hornbills.

Flame-throated Bulbul

White-rumped Shama

Dark-fronted Babbler

Access & Accommodation
Anshi National Park can be reached by rail, road and air. The Karwar railway station, which is 55km away, is the nearest option. The nearest airport is Dharwad, which is 106km from the park. Regular buses from Dandeli to Bangalore, Pune, Panaji, Dharwad, Belgaum, Karwar and Mumbai are available. The nearest railhead to reach the Ganeshgudi area is Londa, which is about 30km away. You can take a taxi or bus from Londa.

The Old Magazine House, a Jungle Lodges property, is about a 10-minute walk from the bus stop. Jungle Lodges also arrange safaris in the Dandeli and Anshi area.

Conservation
Anshi National Park and the adjoining Dandeli Wildlife Sanctuary have many conservation problems arising out of tribal communities residing inside protected areas, the question of their displacement and rehabilitation, and ill-conceived development projects. Another major issue is rapid development and construction in nearby areas.

Lesser Hill Myna

Black-naped Monarch

Malabar Trogon

Hampi

Daroji Sloth Bear Sanctuary

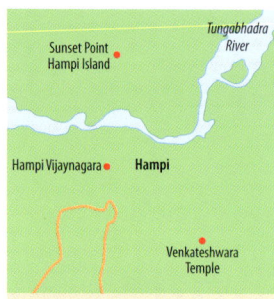

Hampi is a historic temple town and a UNESCO World Heritage Site in northern Karnataka. At its prime in AD 1500–1565, this capital of the great Vijayanagar Empire was one of the richest and largest cities in the world. Hampi was sacked in 1565 by the Sultanate armies, and today the ruins lie scattered over 40km² of craggy hills, open plains and granite outcrops of the Tungabhadra River Basin in central Karnataka's Bellary district. The Hampi area and nearby Daroji Sloth Bear Sanctuary provide an ideal getaway for birdwatchers and wildlife lovers.

Birdwatching Sites

The Hampi area lists more than 200 bird species, of which the main draw is the endemic and threatened Yellow-throated Bulbul. Other important birds here include Painted and Red Spurfowl, Jungle Bush Quail, Barred Buttonquail, Brown Fish Owl, Indian Eagle Owl, Chestnut-bellied and Painted Sandgrouse, Sirkeer Malkoha, Indian Courser, Montagu's Harrier, Eurasian Sparrowhawk, Rufous-tailed Lark, Jerdon's Bushlark, White-browed Bulbul, Yellow-billed Babbler, Hume's Whitethroat, Loten's Sunbird and Grey-necked Bunting.

Daroji Sloth Bear Sanctuary This sanctuary is spread over 83km² and located 20km south of the Hampi ruins. There is a watchtower from which bears can be viewed in the evening. This is also a good place to see Painted Spurfowl and bush quail.

Access & Accommodation

Hampi is about 350km from Bangalore and it takes about six hours to reach it by road. Hubballi Airport (143km) and Jindal Vijaynagar Airport (38km) are two nearby airports, but they are less connected than Bangalore Airport.

KEY FACTS

Nearest Major Towns
Bellary

Habitats
Tropical thorn forest, tropical secondary scrub

Key Species
Yellow-throated Bulbul, Ashy-crowned Sparrow Lark, Black Redstart, Common Sandpiper, Indian Robin, Little Cormorant, Painted Sandgrouse, Rufous-tailed Lark, Sirkeer Malkoha, White-browed Bulbul, Grey Francolin, Plain Prinia, Indian Peafowl

Other Specialities
Sloth Bear, Indian Wolf, Blackbuck, Bonnet Macaque, Indian Grey Mongoose, Southern Plains Grey Langur, Bengal Monitor, Indian Star Tortoise

Best Time to Visit
Throughout the year

Montagu's Harrier

Yellow-throated Bulbul

Painted Spurfowl

The Jungle Lodges resort named the Sloth Bear Heritage Resort (www.junglelodges.com/hampi-heritage-and-wilderness-resort) is a famous place among birdwatchers. For a more luxurious stay option, there is an Orange County Resort (www.evolveback.com/hampi).

Conservation

There are no major immediate conservation threats. Hampi is now included in the UNESCO list of places of global significance. The Karnataka government, together with the Archaeological Survey of India and assistance from UNESCO, the Smithsonian Institution and several other countries, has undertaken the gigantic task of excavation in the 26km^2 city. The principal remains are being mapped and many areas are being excavated to view the remains of temples, tanks and other structures that have lain buried for centuries. Several structural remains and scriptures are being carefully restored, in addition to partial rebuilding and replacement of building elements. In the late 1990s Karnataka's first and southern India's only sanctuary for bears – the Daroji Bear Sanctuary – began taking shape near the Hampi ruins. The final notification of this sanctuary has been completed. It will be of great advantage to the fauna of Hampi, which sometimes encounters disturbance caused by the tourists who visit the site.

Jungle Bush Quail

Rock Bush Quail

Nandi Hills

Nandi Hills

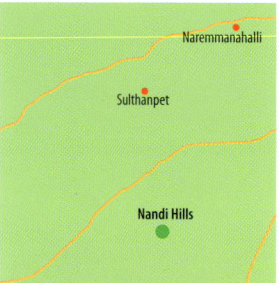

KEY FACTS

Nearest Major Towns
Bengaluru and Chickballapur

Habitats
Tropical dry deciduous forest, tropical secondary scrub

Key Species
White-rumped Vulture, Long-billed Vulture, Nilgiri Wood Pigeon, Yellow-throated Bulbul, Red-headed Vulture, Jungle Bush Quail, Painted Spurfowl, Plum-headed Parakeet, Sirkeer Malkoha, Indian Nightjar, Lesser Golden-backed Woodpecker, Black-rumped Flameback, Ashy-crowned Sparrow Lark, Black-headed Cuckooshrike, Small Minivet, Common Woodshrike, White-browed Bulbul, Indian Robin, Tawny-bellied Babbler, Large Grey Babbler, Jungle Babbler, White-hooded Babbler

Other Specialities
Bonnet Macaque, Southern Plains Langur

Best Time to Visit
Throughout the year

Also referred to as Nandi Durg, Nandi Hills is about 60km north of Bangalore, and is a popular tourist spot. The site lies within the Nandi State Forest, comprising three main hillocks (over 1,400m), with seven peaks in all. Of these, Nandi Hills is the tallest (1,435m). Though Nandi Hills has a general pattern of scrub and deciduous vegetation, altitudinal variations in the floristic composition can be seen. Nandi Hills is a good place in which to look for many Western Ghats endemic birds.

Birdwatching Sites

Nandi Hills is just 60km from Bangalore, but its terrain is quite different. A winding road with many curves leads you to the top of the hill, where vehicles can be parked and you can walk around. There are many trails and you can enjoy birding on foot. Some of the important bird species found in the area are the Nilgiri Wood Pigeon, Yellow-throated Bulbul, Painted Spurfowl, Sirkeer Malkoha, Plum-headed Parakeet, Small Minivet, Common Woodshrike, Indian Paradise-flycatcher, Orange-headed Thrush, Tickell's Blue Flycatcher and Puff-throated Babbler.

Access & Accommodation

Nandi Hills is quite close to Bangalore city, which is very well connected with other major Indian cities. You can hire a taxi or take a private vehicle to reach the site from Bangalore.

Many accommodation options are available in Bangalore, from budget hotels to expensive and luxurious resorts. November–April is a good time to visit for the best birding experience.

Conservation

The habitat of the Nandi Hills gives shelter to many globally

Crested Serpent Eagle

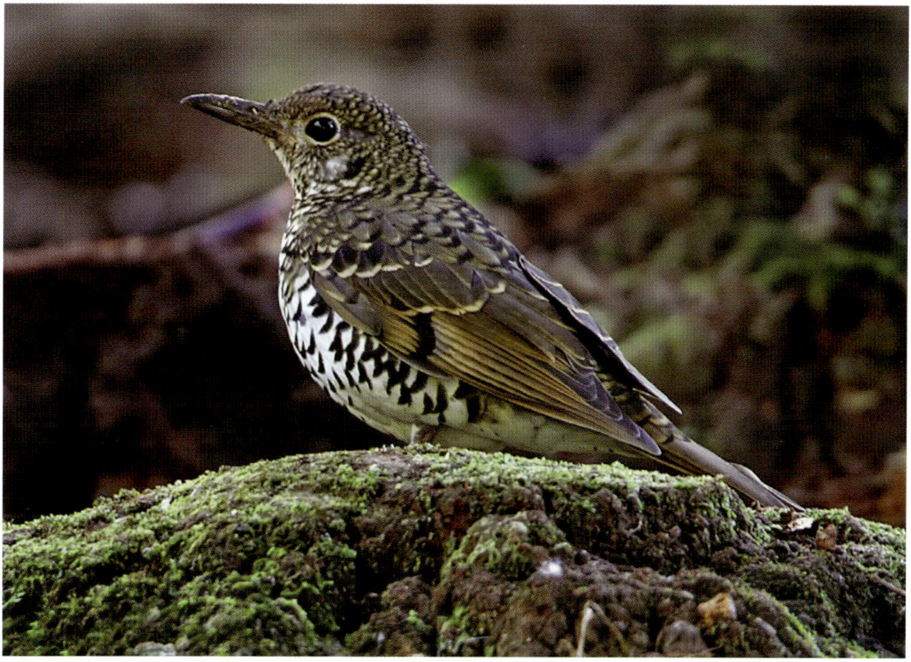

Clement M Francis

Nilgiri Thrush

threatened and biome-restricted bird species – the Yellow-throated Bulbul is one of the vulnerable key species in this IBA. Since the site is well known to tourists, causing disturbance to the birds and their habitat, some specific areas where these birds are normally seen should be completely protected for the long-term conservation of these species.

Clement M Francis

Nilgiri Wood Pigeon

Garima Bhatia

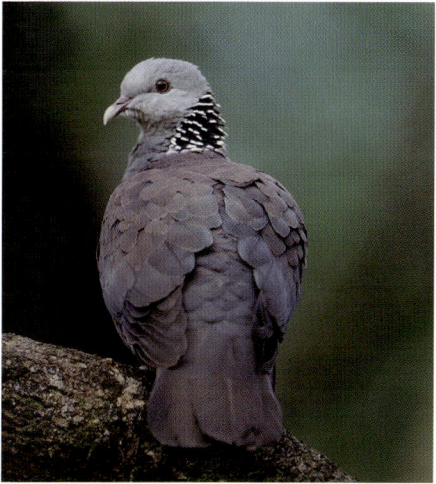

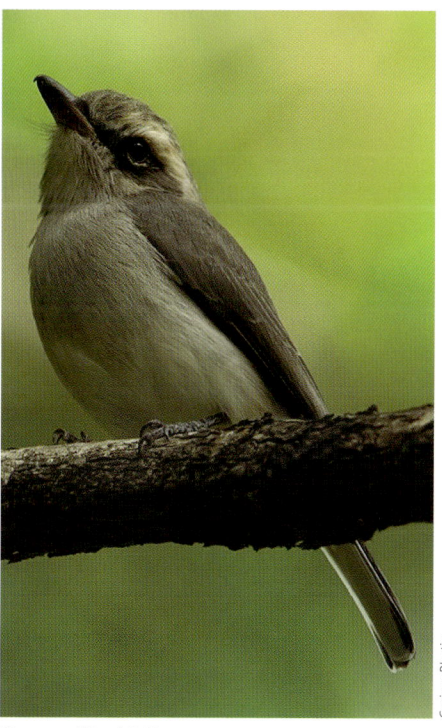

Common Woodshrike

Ranganathittu Bird Sanctuary

Spot-billed Pelican

Located in the Srirangapatna taluka of the Mandya district of Karnataka, this is a small yet important birding area. The sanctuary comprises six islands and six islets in the Kaveri River, and attracts nearly 40,000 birds in winter. The islets are surrounded by the water of a reservoir formed by the construction of a weir across the Kaveri River. In 1940, the area was declared a wildlife sanctuary based on the initiative of Salim Ali (the 'Birdman of India'; an Indian ornithologist and naturalist), who wanted to protect this important nesting site of many species.

Birdwatching Sites

Ranganathittu is a small bird sanctuary, but it is home to many resident as well as migratory birds. The checklist of birds for the area contains more than 200 bird species. The sanctuary is primarily a large nesting heronry where birds like cormorants,

KEY FACTS

Nearset Major Towns
Mysore

Habitats
Riverine vegetation and wetland

Key Species
White-rumped Vulture, Greater Spotted Eagle, Oriental Darter, Painted Stork, Black-headed Ibis, Spot-billed Pelican, Painted Stork, Black-headed Ibis, Eurasian Spoonbill, Great Thick-knee, River Tern, Asian Openbill, Brahminy Kite, Western Marsh Harrier, Changeable Hawk Eagle, Indian Paradise-flycatcher, Indian Pitta, Rosy Starling, Forest Wagtail, Tickell's Blue Flycatcher, Indian Golden Oriole

Other Specialities
Marsh Crocodile, Indian Grey Mongoose, Water Monitor, Bonnet Macaque, Indian Flying Fox

Best Time to Visit
Throughout the year

Forest Wagtail

Little Cormorant

Purple-rumped Sunbird

Asian Openbill

egrets, herons, darters, storks, ibises and pelicans dominate the landscape. Key birds include the Spot-billed Pelican, Painted Stork, Black-headed Ibis, Eurasian Spoonbill, Great Thick-knee, River Tern and Asian Openbill. Raptors are regular and include the Brahminy Kite, Western Marsh Harrier and Changeable Hawk Eagle. Woodland species like the Indian Paradise-flycatcher, Indian Pitta, Rosy Starling, Forest Wagtail, Tickell's Blue Flycatcher and Indian Golden Oriole can be found at waters' edges.

Access & Accommodation

The sanctuary is 18km from Mysuru (Mysore) City, which is very well connected with the rest of India. Mysore offers a variety of accommodation to suit every pocket. If you want to stay closer to the sanctuary, book at the nearby Karnataka State Tourism Development Corporation's Hotel Mayura Riverview (https://kstdc.co/hotels/hotel-mayura-riverview-srirangapatna). The sanctuary is open throughout the year, but the winter season in November–April is considered the best time to visit for birding.

Conservation

Pollution by fertilizers and pesticides drained from the surrounding agricultural land is a major concern. Restricting the recreation and tourism in the nesting area of the sanctuary would help to prevent degradation of the habitat as well as biodiversity. Tourism seems to be the major problem here.

River Tern

Purple Herons

KERALA

1. Periyar Tiger Reserve
2. Vembanad Lake, Kumarakom Bird Sanctuary
3. Thattekad
4. Eravikulam National Park, Munnar

Kerala is one of the smaller states of India and is known as 'God's own country'. Much folklore is linked to the state – according to one Hindu mythological tale, Kerala was created by Lord Parasurama, an incarnation of the great Lord Vishnu, when he threw his axe across the sea to create new land for his devotees to live in peace. The natural beauty of Kerala is unparalleled. Nestled between the pristine waters of the Arabian Sea on the west and the lush mountains of the Western Ghats on the east, an intense network of rivers and lagoons, thick forests, exotic wildlife, tranquil stretches of emerald backwaters and a long shoreline of serene beaches make it a travellers' paradise. The Bharathappuzha, Periyar, Pampa and Chaliyar Rivers are the main rivers of this state. Comprising big rivers along with many smaller ones, they have blessed the state with an abundance of water resources and have supported many forms of flora and fauna. The state has some of the best-known sanctuaries in India, such as Periyar Tiger Reserve, Eravikulam National Park, Parambikulam Wildlife Sanctuary and Thattekad Bird Sanctuary.

Climate

Kerala has an extensive range of physical features that result in a corresponding diversity of climatic features. The high ranges have a cool and bracing climate, while the plains are hot and humid. The temperature ranges from 20 °C to 37 °C. Due to the mountainous nature of the state, it receives heavy rainfall. However, some rain-shadow areas receive less than 1,000mm rainfall. The most important of the rain-shadow areas is Chinnar Wildlife Sanctuary (an IBA) where the rainfall is about 500mm. However, in most other areas, the average annual rainfall is 1,520–4,075mm, with June being the month of the heaviest rainfall. November–April, when the weather is both clear and pleasant, is considered to be the best time for birding.

KEY BIRDS

Top 10 Birds
1. Great Eared Nightjar
2. Red Spurfowl
3. Sri Lanka Bay Owl
4. Indian Swiftlet
5. Chestnut-winged Cuckoo
6. Slaty-legged Crake
7. Caspian Plover
8. Great Knot
9. Long-billed Dowitcher
10. Mountain Hawk Eagle

Access, Transportation & Logistics

Kerala is well connected with the rest of the country via airways, railways and roadways. There are three international airports located in Kerala, which connect the state to major national and international cities. Thiruvananthapuram International Airport is in the southern part of Kerala, and Cochin International Airport is in the central part. Up north, Calicut International Airport is also connected to both Indian and international cities. Kerala is easily accessible by road from other parts of India through national highways. Most of the places in the state are interconnected via railway stations. There are direct trains to Kerala from Delhi, Mumbai, Chennai, Bangalore, Kolkata and other major Indian cities. The backwaters of Kerala not only provide a popular means of transport, but are tourist attractions as well. Presently, these internal water-navigation systems are the single most popular travel product of Kerala, with public ferry services, speedboats and houseboats.

Health & Safety

Kerala is one of the safest states for visitors to India, with no major health concerns. It is a low- to no-risk area when it comes to malaria, but do take precautions especially after the monsoons. Visit a GP or travel clinic 6–8 weeks before departure to make sure you are up to date with any vaccinations. Avoid roadside food and drink bottled water. During the summer months keep hydrated, and carry a stole, scarf or cap to protect yourself from direct sunlight.

Birdwatching Highlights

The birdlife of Kerala is unique and remarkable. More than 525 bird species have been identified in this region, including some extremely rare ones. Among the Critically Endangered species, White-rumped Indian and Red-headed Vultures occur in the state, as does the Nilgiri Laughingthrush, one of the Endangered species. The IBAs and protected areas of Kerala are very important for the long-term survival of many birds like the Ferruginous Duck, Eurasian Wigeon, Painted Bush Quail, Painted Spurfowl, Nilgiri Wood Pigeon, Swinhoe's Storm-petrel, Flesh-footed Shearwater, Streaked Shearwater, Jouanin's Petrel, Great Frigatebird, Curlew Sandpiper, Wood Snipe, Broad-tailed Grassbird, Bristled Grassbird and five species of nightjar, including the Great Eared Nightjar.

Rahul & Khushboo Sharma

Lesser Kestrel

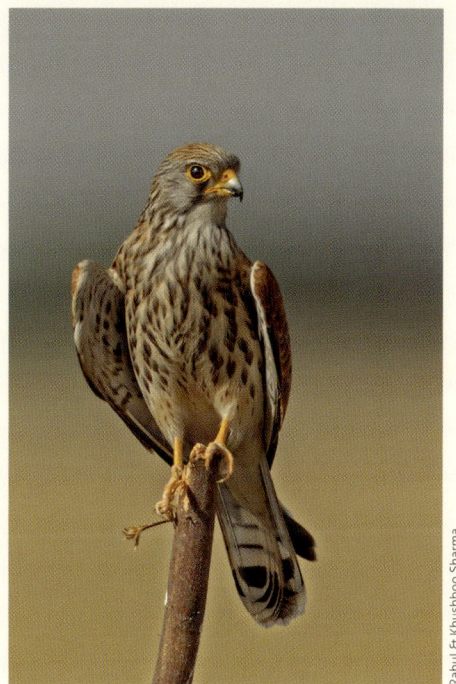

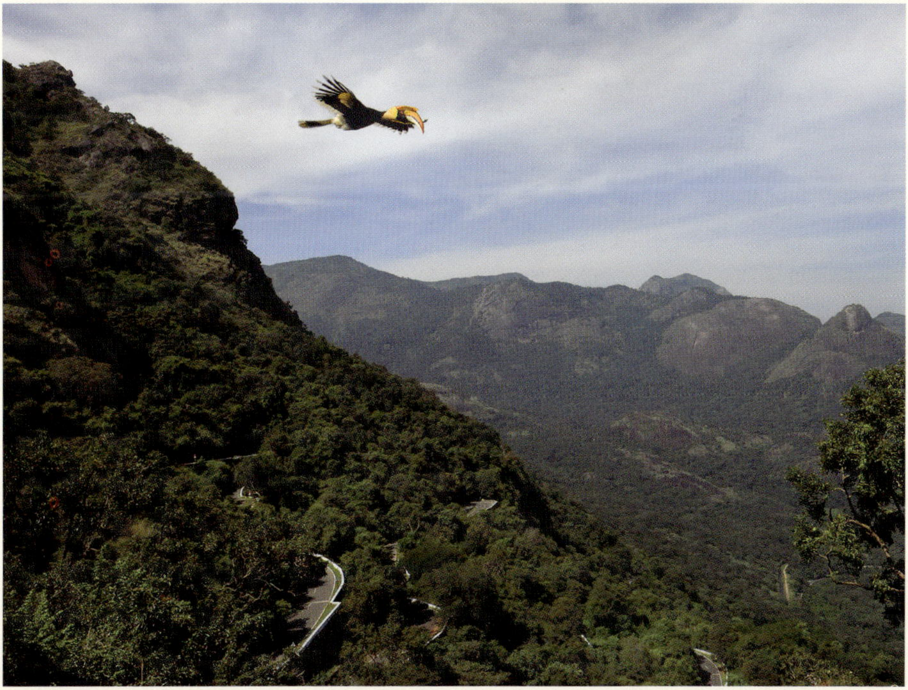

Clement M Francis

Annamalai Hills

Periyar Tiger Reserve

Periyar landscape

KEY FACTS

Nearest Major Towns
Idukki

Habitats
Tropical wet evergreen forest, reservoir

Key Species
White-rumped Vulture, Long-billed Vulture, Greater Spotted Eagle, Wood Snipe, White-bellied Shortwing, Nilgiri Pipit, Black-and-orange Flycatcher, Nilgiri Flycatcher, Nilgiri Wood Pigeon, Malabar Parakeet, Malabar Grey Hornbill, Grey-headed Bulbul, Wayanad Laughingthrush, Rufous Babbler, Broad-tailed Grassbird

Other Specialities
Asian Elephant, Gaur, Tiger, Sambar, Wild Pig, Malabar Giant Squirrel, Travancore Flying Squirrel, Jungle Cat, Nilgiri Tahr, Lion-tailed Macaque, Nilgiri Langur, Nilgiri Marten

Best Time to Visit
October to March

This is one of the most famous Tiger Reserves in India. Named after the Periyar River, the park is bounded by the Madurai and Ramanathapuram districts in the east, the Kottayam district in the west and the Pathanamthitta district in the south. It was one of the first Project Tiger areas in the country. Though the main attraction of the park is the Tiger, the area is very rich in birdlife, hence the park is categorized as an IBA. The vegetation in the reserve is mainly composed of tropical evergreen forests and semi-evergreen forests; moist deciduous forests and grassland predominate in the central part of the sanctuary.

Birdwatching Sites

Periyar is one of the most visited places in South India. Birdwatchers visit to see the Western Ghats endemics and forest birds. More than 300 bird species have been recorded in this area, including almost all Western Ghats endemic species. Periyar is also an important wintering site for many long-distance migrants such as Tickell's, Large-billed and Western Crowned Leaf Warblers, Rusty-tailed Flycatcher, Pied Thrush, Painted Bush Quail and Red Spurfowl. Western Ghats endemics include the Nilgiri Wood Pigeon, Malabar Grey Hornbill, Malabar and White-cheeked Barbets, Malabar Parakeet, White-bellied Treepie, Crimson-backed Sunbird and Nilgiri Pipit.

Boating is a preferred way of birding in the reserve, providing ample opportunities for photography. There are also some birding trails. Visitors need to take precautions as most of the trails are infested with leeches, especially just after the rainy season. Regular boat, jeep and elephant safaris are organized by the Forest Department within the park.

Black Bittern

Crested Treeswift

Black-hooded Oriole

Access & Accommodation

Periyar Tiger Reserve is accessible from Cochin International Airport (about 200km away) and Madurai Airport in Tamil Nadu (about 140km). Kottayam Railway Station, about 114km away, serves as the nearest railway station to reach the park. Following arrival in Kottayam, visitors can either opt for regular buses or hire independent cabs or taxis to reach the park.

There are accommodation options inside the park managed by the Kerala Tourism Development Corporation (Aranya Nivas and Periyar House). Lake House is one of the prime properties managed by the corporation. Bookings for its resorts can be made online (www.ktdc.com). There are many private properties outside the park, which visitors may opt for depending upon their budgets and requirements.

Conservation

Some of the common conservation issues faced by the park are grazing, poaching of wild mammals and birds, illegal cutting of trees and firewood collection. There have been many incidents of elephant poaching, which has left only a handful of tuskers in the park.

Cormorants

Great Eared Nightjar

Vembanad Lake, Kumarakom Bird Sanctuary

Kerala

Vembanad Lake

Vembanad Lake is a coastal lagoon – one of the richest wetland habitats and the largest on the southwestern coast of India. It was declared a Ramsar Site of international importance in 2002. The town of Kumarakom, bordered by the Vembanad Lake, is known for backwater tourism in Kerala. Set amid the Vembanad Lake, the Kumarakom Bird Sanctuary is home to a number of birds. You can see the resident birds as well as migratory species flocking around the sanctuary. It has the largest number of heron species in the state, and during the 1970s it was the only breeding place for night herons.

KEY FACTS

Nearest Major Towns
Kochi, Alleppey and Kottayam

Habitats
Wetlands

Key Species
Lesser Whistling Duck, Cotton Pygmy-goose, Oriental Turtle Dove, Asian Palm Swift, Greater Coucal, Common Hawk Cuckoo, Slaty-legged Crake, Ruddy-breasted Crake, White-breasted Waterhen, Common Moorhen, Asian Openbill, Spot-billed Pelican, Yellow Bittern, Striated Heron, Purple Heron, Intermediate Egret, Black-headed Ibis, Little Cormorant, Oriental Darter, Pacific Golden Plover, Greater Painted-snipe, Pheasant-tailed Jacana, Common Redshank, Pallas's Gull, Sooty Tern, Caspian Tern, Western Marsh Harrier, Orange-headed Thrush, Pied Bushchat, Common Stonechat, Bluethroat, Oriental Magpie Robin, Indian Robin, Jungle Myna, Common Myna, Chestnut-tailed Starling, Rosy Starling, Brahminy Starling

Other Specialities
Smooth-coated Otter, Wild Pig, Malabar Giant Squirrel

Best Time to Visit
October to March

Intermediate Egrets

Grey-headed Swamphen

Cinnamon Bittern

Whiskered Tern

Birdwatching Sites

More than 180 bird species have been recorded in this area, including over 90 migratory ones. You can see several waterbird species at the sanctuary as it is bordered by waterbodies. Some of the common resident birds in the sanctuary are egrets, Brahminy Kites, cormorants, cuckoos and many waterfowl. Apart from these, some migratory birds like cranes, flycatchers, Western Marsh Harriers, Ospreys and Steppe Eagles are seen here. The best way to birdwatch is via a boat ride – many local boats are available near the lake. If you prefer to walk, an early morning trip along the 1.5km-long walkway into the sanctuary is the best option. June–August is the best time of the year for birdwatching. Migratory birds in huge numbers visit the sanctuary in November–February.

Access & Accommodation

Kumarakom Bird Sanctuary is well connected to many major cities in India by air, railways and roads. Kottayam is the closest railway station and is 16km away. A cab can be taken to Kottayam. Cochin International Airport is the nearest airport to Kumarakom.

There are various accommodation options available near the sanctuary. There are also some houseboats available in Vembanad Lake, which offer a unique experience.

Conservation

Vembanad Lake is one of the tourist attractions in Kerala facing increasing threats of excessive tourism. Due to increasing numbers of speedboats and houseboats, many migrant ducks that used to congregate here have disappeared as a result of the disturbance. There is an urgent need to control the number of motorboats and unregulated developments.

Pheasant-tailed Jacana

Thattekad

White-bellied Treepie

Malabar Pied Hornbill

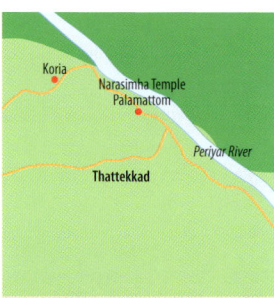

KEY FACTS

Nearest Major Towns
Idukki

Habitats
Tropical wet evergreen, tropical semi evergreen and tropical moist-deciduous forests

Key Species
White-bellied Shortwing, Broad-tailed Grassbird, Oriental Darter, Grey-headed Fish Eagle, Pallid Harrier, Malabar Pied Hornbill, Great Hornbill, Malabar Parakeet, Malabar Grey Hornbill, Grey-headed Bulbul, Wayanad Laughingthrush, Rufous Babbler, White-bellied Blue Flycatcher, Crimson-backed Sunbird, White-bellied Treepie, Sri Lanka Frogmouth, Malabar Trogon

Other Specialities
Asian Elephant, Indian Porcupine, Sloth Bear

Best Time to Visit
October to March

Also known as Salim Ali Bird Sanctuary, Thattekad is located in south-west India, in the Ernakulam district. This bird sanctuary was established in 1983. Situated between two branches of the Periyar River, Thattekad is a hotspot of endemism and is famed for its bird diversity. The total area of the sanctuary comprises roughly 25km^2 of undulating terrain, with a peak altitude of 488m. Several marshy areas (called *vayals*) occur in areas bordering the Periyar and Idamalayar Rivers.

Birdwatching Sites

Thattekad is a birdwatchers' paradise and one of the prime areas for birding in South India. Internationally known ornithologist Dr Salim Ali described this place as the richest bird habitat in peninsular India, comparable only with the Eastern Himalayas.

More than 260 bird species have been recorded in the area. Some of the key species among them include the Spot-bellied Eagle Owl, Sri Lanka Frogmouth, Red Spurfowl, Sri Lanka Bay Owl, Rufous Babbler, Blue-faced Malkoha, Brown-breasted, White-bellied Blue and Rusty-tailed Flycatchers, Grey-headed Bulbul and Wayanad Laughingthrush. Thattekad is ideal for seeing a good number of the endemics of the Western Ghats. The Urulanthanni area is good for some lowland birding. Interesting species that can be found here include the Wyanad Laughingthrush, Dark-fronted Babbler, White-bellied Woodpecker, Spot-bellied Eagle Owl, Oriental Sri Lanka Bay Owl, Crested Goshawk, Mountain and Green Imperial Pigeons, Scarlet Minivet, White-bellied Blue Flycatcher, Black-naped Monarch, Grey-headed Bulbul, Brown Fish Owl, Brown-cheeked Fulvetta, Indian Scimitar Babbler. Edamalayar Power Plant land and the south side of the river area are good for the owls, and the Streak-throated Woodpecker and Yellow-billed Babbler.

Sri Lanka Bay Owl

Access & Accommodation

Thattekad is located 60km towards the north-east of Kochi in the Kothamangalam taluka of the Ernakulam district. The nearest city is Kochi, and the nearest railway station is Aluva, which is a few kilometres away. It is well connected to other cities in the state and the rest of South India by road and rail. The nearest airport is Cochin International Airport.

There are several places to stay in and around Thattekad. A Forest Department Inspection Bungalow can be booked with advance notice. There are also many simple homestays, as well as luxurious tented facilities. The Hornbill Camp (www.thehornbillcamp.com) and Eldhose Birding Camp (https://birdingsouthindia.com) are quite famous among birdwatchers. These camps can arrange birding trips in nearby areas.

Conservation

There are no major conservation threats and issues. Some of the common issues are illegal use of land for agriculture, illegal tree cutting, and grazing of livestock inside the sanctuary. There are small villages at the fringes of the park, and the main occupation of the villagers is agriculture. To meet their various needs, they enter the forest illegally.

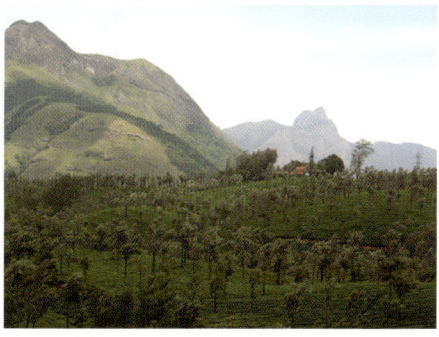

Thattekad landscape

Rahul & Khushboo Sharma

Indian Blue Robin

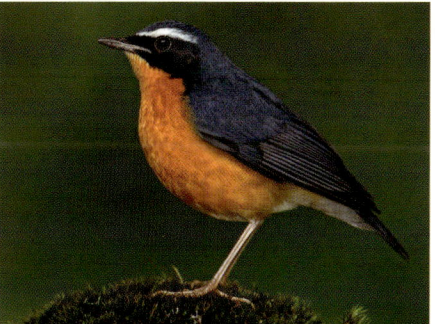

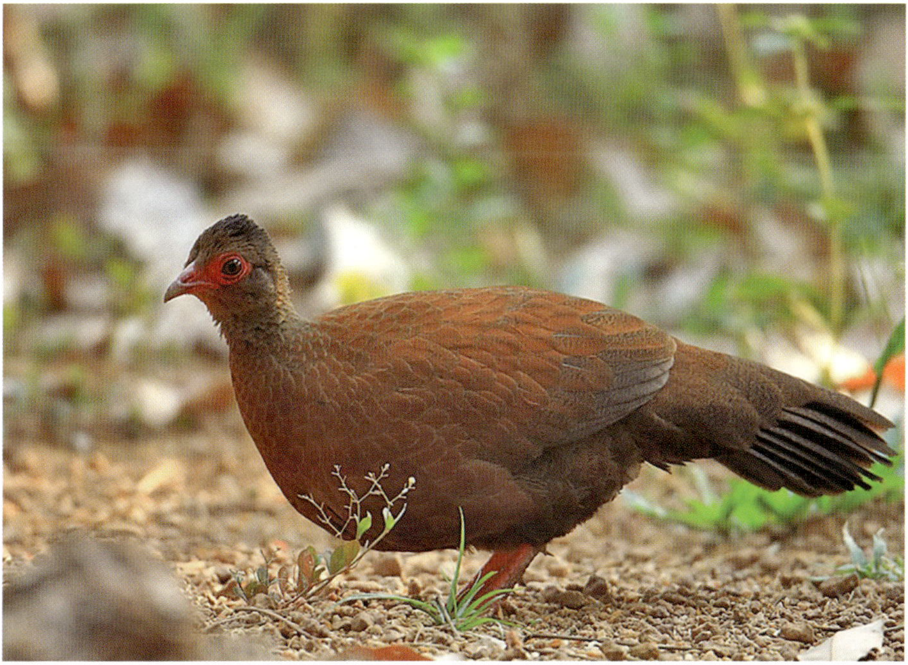

Ramki Sreenivasan

Red Spurfowl

Kerala

Eravikulam National Park, Munnar

Nilgiri Tahr

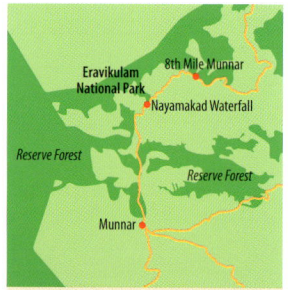

KEY FACTS

Nearest Major Towns
Idukki

Habitats
Subtropical broadleaved hill forest, montane grassy slopes

Key Species
White-bellied Shortwing, Broad-tailed Grassbird, Nilgiri Pipit, Palani Laughingthrush, Black-and-orange Flycatcher, Nilgiri Flycatcher, Nilgiri Wood Pigeon, Malabar Parakeet, Malabar Grey Hornbill, Grey-headed Bulbul, Rufous Babbler, White-bellied Blue Flycatcher, Crimson-backed Sunbird, White-bellied Treepie, Malabar Trogon, Edible-nest Swiftlet, White-cheeked Barbet, Yellow-browed Bulbul, Malabar Whistling Thrush, Indian Scimitar Babbler, Dark-fronted Babbler

Other Specialities
Nilgiri Tahr, Lion-tailed Macaque, Gaur, Indian Chevrotain, Nilgiri Marten

Best Time to Visit
October to March

Eravikulam National Park is located in the High Ranges (Kannan Devan Hills) of the Southern Western Ghats in the Devikulam taluka of Idukki District. It was declared a national park in 1978, mainly to protect the endemic Nilgiri Tahr. The geographical area of the park mostly consists of high-altitude grassland that is interspersed with sholas. The main body of the park comprises a high, rolling plateau with a base elevation of about 2,000m.

Birdwatching Sites

The park is an important habitat for many bird species, such as the Palani Laughingthrush, Nilgiri and Black-and-orange Flycatchers, Nilgiri Wood Pigeon and Nilgiri Pipit, which are endemic to the Western Ghats. There are also some records of White-bellied Shortwings from this area. More than 150 bird species have been recorded here. The park is also an important site for winter migrants from the Himalayas and beyond. For example, Large-crowned and Large-billed Leaf-warblers, and Rusty-tailed Flycatchers, are some of the birds that winter here in large numbers. Along with its rich avifaunal diversity, Eravikulam is famous for the Nilgiri Tahr, which is the park's star attraction.

Access & Accommodation

Eravikulam National Park is accessible from Kochi (Kerala) and Coimbatore (Tamil Nadu) Airports, which are about 148km and 175km away respectively. Munnar, the nearest town (13km) is well connected by roads from Kerala and Tamil Nadu. The nearest railway station in Kerala is Alwaye (120km).

Munnar is a well-known tourist destination so there are various accommodation options. Visitors can book

Common Tailorbird

White-browed Bulbul

accommodation depending on their budgets and needs. Mistletoe Munnar (www.mistletoemunnar.com/index.php) is a private property near Munnar that organizes birding and sightseeing trips. There are also many homestay options available. Advance booking is recommended. There are no organized safaris inside the park, and all movements in the park take place on foot along trails and footpaths.

Conservation

There are no major immediate conservation threats. From a long-term perspective, there are some minor issues like the grazing of livestock inside the park and collecting firewood. A few hill tribes stay within the park and are dependent on it for their livelihoods. They engage in cultivating lemongrass and collecting minor forest produce from the adjoining forest areas. Communities around the park are already motivated for participatory conservation. Eco-development best practices are implemented to reduce the threats of the unsustainable harvesting of forest resources.

Nilgiri Pipit

Broad-tailed Grassbird

LAKSHADWEEP

The Lakshadweep archipelago is the smallest union territory of India, with a geographical area of only 32km². It comprises a group of 36 coral islands covering 12 atolls, three reefs and sandbanks that are submerged at high tide. Only 11 islands are inhabited, of which Agatti is the most populated. These islands are irregularly scattered in the south Arabian Sea, about 280km to 480km west of Kochi on the Kerala coast. The Lakshadweep, Maldive and Chagos archipelagos form a contiguous mountain ridge in the ocean. The ridge is believed to be a continuation of the Aravalli mountain range of Rajasthan and Gujarat from late tertiary times. The terrestrial flora and fauna of the archipelago are poor, and no endemic plant species have been reported. A large number of crop plants (rice, vegetables, fruits, tubers, spices, sugar and areca nuts) were introduced from the Indian mainland, along with domestic livestock such as cattle, goats and poultry chickens, and domestic cats.

1. Pitti Island

Climate

Lakshadweep has an agreeable tropical climate with summer temperatures of 22–33 °C and winter temperatures of 20–32 °C. The south-west monsoons in June–September bring plenty of rainfall to the islands. October–May are pleasant, with a slight rise in temperature in March–April. Cool sea breezes and abundant shade provided by the canopy of coconut palms make the climate enjoyable even during peak summer.

Access, Transportation & Logistics

Lakshadweep Island can be reached by ships and flights operated from Kochi. For all tourist purposes, Kochi is the gateway to Lakshadweep. Agatti and Bangaram Islands can be reached by flights from Kochi operated by Indian Airlines. Onward flights from Kochi are available to most airports in India and abroad. From Agatti boats are available to Kavaratti and Kadmat during the fair season in October–May. Helicopter transfer is possible from Agatti to Kavaratti during the monsoon season, subject to the availability of a helicopter. The flight from Cochin to Agatti takes approximately an hour and a half.

Seven passenger ships – MV *Kavaratti*, MV *Arabian Sea*, MV *Lakshadweep Sea*, MV *Lagoon*, MV *Corals*,

KEY BIRDS
Top 10 Birds
1. Red-billed Tropicbird
2. Swinhoe's Storm-petrel
3. Flesh-footed Shearwater
4. Great Frigatebird
5. Great Knot
6. Arctic Skua
7. Black Noddy
8. Sooty Tern
9. White-cheeked Tern
10. Pacific Golden Plover

Lesser Noddy

Western Reef Egret

or motion sickness, consult a doctor and carry your medicine with you. Avoid roadside food and drink bottled water. During the summer months keep hydrated, and carry a stole, scarf or cap to protect yourself from direct sunlight.

Birdwatching Highlights
Only one site in Lakshadweep has been selected as an IBA, known as Pitti Island. Its selection was mainly on the basis of the congregation of nesting seabirds. Though some migratory waders are seen on Pitti, terns are the only breeding birds on the island. The tern species that nest on and feed around Pitti are Sooty, Greater Crested and Bridled Terns, and the Brown Noddy. Waders reported from Pitti include the Ruddy Turnstone, Eurasian Curlew and Lesser Sand Plover. The island is devoid of any vegetation

MV *Amindivi* and MV *Minicoy* – operate between Cochin and Lakshadweep Islands. The passage takes 14–18 hours, depending on the island chosen for the journey. During the fair season, high-speed vessels operate between islands.

Health & Safety
Lakshadweep is one of the safest places for visitors in India, with no major health concerns. It is a low- to no- risk area of India when it comes to malaria, but do take precautions especially after the monsoon. Visit a GP or travel clinic 6–8 weeks before departure to make sure you are up to date with any vaccinations. If you are prone to seasickness

Lesser Crested Tern

Lakshadweep coastal scene

Brown Skua

Pitti Island

Brown Noddy

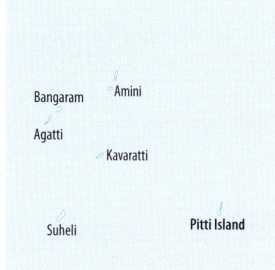

KEY FACTS

Nearest Major Towns
Kavaratti

Habitats
Coastal

Key Species
Sooty Tern, Red-billed Tropicbird, Pacific Golden Plover, Common Sandpiper, Bar-tailed Godwit, Lesser Crested Tern, Bridled Tern, Brown Noddy, Greater Crested Tern, Saunder's Tern, Ruddy Turnstone, Grey Plover, Whimbrel, Caspian Tern, Swinhoe's Storm-petrel, Flesh-footed Shearwater, Great Frigatebird, Great Knot, Arctic Skua, Black Noddy, White-cheeked Tern, Eurasian Curlew, Lesser Sand Plover, Wilson's Storm-petrel, Black-bellied Storm-petrel, Lesser Frigatebird

Other Specialities
Butterfly fish, Bottlenose Dolphin, Bryde's Whale, Sperm Whale

Best Time to Visit
October to March

Pitti Island is an uninhabited, barren reef with a sandbank, located in the Lakshadweep archipelago. The island is mostly made up of broken coral that has been pounded to a rough sand-like consistency, with large boulders and loose stones. The island is a low plateau rising 2m above mean sea level at high tide, with sloping beaches all around. It was formed by the accumulation of coral sand in the form of a sandbank with the action of wind waves and currents.

Birdwatching Sites

The island harbours many varieties of tern. It is the breeding ground for four tern species, Sooty, Great Crested and Bridled Terns, and the Brown Noddy. It is of great importance for breeding colonies of pelagic birds in the Indian territory. Though some migratory waders are seen on Pitti, terns are the only breeding birds on the island. Waders reported from Pitti include the Ruddy Turnstone, Eurasian Curlew and Lesser Sand Plover.

Access & Accommodation

The only way to reach Pitti Island is via Agatti Airport. From Agatti, you need to take a boat to reach the island. Pitti is very close to Kalpeni Island, which is reachable by small boat – it takes only 5-10 minutes to reach it from Kalpeni.

Visitors can book accommodation via the Lakshadweep tourism portal (file://localhost/(https/::samudram.utl.gov.in:sprt_Home.aspx). Lakshadweep Tourism also arranges packaged tours.

Ruddy Turnstone

Bridled Terns

Conservation

Although Pitti Island is uninhabited, fishermen from other islands in Lakshadweep visit it for fishing, and to collect shells and tern eggs. This poses a severe threat to the colony throughout the year, except during the monsoon when the island becomes inaccessible to humans. Heavy rains, however, take their toll on the chicks, and mortality is very high. Predation by crabs is another natural threat to the birds' eggs. Introducing vegetation to the island has been suggested, but this would be disastrous for the nesting colony of terns.

Sooty Tern

Caspian Tern

Greater Sand Plover

DELHI & NCR

India's capital territory, Delhi is a massive metropolitan area overflowing with modern life. This northern state of India is divided into two main ecological zones, the Aravalli Hills and the Plains. The Yamuna River, where a large number of waterbirds congregates during winter, is the main source of water. Despite the dense human population, Delhi does have some natural forests, especially on what is known as the Ridge. The most prominent forest type of this region is tropical dry deciduous forests.

1. Okhla Bird Sanctuary
2. Sultanpur National Park

Climate

Delhi experiences extreme temperatures. It is very hot in summer (April–July) and cold in winter (December–January). Winter temperatures can fall to as low as 2 °C, while the mercury soars to an uncomfortable 50 °C in summer. Delhi is located in a semi-arid zone, so the rainfall is low, reaching a maximum of 600mm. Autumn (September–November) and spring (February–March) are the pleasant months for visitors. Winter is considered the best time for birdwatching.

Access, Transportation & Logistics

Delhi is well connected to all the major cities within and outside India via airways, railways and roadways. Almost all the major airlines have their flights operating from Indira Gandhi International Airport in New Delhi. The three important railway stations in Delhi are New Delhi Railway Station, Old Delhi Railway Station and Hazrat Nizamuddin Railway Station. Delhi is well connected with all of India's major cities by a network of roads and national highways. The three major bus stands in Delhi are the Inter-State Bus Terminus at Kashmiri Gate, Sarai Kale-Khan Bus Terminus and Anand Vihar Bus Terminus. Both government and private transport providers provide frequent bus services. You can also get government as well as private taxis here.

Health & Safety

There are no major health concerns in Delhi. As a precaution visitors may take vaccinations for hepatitis A and B, and typhoid. Malaria is prevalent in remote areas and prophylaxis should be taken, although this is less of a problem during the dry

Spotted Dove

KEY BIRDS

Top 10 Birds
1. Marbled Duck
2. Great Bittern
3. Dusky Eagle Owl
4. Sarus Crane
5. Indian Courser
6. Sind Sparrow
7. White-bellied Minivet
8. Eurasian Roller
9. Brooks's Leaf Warbler
10. Bonelli's Eagle

season. Dengue fever is also present and appropriate precautions should be taken. Avoid roadside food as the vendors may not follow food-safety norms. Visit a GP or travel clinic 6–8 weeks before departure to make sure you are up to date with any vaccinations. In summer adequate precautions need to be taken to avoid the intense heat, such as wearing light cotton clothing, wearing a hat or using a sunshade while going outdoors, and drinking plenty of liquids.

White-breasted Waterhen

Yellow-crowned Woodpecker

Birdwatching Highlights

Despite its dense human population, Delhi is a birdwatchers' paradise, due to its old avenue trees, large number of parks, historical monuments with gardens, colonial bungalows with large lawns and the famous Delhi Ridge. More than 400 bird species have been identified here, including some rare ones such as the Spot-billed Pelican, Lesser Adjutant, Greater Spotted Eagle, Sarus Crane and Indian Skimmer. Some other important birds recorded here are the Oriental Darter, Painted and Black-necked Storks, Lesser Grey-headed Fish Eagle and Black-headed Ibis.

Bronze-winged Jacana

Okhla Bird Sanctuary

Okhla Bird Sanctuary

Delhi and the neighbouring state of Uttar Pradesh harbour a huge wetland refuge for birds. The site is located at the point where the Yamuna River leaves the territory of Delhi and enters Uttar Pradesh. The eastern side of Delhi along the Yamuna River offers excellent birding in winter and during migration, when there are large numbers of waterbirds. The most prominent feature of the sanctuary is a large lake created by damming the river, which lies sandwiched between Okhla village towards the west and Gautam Budh Nagar towards the east.

Birdwatching Sites

Okhla is arguably one of the finest sites for birdwatching in Delhi, with more than 300 bird species having been recorded in the area. The sanctuary is located just 10km south-east of Connaught Place, a business and financial hub of Delhi. It has been identified as an IBA because of its birdlife and the presence

KEY FACTS

Nearest Major Towns
New Delhi

Habitats
Urban, wetland, freshwater swamp, riverine forest

Key Species
White-rumped Vulture, Long-billed Vulture, Spot-billed Pelican, Lesser Adjutant, Baikal Teal, Baer's Pochard, Pallas's Fish Eagle, Sarus Crane, Greater Spotted Eagle, Sociable Lapwing, Indian Skimmer, Bristled Grassbird, Finn's Weaver, Dalmatian Pelican, Oriental Darter, Painted Stork, Black-necked Stork, Black-headed Ibis, Ferruginous Pochard, Grey-headed Fish Eagle, Black-bellied Tern

Other Specialities
Nilgai, Indian Jackal, Rhesus Macaque, Indian Grey Mongoose, Indian Porcupine, Black-naped Hare

Best Time to Visit
Throughout the year

Bristled Grassbird

Common Shelduck

Baikal Teal

Zitting Cisticola

of globally threatened species such as the Sarus Crane, Bristled Grassbird, Greater Spotted Eagle and Indian Skimmer.

Birding in the area is good and relatively undisturbed. Visiting is recommended in the morning, when the birds are most active. Some of the important birds found here are Spot-billed and Dalmatian Pelicans, Lesser Adjutant, Baikal Teal, Oriental Darter, Painted and Black-necked Storks, and Ferruginous Pochard.

Access & Accommodation

Okhla is just a few kilometres away from Delhi city centre, so accessibility and accommodation is never a problem here. For convenience and to maximize the birding time, it is best to hire a vehicle for your disposal. There are plenty of options for accommodation in Delhi. From the most luxurious hotels and resorts, to budget lodges, all are available within a range of only a few kilometres.

Conservation

Okhla Wildlife Sanctuary has great potential to attract national and international birdwatchers. Its greatest advantages are its easy accessibility and its congregation of waterfowl. However, the sanctuary is not properly protected. Pollution and solid waste are the biggest problems of Okhla wetland. Encroachment on forest land and illegal construction are common and ongoing threats.

River Lapwing

Red Avadavat

Sultanpur National Park

Gadwall

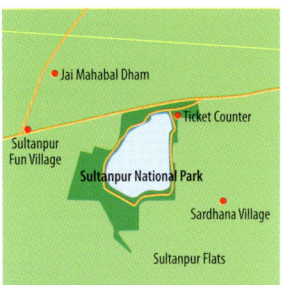

KEY FACTS

Nearest Major Towns
Gurugram

Habitats
Urban plains, wetlands, scrub

Key Species
Indian Peafowl, Black Francolin, Grey Francolin, Common Pigeon, Eurasian Collared Dove, Red Collared Dove, Yellow-wattled Lapwing, Red-wattled Lapwing, Barred Buttonquail, Oriental Pratincole, Cattle Egret, Indian Black Ibis, Black-winged Kite, Black Kite, Spotted Owlet, Common Hoopoe, White-throated Kingfisher, Green Bee-eater, Blue-cheeked Bee-eater

Other Specialities
Nilgai, Indian Jackal, Rhesus Macaque, Indian Grey Mongoose, Indian Porcupine, Black-naped Hare

Best Time to Visit
Throughout the year

The Sultanpur Bird Sanctuary and National Park is located in the Gurugram district of Haryana. Sultanpur National Park was formerly known as Sultanpur Bird Sanctuary, which was established in 1971. It was declared a national park in 1991. The national park is categorized as an IBA due to the huge congregation of waterbirds there. It is a bird paradise for birdwatchers, famous for its migratory as well as its resident birds. Migratory birds start arriving in the park in September. They then use the park as a resting place until the following March–April.

Ruddy-breasted Crake

Stoliczka's Bushchat

Barred Buttonquail

Birdwatching Sites

Sultanpur National Park is located just off the Gurugram-Farrukhnagar road, 45km south-west of Delhi. Sultanpur Lake forms the core area. This shallow lake is fed by the overflow from neighbouring canals and agricultural fields, and replenished by saline groundwater. The park has seasonal aquatic vegetation and open grassland, dotted with artificial islands planted with Acacia nilotica. It experiences extreme weather conditions. There are four watchtowers located at different points. The Educational Interpretation Centre offers guidance to visitors.

More than 300 species of bird have been recorded in the park. Some are resident, while others come from distant regions like Siberia, Europe and Afghanistan. Some of the common resident bird species are the Common Hoopoe, Purple Sunbird, Black Francolin, Little Cormorant, Indian Cormorant, Paddyfield Pipit, Eurasian Spoonbill, Grey Francolin and Indian Roller. Among migratory birds important species are the Common Crane, Lesser Flamingo, Ruff, Black-winged Stilt, Common Teal, Common Greenshank, Northern Pintail, Western Yellow Wagtail, White Wagtail, Northern Shoveler, Great White Pelican and Gadwall.

Access & Accommodation

The nearest airport is Indira Gandhi International Airport, about 35km from Sultanpur National Park. The airport is well connected by road to the park. The nearest railway station is Gurugram, which is about 15km from the park and well connected to it. The park is also well connected to major cities by the road network. There are a number of government and privately operated vehicles that go to the park at frequent intervals. Since this place is quite near Gurugram and Delhi, most visitors prefer to stay in these cities and visit the park as a day trip. Both cities have a wide range of accommodation options.

An entry permit can be obtained from the Divisional Wildlife Officer, Gurugram (0124-2222272). Obtaining the permit before travelling is recommended.

Conservation

The climate and soil in the environs of Sultanpur Lake are not conducive to supporting permanent waterbodies. In recent years of low rainfall, only a few pools remain by midwinter and the lake is predominantly dry. Land use practices in the catchment areas have impeded the natural flow of water into the lake. While the core area of the lake is under the jurisdiction of the Haryana Forest Department, the buffer zone is under private or Gram Panchayat or village council control. These areas are cultivated, heavily grazed and for the most part heavily degraded. Encroachment on the forest land and illegal constructions are common and ongoing threats. Tourist activity causes a high level of disturbance. Siltation due to soil erosion in the catchment area and windblown dust is on the increase. Mining of sand for the nearby brick and quick lime industries is also a major concern.

White-tailed Lapwing

Black-breasted Weaver

MADHYA PRADESH

1. Bandhavgarh National Park
2. Kanha National Park
3. Panna National Park
4. Pench National Park

Madhya Pradesh is one of the larger states in central India. It is bounded by Chhattisgarh in the east, Rajasthan and Gujarat in the west, Uttar Pradesh in the north and Maharashtra in the south. Bhopal, one of the best-known cities of India, is the capital of the state. Madhya Pradesh comprises three main sections of the Deccan Plateau, namely the Central Highland, the Satpura Maikal ranges and the Eastern plateau. The state is blessed with many rivers, such as the Chambal, Narmada, Tapti, Betwa, Ken, Sone and Jamner. The Tropic of Cancer crosses Madhya Pradesh.

Climate
The state has three main seasons – winter, summer and monsoon. The rainfall decreases from the south-east and east to the north-west and west. The average annual rainfall varies from 500mm to

Common Kestrel

KEY BIRDS
Top 10 Birds
1. Indian Scops Owl
2. Indian Pitta
3. Rufous Treepie
4. White-eyed Buzzard
5. Bay-backed Shrike
6. Red-headed Vulture
7. Black-necked Stork
8. Painted Spurfowl
9. Jacobin Cuckoo
10. Indian Grey Hornbill

Lesser Adjutant

Health & Safety
There are no major health concerns, and the state is a low- to no-risk area of India when it comes to malaria, but do take precautions, especially after the monsoon. Visit a GP or travel clinic 6–8 weeks before departure to make sure you are up to date with any vaccinations. Avoid roadside food and drink bottled water. During the summer months keep hydrated, and carry a stole, scarf or cap for protection from direct sunlight.

Birdwatching Highlights
The natural splendour of Madhya Pradesh includes a wide spectrum of wildlife, ranging from Tigers, Indian Leopards, antelopes and gazelles, to other mammals and reptiles, and an abundance of birdlife. More than 570 bird species have been recorded in the state. They include some Critically Endangered birds like White-rumped and Long-billed Vultures, some Endangered species like the Lesser Florican and Lesser Adjutant, and many Vulnerable ones such as the Common Pochard, Marbled Duck, Pale-capped Pigeon, Sarus Crane, Lesser Adjutant, and Indian and Greater Spotted Eagles.

3,000mm, and the temperatures range from 1 ºC to 48 ºC. There are four forest types, namely tropical moist deciduous, tropical dry deciduous, tropical thorn and subtropical broadleaved hill forests. The central, southern and eastern parts of the state have a better forest cover than the northern and western parts. Teak and sal forests are the two most important forest formations in the state.

Access, Transportation & Logistics
The state is well connected with the rest of India via air and roads. Its major cities, including Indore, Bhopal, Jabalpur and Ujjain, all enjoy good connectivity by regular flights and long-distance trains with the rest of India. The state has five air terminals, numerous railway junctions and an extensive network of national highways. Devi Ahilyabai Holkar Airport (IDR), located in Indore, is the largest airport in Madhya Pradesh. It is served by direct flights from all major cities in India. Additionally, Raja Bhoj International Airport (BHO) in Bhopal, Dumna Airport (JLR) in Jabalpur, Gwalior Airport (GWL) and Khajuraho Airport (HJR) have scheduled flights to major cities in India. The state has a total of 20 major railway junctions, including stations at Ratlam, Indore, Bhopal, Jabalpur, Gwalior and Khajuraho. There are direct trains from all major stations in India. There is also a huge network of national and state highways connecting the state to the rest of India.

Indian Peafowl

Bandhavgarh National Park

Bandhavgarh landscape

This park is located in the Umaria and Shahdol districts of northeastern Madhya Pradesh. It became a national park in 1965, and before this was the personal hunting ground of the Maharajas of Rewa. The area under the national park is rugged and marked by sharp-crested hills, sal forests and grassy pastures. The altitude varies at 440–811m, and the impressive Bandhavgarh Hill is the highest place in the reserve. Bandhavgarh has strategic importance for wildlife and conservation in India, as it is not an isolated and fragmented patch of forest, and forms part of a larger forest block. Apart from the Panpatha Wildlife Sanctuary that is connected to the park in the north, there are several smaller pockets of protected and reserve forest, interspersed with small agricultural communities, which act as one of the most important wildlife corridors.

KEY FACTS

Nearest Major Towns
Umaria, Satna, Jabalpur and Shahdol

Habitats
Tropical moist deciduous forest and tropical grassland

Key Species
White-rumped Vulture, Long-billed Vulture, Lesser Adjutant, Sarus Crane, Oriental Darter, Black-headed Ibis, Grey-headed Fish Eagle, Malabar Pied Hornbill, White-eyed Buzzard, Yellow-legged Green Pigeon, Indian Grey Hornbill, Lesser Golden-backed Woodpecker, Small Minivet, Indian Robin, Large Grey Babbler, Jungle Babbler, Ashy Prinia, White-browed Fantail, Brahminy Starling, White-bellied Drongo

Other Specialities
Tiger, Indian Leopard, Sambar, Spotted Deer, Indian Chevrotain, Sloth Bear, Asiatic Wild Dog, Northern Plains Langur

Best Time to Visit
October to March

Chestnut-headed Bee-eaters

Eurasian Cuckoo

Orange-headed Thrush

Rose-ringed Parakeet

Birdwatching Sites

Bandhavgarh supports a variety of bird species that occur in central India – more than 300 species have been listed here. The park is an important area for the conservation of the severely endangered vulture species, including Indian, White-rumped, Red-headed and Egyptian Vultures – many of which breed here. Birdwatching in Bandhavgarh is a pleasure, with significant bird density. Some of the sought-after birds are the Rufous-tailed Lark, Common Pochard, Black-bellied Tern, White-naped Woodpecker, Painted Spurfowl, Steppe and Indian Spotted Eagles, Plum-headed Parakeet, Lesser Adjutant, Mottled Wood Owl and many varieties of flycatcher. Most of the birding is done easily from the comfort of a safari vehicle. Since the park is a Tiger Reserve, the birds are often overlooked, but birders will have a rewarding time here if they look for them.

Tala Gate is the main entry gate of Bandhavgarh. The area is rich in water and food resources, and harbours a great variety of wildlife, including avian life. Jeep safaris are available in the morning and evening, when the animals are most active. A Forest Department guide accompanies visitors. The meadows of Chakradhara, Bhaitari Bah, Raj Bahera and Sehra are some of the nearby areas known for birdwatching.

Access & Accommodation

The nearest railheads for Bandhavgarh are Jabalpur (170km), Katni (102km) and Satna (112km) on the central railway, and Umaria (35km) on the Katni Bilaspur route. From the Umaria railway station, it is an hour's drive to Tala. There are regular buses and taxis available between Katni and Umaria, and from Satna and Rewa to Tala. Jabalpur and Khajuraho are the nearest airports, from where it is an approximately five-hour drive to Bandhavgarh.

Various accommodation options are available in and around Bandhavgarh. Samode Safari Lodge (www.samode.com/safarilodge) and Forest Hills (www.foresthillstala.com) are two of the premium properties and are well known among birdwatchers. White Tiger Lodge is a Madhya Pradesh Tourism lodge near Tala. There is also a forest rest house in Tala. The ideal time to visit is November–June. The park is closed in July–October.

Conservation

Some of the common conservation issues faced by the park are grazing, poaching of wild animals including birds, illegal cutting of trees and firewood collection. One of the main problems is firewood and timber collection, and a huge area has been degraded solely because of deforestation and clearance of woods and grassland for grazing. Poaching of wild animals takes place regularly in many areas. There have been many incidents of human-animal conflict, mainly due to crop damage by herbivores and the lifting of cattle by large carnivores.

Kanha National Park

Kanha forest

KEY FACTS

Nearest Major Towns
Jabalpur, Mandla and Balaghat

Habitats
Tropical moist deciduous forest, tropical dry deciduous forests, tropical grasslands

Key Species
White-rumped Vulture, Long-billed Vulture, Lesser Florican, Lesser Adjutant, Sarus Crane, Painted Stork, Black-headed Ibis, Grey-headed Fish Eagle, Red-headed Vulture, Pallid Harrier, Malabar Pied Hornbill, Orange-headed Thrush, Grey Bushchat, Blue Rock Thrush, Ultramarine Flycatcher, Bluethroat, Verditer Flycatcher, White-rumped Shama, Oriental Magpie Robin, Jungle Myna, Chestnut-tailed Starling, Jungle Babbler

Other Specialities
Tiger, Indian Leopard, Swamp Deer, Sambar, Indian Chevrotain, Wild Pig, Gaur, Chinkara, Bengal Fox, Sloth Bear, Jungle Cat, Asiatic Wild Dog, Four-horned Antelope, Striped Hyena, Indian Grey Mongoose, Palm Civet, Small Indian Civet, Lesser Bandicoot

Best Time to Visit
October to March

This Tiger Reserve in the central Indian highlands is one of the most picturesque national parks in India. The park is known not only for its large mammals, especially the Tiger, but also for its avifaunal diversity. Two river valleys are prominent features of the park's topography: the Banjar in the west and the Halon in the east, both tributaries of the Narmada River. Kanha is one of the largest and best-protected areas, with well-preserved forests. As a consequence, the birdlife is rich and varied, containing most of the representative species of the Indo-Malayan Tropical Dry Zone. The park was originally established to protect the Swamp Deer (Barasingha). Four principal vegetation types have been identified in Kanha: moist deciduous forests, dry deciduous forests, valley meadows and plateau meadows. The vegetation is chiefly made up of sal and bamboo forests, and grassland.

Birdwatching Sites

The rich birdlife of Kanha offers great opportunities for birdwatchers and photographers. Morning and evening jeep safaris are organized by the Forest Department. Tourists are not allowed to get down from a vehicle inside the park, so most of the birding is done only from the jeep. There are four zones in Kanha: Kisli, Kanha, Mukki and Sarhi. Hiring a local birding guide is recommended for the best birding experience. Along with jungle safaris, there are also nature trails for those who want to explore the jungle closely on foot. The nature trails also offer some good birdwatching experiences.

More than 300 bird species have been recorded in Kanha and its surroundings. Some of the commonly seen birds are the Indian Roller, Indian Grey Hornbill, Cattle Egret, Blue-bearded Bee-eater, Black Drongo, Indian Black Ibis, Common Teal, Crested Serpent Eagle, Asian Paradise-flycatcher, Indian

Great Thick-knee

Mottled Wood Owl

Black Francolin

Peafowl, White-eyed Buzzard, Stork-billed Kingfisher, Lesser Adjutant, Lesser Whistling Duck, Steppe Eagle, Little Grebe and many more. The Critically Endangered White-rumped and Long-billed Vultures, and the highly endangered Lesser Florican, are also occasionally seen in the grassland. The winter months (November–February) are good for birdwatching in Kanha – this is when numerous migratory birds, along with endemic birds, inhabit the park.

Access & Accommodation

Jabalpur is the nearest airport to Kanha (170km, about three and a half hours). The second best option is Raipur Airport (225km, about four and a half hours). Another way to reach Kanha is from Nagpur Airport (27 km, about five and a half hours). Jabalpur and Raipur are better options for travellers boarding from Delhi, and Nagpur is a good option for Mumbai, Pune and Bangalore based visitors. Jabalpur is the nearest and most convenient railhead for visits to Kanha. Gondia is another railhead that is more convenient for Mumbai and Kolkata wildlifers.

The park is well connected by road to the rest of India.

Kanha is one of the most famous national parks in India, and there is a large variety of options available for accommodation, depending on budgets and needs. Kanha Earth Lodge by Pugdundee Safari (www.pugdundeesafaris.com/kanhaearthlodge-kanha.php) is popular for its eco-friendly architecture. Banjaar Tola by Taj is another premium and well-known property. Safari tickets can be booked online (mponline.gov.in). Most of the resorts and hotels also help in booking safari tickets and safari vehicles.

Conservation

Due to strict protection measures undertaken by the Project Tiger management, illicit felling, poaching and encroachment are well under control in Kanha. There are some common minor conservation issues like human habitation, illegal cutting of trees and cattle grazing in the buffer zone. Excessive tourism is creating a disturbance to wild animals and birds. There are occasional incidents of wild animal poaching.

Panna National Park

Ken River

KEY FACTS

Nearest Major Towns
Khajuraho

Habitats
Tropical dry deciduous and mixed forests

Key Species
White-rumped Vulture, Long-billed Vulture, Lesser Adjutant, Oriental Darter, Painted Stork, Black-necked Stork, Black-headed Ibis, Ferruginous Pochard, Red-headed Vulture, Painted Spurfowl, Little Grebe, Painted Francolin, Northern Pintail, Indian Spot-billed Duck, Tufted Duck, Common Moorhen

Other Specialities
Tiger, Indian Leopard, Asiatic Wild Dog, Spotted Deer, Indian Fox, Indian Wolf, Jungle Cat, Chinkara, Striped Hyena, Sambar, Indian Jackal, Nilgai

Best Time to Visit
October to March

This park is located in the northern part of Madhya Pradesh and spreads over two districts, Panna and Chhatarpur. It was constituted in 1981 and declared a Tiger Reserve in 1994. The forest of Panna was the game reserve of the erstwhile princely states of Bijawar, Chhatarpur and Panna. The main forest types here are southern tropical dry teak forests and northern tropical dry deciduous mixed forests, and the reserve is rich in fodder grasses. The lifeline of the park is the Ken River, which meanders for about 55km through the Tiger Reserve from south to north. Khajuraho, a World Heritage Site famous for its beautifully sculpted temples, is a nearby attraction of Panna.

Birdwatching Sites

Panna is perhaps the best and most extensive forest left in north-central Madhya Pradesh in the Bundelkhand area. It is located at the junction of the Deccan and Indo-Gangetic Plains, so the birdlife is very rich.

More than 250 bird species have been recorded in Panna, including some that are Vulnerable and Endangered, like the Common Pochard, Ferruginous Duck, Lesser Adjutant, Painted and Black-necked Storks, Black-headed Ibis, Oriental Darter, Greater Thick-knee, River Lapwing and Egyptian Vulture. Panna is a breeding place for the Critically Endangered Long-billed, Red-headed and White-rumped Vultures. A jungle safari provides the best way of birdwatching in Panna. Madna and Hinauta are two entry gates, and regular jungle safaris are organized by the Forest Department from these gates. Private vehicles are allowed but due to the rough terrain, unmetalled roads and steep inclines, a four-wheel-drive petrol vehicle is recommended for the best bird- and general wildlife-watching experience. Private vehicles can be hired at Khajuraho and Panna.

Egyptian Vulture

Red-headed Vulture

Painted Francolin

Access & Accommodation
The nearest airport to Panna is Khajuraho, which is just 25km from the park. Khajuraho is a well-known tourist spot, and there are regular buses and taxis available from here to Panna. Satna is the nearest railhead, about 90km from Panna. Jhansi is a bigger and busier railhead, about 180km from the park.

There are many accommodation options near the park. There is an eco complex at Hinauta and Madla run by the Forest Department. Bookings can be made online through the Madhya Pradesh Ecotourism Portal (https://ecotourism.mponline.gov.in). There are also many private accommodation options available near the park. Ken River Lodge (+91-07732-275235), Panna Tiger Resort (+91-07732-275248), and Pashangarh, Taj Safaris (+91-8959904701/02/03/04) are some of the best known among birdwatchers.

Conservation
The park has a special place in north-central Madhya Pradesh from an ecology viewpoint, and its vegetation and culture, and is a true representative of the Bundelkhand region. The biggest conservation threat to Panna at the present time is the Ken-Betwa river-linking project, which involves diversion of about 60km^2 of forests, mostly from the Panna Tiger Reserve. The project will threaten endangered species such as the Tiger, Garial and several kinds of vulture.

Plum-headed Parakeets

Pench National Park

Pench landscape

KEY FACTS

Nearest Major Towns
Seoni and Chhindwara

Habitats
Tropical dry deciduous forest

Key Species
White-rumped Vulture, Long-billed Vulture, Greater Spotted Eagle, Eastern Imperial Eagle, Lesser Kestrel, Green Avadavat, Oriental Darter, Painted Stork, Black-necked Stork, Black-headed Ibis, Ferruginous Pochard, Grey-headed Fish Eagle, Cinereous Vulture, Malabar Pied Hornbill, Red-headed Vulture

Other Specialities
Tiger, Indian Leopard, Jungle Cat, Leopard Cat, Indian Fox, Small Indian Civet, Indian Palm Civet, Sloth Bear, Striped Hyena, Indian Grey Mongoose, Indian Flying Fox

Best Time to Visit
October to March

Also known as Pench Tiger Reserve, Pench National Park is named after the Pench River, which flows from north to south through the park. Pench is located in the southern reaches of the Satpura Hills in the Seoni and Chhindwara districts of Madhya Pradesh, and continues in the Nagpur district in Maharashtra as a separate sanctuary. Renowned British author Rudyard Kipling mentioned this park in *The Jungle Book*, which made it one of the most popular national parks in the world and also a prominent tourist attraction in India. The geographical area of Pench is crisscrossed by numerous seasonal streams and nullahs. The Pench River, which flows through the centre of the reserve, is dry by the end of April, but several pools, locally known as *doh*s, remain, serving as waterholes for wildlife. The Pench Reservoir at the centre of the reserve is the only major water source during the summer period.

Birdwatching Sites

Pench is quite well known among birdwatchers for its rich birdlife. More than 300 bird species have been recorded in the park and its surrounding area. The Pench Reservoir is a major attraction to migratory waterfowl. The dead trees scattered amid the reservoir are good nesting sites for cormorants, egrets, herons and storks. White-necked and Painted Storks, and the Asian Openbill, Black-headed Ibis and Purple Heron are some of the birds that breed around the reservoir. A jungle safari provides the best way of birdwatching in Pench. Turia, Karmajhiri and Jamtara are three entry gates from which regular jungle safaris are organized by the Forest Department. Private vehicles are not allowed. Only registered four-by-four Maruti Gypsy vehicles are allowed. It is recommended that visitors book a vehicle in advance. The open jeeps provide an

Indian Pitta

Asiatic Wild Dogs

Indian Gaur

easy and great birdwatching and photography experience.

Access & Accommodation

Dr Babasaheb Ambedkar International Airport in Nagpur and Jabalpur Airport are the two options for travelling to Pench. Nagpur Airport has better connectivity from the rest of India, and is about 130km from the park. Seoni railway station is the nearest railway station to Pench.

There are many accommodation options available near the park gate. Pench Jungle Camp (www.penchjunglecamp.com), Vannraj Resorts (https://vannrajresorts.com) and Mahua Van (www.vresorts.in/resorts/v-resorts-mahua-vann-pench) are some of the private properties that are quite well known to birdwatchers. There are also forest rest houses and eco resorts. Jungle safaris can be booked online through the Madhya Pradesh Forest Department portal (https://forest.mponline.gov.in).

Conservation

Human settlements in the vicinity of Pench are dependent on the park for firewood, fodder and to some extent timber, which they use for construction. Some of the common conservation issues in Pench are illegal fishing and the grazing of cattle in the buffer zone. There have been frequent incidents of human-animal conflicts and poaching of wild animals.

Sarus Cranes

MAHARASHTRA

1. Tadoba National Park
2. Melghat Tiger Reserve
3. Sanjay Gandhi National Park
4. Koyna Wildlife Sanctuary & Chiplun
5. Pune

Maharashtra, a western-central state, is India's third largest state by area. It is bounded by the Arabian Sea to the west, and the Indian states of Karnataka, Telangana, Goa, Gujarat, Chhattisgarh and Madhya Pradesh, and union territory of Dadra and Nagar Haveli. Mumbai (formerly known as Bombay), a sprawling metropolitan city, is the capital of the state. It is world famous for its Bollywood industry and monuments. The Western Ghats (also known as Sahyadri Mountain), spread from north to south, has divided Maharashtra into Konkan (coastal region) and Desh (Deccan Plateau region), its two main natural zones. The Western Ghats is said to be the backbone of Maharashtra as well as the whole of southern India.

Climate

The state has three well-defined seasons, monsoon, winter and summer. The monsoon lasts from mid-June to the end of September, winter is in October–January, and summer is in February–May to June. Rainfall varies according to the topography of the region. The average annual rainfall in the Western Ghats is 2,000mm, but in some areas it reaches up to 3,500mm. Many districts, like Nashik, Pune, Ahmednagar, Nandurbar, Jalgaon, Beed, Akola, Satara, Sangli, Solapur and some parts of Kolhapur, lie in the rain shadow of the Ghats and have a mean annual rainfall of about 600mm.

Access, Transportation & Logistics

Maharashtra state is well connected to the rest of India via airways, railways and roadways. Its major cities, including Mumbai, Pune and Nagpur, have good connectivity by regular flights and long-distance trains with the rest of the country. The state has many air terminals, railway junctions and an extensive network of national highways. One of India's major airports is located in Mumbai. Chhatrapati Shivaji International Airport of Mumbai is connected to several cities around the world. The state's other international airports are in Nagpur and Pune, which are also connected to various Indian

KEY BIRDS

Top 10 Birds
1. Falcated Teal
2. Painted Francolin
3. Jerdon's Nightjar
4. White-throated Needletail
5. Heart-spotted Woodpecker
6. Tropical Shearwater
7. Masked Booby
8. Pacific Golden Plover
9. Long-billed Dowitcher
10. Wood Snipe

cities. The major railway station, Chhatrapati Shivaji Terminus, serves as the headquarters of the Central Railways of India and is connected to various cities across the country. There are also railway stations in Pune, Nashik, Nagpur, Kolhapur, Aurangabad and Amravati, which have great connectivity. The state is well connected by road to all important cities in India.

Health & Safety
There are no major health concerns in Maharashtra. The state is a low- to no-risk area when it comes to malaria, but do take precautions, especially after the monsoon. Visit a GP or travel clinic 6–8 weeks before departure to make sure you are up to date with any vaccinations. Avoid roadside food and drink bottled water. During the summer months keep hydrated, and carry a stole, scarf or cap for protection against direct sunlight. The Vidarbha and Marathwada regions of Maharashtra experience excessive heat during summer, when the maximum temperature rises to 50 °C; visitors travelling to Tadoba Tiger Reserve or nearby areas during summer should take extra precautions.

Birdwatching Highlights
The state has a very rich avifauna, with more than 636 species recorded so far. The main birding habitats can be broadly classified into forests, scrub and grass country, freshwater bodies, seashores, cultivated country and urban areas. Tropical deciduous forest is the dominant vegetation type, found all over the state in all divisions. Small pockets in the Western Ghats – like Bhimashankar, Matheran, Mahabaleshwar and Chandol – support evergreen and semi-evergreen vegetation. Tropical thorn forest is the dominant type of vegetation throughout the plateau area, and the Great Indian Bustard Sanctuary is located within this belt. There are some narrow mangrove patches and many creeks in the coastal belt.

The state has more reservoirs than any other in India. These waterbodies are full of migratory birds in the winter months. Nandur-Madhmeshwar (Nashik), Jayakwadi (Aurangabad), Ujani (Pune-Solapur), Mayani (Satara) and Navegaon (Bhandara) are some of the popular birding spots for wetland birds. Despite Maharashtra being the most urbanized state, its cities support high bird diversity. The Pune city bird checklist includes more than 400 bird species, and Mumbai, Nagpur, Nashik and Aurangabad support equally high bird diversity. Along the coast, the sandy beaches are full of waders in the winter months. Thane, Kihim (Raigad), Guhagar and Malvan-Tarkarli (Sindhudurg) are some good shore-birding spots.

Lesser Flamingos, Mumbai

Tadoba National Park

Maharashtra

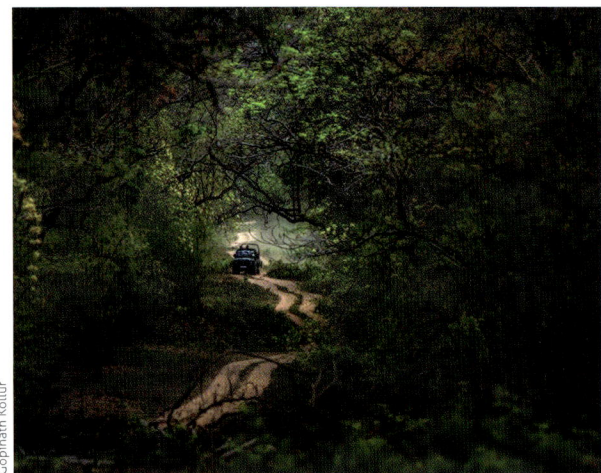

Tadoba forest

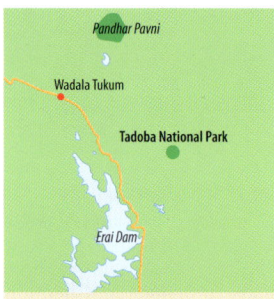

Also known as Tadoba Andhari Tiger Reserve, Tadoba National Park is one of the oldest national parks in India. It was declared a national park in 1955, and over a period of time has become one of the most visited national parks in India due to its increasing populations of the Tiger, other wild animals and birds.
The habitat consists of southern tropical dry deciduous forests interspersed with several large meadows, providing a good herbivore density for large cats. The forest is typical southern tropical dry deciduous forest, dominated by teak and bamboo.

Birdwatching Sites

The park is divided into three separate forest ranges: Tadoba north range, Kolsa south range and Morhurli range, which is sandwiched between the first two. There are two lakes and one river in the park, which fill up during every monsoon: Tadoba Lake, Kolsa Lake and Tadoba River. These lakes and the river provide essential ingredients needed to sustain the park's life. Tadoba Lake is visited by migratory waterfowl in winter, and other lakes and waterbodies are also visited by winter migrants.

The existence of the Andhari River inside the park encourages a wide diversity of waterbirds and raptors. More than 200 bird species have been recorded in the park, including three endangered species, the Grey-headed Fish Eagle, Crested Serpent Eagle and Changeable Hawk Eagle. Some of the other interesting bird species seen in this area are the Orange-headed Thrush, Indian Pitta, Crested Treeswift, Indian Thick-knee, Crested and Oriental Honey Buzzards, Asian Paradise-flycatcher, Bronze-winged Jacana, Lesser Golden-backed Woodpecker, Black-naped Monarch, White-eyed Buzzard, Eurasian Sparrowhawk, Shikra, Short-toed Snake Eagle, Bonelli's Eagle, Common Kestrel, Asian Openbill, Indian Black Ibis, Bar-headed Goose, Black Stork,

KEY FACTS

Nearest Major Towns
Chandrapur

Habitats
Tropical dry deciduous forest, tropical wet evergreen forest, freshwater swamp

Key Species
White-rumped Vulture, Lesser Adjutant, Greater Spotted Eagle, Sarus Crane, Green Avadavat, Indian Black Ibis, White-eyed Buzzard, Painted Francolin, Rain Quail, Jungle Bush Quail, Indian Peafowl, Yellow-legged Green Pigeon, Plum-headed Parakeet, Indian Nightjar, Brown-headed Barbet, Yellow-crowned Woodpecker, Lesser Golden-backed Woodpecker, Indian Robin, Brown Rock Chat, Jungle Babbler, Jungle Prinia, Ashy Prinia, Brahminy Starling, White-bellied Drongo

Other Specialities
Tiger, Indian Leopard, Leopard Cat, Rusty-spotted Cat, Jungle Cat, Small Indian Civet, Common Palm Civet, Indian Grey Mongoose, Ruddy Mongoose, Indian Jackal, Asiatic Wild Dog, Sloth Bear

Best time to Visit
October to March

Red-wattled Lapwing

Eurasian Wryneck

Jerdon's Leafbird

Lesser Adjutant, Brahminy and Comb Ducks, Little Grebe, Grey and Purple Herons, Banded Bay Cuckoo, Painted Sandgrouse, Jungle Bush Quail, Indian Peafowl, Spotted Owlet, Forest Wagtail, Indian Robin, Indian Roller, Himalayan Rubythroat, Bluethroat, Black Redstart, Cinereous Tit and Common Stonechat.

Access & Accommodation

Tadoba National Park is 140km from Dr Babasaheb Ambedkar International Airport, Nagpur. There are regular flights to Nagpur from all of India's major cities. Visitors can hire taxis or cabs from the airport to reach the park. The nearest railhead from the park is located in Chandrapur, which is 45km away. Chandrapur railhead is well connected to other major cities, and there are regular buses and taxis from Chandrapur to Tadoba, which is also well connected by the road network from other major cities of India.

Tadoba has a heavy influx of visitors, and to cater for them there are numerous hotels and resorts. There is a Maharashtra Tourism Development Corporation resort near Moharli gate. Booking can be made online (www.maharashtratourism.gov.in/properties/property/mtdc-tadoba). There are also many private properties near the park. Irai Safari Retreat (http://iraisafariretreat.com) and Camp Serai (www.seraitiger.com) are well known among birdwatchers and wildlife enthusiasts.

Conservation

Tadoba National Park is surrounded by big villages. Some of the common conservation issues faced by the park are grazing, poaching of wild animals including birds, illegal cutting of trees and firewood collection. One of the main problems is firewood and timber collection, and a huge area has been degraded solely due to deforestation and clearance of woods and grassland for grazing. Poaching of wild animals takes place regularly in many areas. There have been numerous incidents of human-animal conflict, mainly due to crop damage by herbivores and the lifting of cattle by large carnivores.

Black-naped Monarch

Melghat Tiger Reserve

Melghat landscape

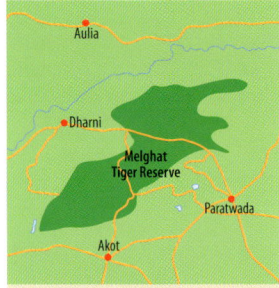

KEY FACTS

Nearest Major Towns
Amravati

Habitats
Tropical dry deciduous forest, tropical grassland

Key Species
White-rumped Vulture, Lesser Kestrel, Green Avadavat, Forest Owlet, Indian Black Ibis, White-eyed Buzzard, Painted Francolin, Rain Quail, Jungle Bush Quail, Indian Peafowl, Yellow-wattled Lapwing, Indian Courser, Yellow-legged Green Pigeon, Plum-headed Parakeet, Indian Nightjar, Indian Grey Hornbill, Yellow-crowned Woodpecker, Lesser Golden-backed Woodpecker, Black-rumped Flameback, Ashy-crowned Sparrow Lark, Black-headed Cuckooshrike, Small Minivet, Common Woodshrike, Indian Robin, Jungle Babbler, Jungle Prinia, White-browed Fantail, Chestnut-tailed Starling, Brahminy Starling, White-bellied Drongo

Other Specialities
Tiger, Indian Leopard, Asiatic Wild Dog, Indian Wolf

Best Time to visit
October to March

This reserve is situated in the Satpura hill ranges of central India. It lies in the Melghat forests of the Amravati, Akola and Buldhana districts in the Vidarbha region of Maharashtra, bordering Madhya Pradesh in the north and east. The name Melghat means 'the place where the ghats meet'. As the name suggests, the reserve is well known for its unique topographical diversity, dotted by several high hills and deep valleys. The high ridge running east–west forms the southwestern boundary of the reserve, and the northeastern boundary is marked by the Tapti River. The forest is tropical dry deciduous, dominated by teak. It is a prime habitat of the Tiger and several other wild mammals and birds.

Birdwatching Sites

Melghat is known for its diverse birdlife and is one of the best places to see species of the Indo-Malayan Tropical Dry Zone. More than 300 bird species have been recorded in the reserve, including some that are Critically Endangered, like the Forest Owlet and White-rumped Vulture. There are regular safaris from Kolkas, Vairat and Harisal gates. Along with jungle safaris, there are nature trails for those who want to explore the jungle more closely on foot. These also offer some good birdwatching experiences.

Some of the commonly seen bird species of the area are the Green Avadavat, Yellow-legged Green Pigeon, Jungle Bush Quail, Rain Quail, Painted Francolin, Indian Roller, Indian Grey Hornbill, Cattle Egret, Green Bee-eater, Black Drongo, Common Teal, Crested Serpent Eagle, Asian Paradise-flycatcher, Indian Peafowl, White-eyed Buzzard, Common Kingfisher and Little Grebe. The winter months (November–February), when many migratory birds along with endemic birds inhabit the park, are good for birdwatching. Most of the park is inaccessible during the monsoon season.

Bonelli's Eagle

Pallid Harrier

Short-toed Snake Eagle

Access & Accommodation
The nearest airport to Melghat is Nagpur, which is about 255km away. Visitors can hire a taxi from Nagpur to reach the park. The nearest railway station is Badnera, located on the central railway zone on the Mumbai–Kolkata route. Badnera is about 110km from the park.

A couple of ecotourism lodges are run by the Forest Department. Kolkas Ecotourism Complex is an affordable and comfortable accommodation option near the reserve. Chikhaldara, a hill station and well-known tourist place in the Vidarbha region of Maharashtra, is another important gateway to the reserve. It has many privately owned hotels and resorts, and a jungle safari gate commonly known as Vairat.

Conservation
Due to its unique flora and rich biodiversity, the Melghat Tiger Reserve has a very significant place in the conservation history of Maharashtra. Some of the major threats and conservation issues are encroachments on wildlife habitats and forest land for agricultural purposes by local people, illicit cutting of trees for local needs and commercial purposes, illegal cattle grazing in the buffer zone, illegal removal of forest produce and valuable medicinal plants, soil, boulders, rock, sand, stones, including minerals, and so on. Poaching and hunting of wild animals for local as well as commercial purposes is another major threat.

Greater Painted Snipe

Forest Owlet

Maharashtra

Sanjay Gandhi National Park

Sanjay Gandhi landscape

This park is located in the Sahyadri Range in the northernmost part of the Western Ghats. It is a unique national park of India, being located within the mega-metropolitan Mumbai. The park's forest is tropical dry deciduous or southern dry deciduous, and is dominated by teak and bamboo. The geographical area of the park constitutes the prime catchment area of two freshwater lakes, Tulsi and Vihar, which supply water to Mumbai city. These two lakes support a variety of aquatic fauna and flora. The existence of the old Buddhist Kanheri Caves at the centre of the park makes the area a place of great historical importance.

KEY FACTS

Nearest Major Towns
Mumbai

Habitats
Urban, tropical dry deciduous forest, tropical dry evergreen forest

Key Species
White-rumped Vulture, Long-billed Vulture, Lesser Adjutant, Pallas's Fish Eagle, Greater Spotted Eagle, Indian Skimmer, Nilgiri Wood Pigeon, Malabar Trogon, Malabar Whistling Thrush, Indian Scimitar Babbler, Loten's Sunbird

Other Specialities
Black-naped Hare, Indian Chevrotain, Palm Civet, Indian Porcupine, Rhesus Macaque, Indian Flying Fox, Mouse Deer, Bonnet Macaque

Best Time to Visit
Throughout the year

Lesser Flamingos

Peregrine Falcon

Oriental Honey Buzzard

Peregrine Falcon

Birdwatching Sites
Sanjay Gandhi National Park is a well-known tourist spot within Mumbai city and attracts a great number of visitors from across the world. With its lush green hideouts and nature trails, the park offers a great birding experience. There are many well-known trails, like the Kanheri Caves, Bamboo Hut and Shilonda Trails, which can yield very good birding results in just one day.

More than 300 bird species have been recorded in this area, including some rarities like White-rumped and Long-billed Vultures, Greater Spotted Eagle and Nilgiri Wood Pigeon. Some of the other important bird species found here are the Asian Emerald Dove, Malabar Trogon, Oriental Dwarf or Three-toed Kingfisher, Crimson-backed Sunbird, Indian Scimitar Babbler, Loten's Sunbird, White-bellied Sea Eagle, Peregrine Falcon, Pallid and Western Marsh Harriers, Puff-throated Babbler and Black-naped Monarch.

Access & Accommodation
The park is situated in Mumbai, one of the busiest cities of India. Mumbai is very well connected with all major national as well as international cities. From Mumbai Airport or railway station, you can take the local train to Borivali or hire a taxi.

There are many Forest Department rest houses in the park, but due to high demand they cannot be booked at short notice. There are numerous accommodation options available in Mumbai city, from the most basic to the most luxurious.

Conservation
The park is surrounded by the Mumbai and Thane districts. Encroachment of slum colonies into the park, the smuggling of timber, firewood collection, poaching, other anti-social activities and human-animal conflicts are some of the major conservation issues. Frequent human-animal conflict indicates the increasing pressure of these negative anthropogenic activities on the natural habitat of wild animals.

Rain Quail

Koyna Wildlife Sanctuary & Chiplun

Koyna Lake

This sanctuary is located in the Satara district in western Maharashtra. It includes the eastern and western catchments of the Koyna Dam, a major hydroelectric project in western Maharashtra. At the centre of the sanctuary lies the historical Vasota Fort, constructed in 1178–1193, which was taken over by the great Maratha warrior Chhatrapati Shivaji in 1655 and used as a prison. The vegetation of the sanctuary consists of southern tropical evergreen forests and southern moist mixed deciduous forests. It is in the tropical monsoon belt, and consequently receives excessive rainfall. Additionally, thunderstorms wreak havoc in the sanctuary. Due to the rainfall, the sanctuary remains emerald green throughout the year.

KEY FACTS

Nearest Major Towns
Satara and Kolhapur

Habitats
Tropical evergreen forest, southern moist mixed deciduous forest

Key Species
White-rumped Vulture, Long-billed Vulture, Nilgiri Wood Pigeon, Malabar Grey Hornbill, Crimson-backed Sunbird, Small Minivet, Green Bee-eater, Eurasian Crag Martin, Brahminy Kite, Little Cormorant, Oriental Dwarf Kingfisher, Blue-eared Kingfisher, Indian Pitta, Asian Fairy-bluebird, Malabar Grey Hornbill

Other Specialities
King Cobra, Tiger, Indian Leopard, Sambar, Gaur, Sloth Bear, Mouse Deer, Indian Chevrotain, Malabar Giant Squirrel, Smooth-coated Otter

Best Time to Visit
Throughout the year

Western Ghats landscape

Oriental Dwarf Kingfisher

Common Snipe

Birdwatching Sites

The sanctuary is not very well studied, but it has great potential for birdwatching and is categorized as an IBA. Some of the commonly seen birds are the Heart-spotted Woodpecker, Crested Goshawk, Asian Fairy-bluebird, Large-tailed Nightjar, Malabar Grey Hornbill, many varieties of sunbird and the endangered Nilgiri Wood Pigeon. Regular safaris are organized in the sanctuary. The resort of your stay can usually organize a safari for you, and a safari runs between 7 a.m. and 6 p.m. If you are travelling by private car, you can enter the park in your vehicle for an extra fee.

Chiplun

A small place in the Ratnagiri district of Maharashtra, Chiplun has established itself as a great birding destination due to its star bird, the Oriental Dwarf Kingfisher. This small, very colourful bird breeds in the region during the monsoon season and attracts birders from all over India. Local people have established hides, which provide an opportunity to get a closer look at the bird without much difficulty. Chiplun also offers good sightings of the Blue-eared Kingfisher, Indian Pitta, Black-hooded Oriole and many varieties of sunbird. Koyna Wildlife Sanctuary is about four hours' drive from Chiplun and a visit here can incorporate a birding trip to the sanctuary. There are some homestay options available in Chiplun.

Access & Accommodation

The sanctuary is in the Satara district. It lies 180km from Pune, and access from Pune is quite easy by private vehicle. The Pune-Satara highway is well paved and scenic, and visitors generally prefer to travel by road to reach the sanctuary.

There are many resorts in the sanctuary. Some are relatively cost-effective and offer good services, like Nisarga Agro Tourism, the Riverview Resort and the Forest County Resort. There is one Maharashtra Tourism Development Corporation resort near Koyna Dam.

Conservation

Koyna Wildlife Sanctuary is honeycombed with privately owned forests and patches of agricultural land. At the periphery of the sanctuary the area is being actively promoted as a tourism zone by the government of Maharashtra. This is creating serious environmental problems in the form of increased tourist traffic, water pollution, littering of non-degradable waste and general disturbance. The topography of the sanctuary is conducive to high wind velocity, and it is therefore likely to be considered for the exploitation of non-conventional energy. The windmills that dot the landscape are currently located away from the boundary of the sanctuary, but are likely to also be erected inside it.

Pune

Pune landscape

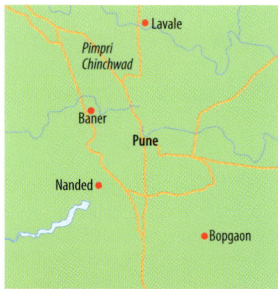

KEY FACTS

Nearest Major Towns
Pune

Habitats
Urban plains, wetlands, scrub

Key Species
Little Swift, Black Kite, Dusky Crag Martin, Red-vented Bulbul, Oriental White-eye, Common Myna, Thick-billed Flowerpecker, Indian Peafowl, Common Pigeon, Laughing Dove, Greater Coucal, Grey-bellied Cuckoo, Crested Treeswift, Cattle Egret, Shikra, Common Hoopoe, Green Bee-eater, Common Iora, Long-tailed Shrike, Black Drongo, Glossy Ibis, Barn Swallow, Nilgiri Flowerpecker, Brahminy Kite, White-bellied Drongo, Common Babbler, Asian Brown Flycatcher, Indian Vulture

Other Specialities
Wild Pig, Malabar Giant Squirrel, Malabar Large-spotted Civet, Indian Grey Mongoose

Best Time to Visit
Thoughout the year

Pune is the second largest city in Maharashtra, after Mumbai. The city is world famous for its sprawling information technology and manufacturing industry. It was the base of the Peshwas (prime ministers) of the Maratha Empire. Pune comes into the rain-shadow region of the Sahyadri mountain range. The city is surrounded by hillocks, lakes, dams, backwaters and open grassland, which harbour a great variety of resident birds and attract a lot of migratory species during the winter.

Birdwatching Sites

Pune is one of the most famous cities for birding, and more than 400 different resident and migratory birds have been recorded from this region. Pune has many hill forts and lakes that make great birding destinations. Most of the birding destinations are within a few kilometres from the city, so they can provide a comfortable birding experience for birdwatchers.

Sinhagad Valley Sinhagad is a hill fortress about 36km southwest of Pune city. It is quite well known among birdwatchers, and attracts huge crowds during weekends. Visiting early in the morning, preferably on weekdays, is recommended for some quiet birding hours. Sinhagad Valley is a great place in which to look for flycatchers like the Indian Paradise-flycatcher, Black-naped Monarch, White-spotted Fantail, and Tickell's Blue, Verditer and Ultramarine Flycatchers. During the early morning hours you may encounter some surprises, like Crested Treeswifts.

Some of the common birds found here are the Oriental White-eye, Scaly-breasted Munia, Crested Serpent Eagle, Shikra, Black-winged Kite, Jungle Bush Quail, Orange-headed Thrush, Rufous Treepie, Coppersmith and White-cheeked Barbets, Black and

Common Stonechat

White-bellied Drongos, Jungle and Brahminy Starlings, Indian Yellow and Cinereous Tits, Common Tailorbird, Jungle, Tawny-bellied and Puff-throated Babblers, Common Iora, Indian Robin, Oriental Magpie Robin, White-throated Kingfisher and Indian Blackbird.

Bhigwan This is a small town on the border of the Pune and Solapur districts. It is located on the Pune-Solapur Highway, about 105km from Pune and Bhigwan. It is well connected by road and the recommended mode of transport is private vehicle, as it will give you the most flexibility to travel around. The area attracts many migratory birds during the winter season.

Diksal and Kumbhargaon are the main birding locations in Bhigwan. Kumbhargaon is well known for flamingos, and you can find a huge flock here in December–March. Some other important birds of the region are Painted Stork and Asian Openbill, Bar-headed Goose, Common Teal, Northern Shoveler, Ruddy Shelduck, Western Marsh Harrier, Greater Spotted Eagle, Pallas's and Brown-headed Gulls, Grey and Purple Herons, and Osprey.

Veer Dam This is about 60km south of Pune, off the Pune-Bangalore highway. It is slightly quieter and less well known than other birding spots. Due to fewer crowds, it will reward you with many surprises if you visit early in the morning. Waders, flycatchers and raptors like the Western Marsh Harrier and Common Kestrel are commonly seen, although the chief draws are migratory visitors like the Bar-headed Goose and Demoiselle Crane, which winter here.

Dive Ghat and Bopdeo Ghat Dive Ghat is a hillock region on the Pune-Saswad road. It is a well-known birding area at the southeastern end of the city. Some of the common birds that you can easily find here are Long-tailed, Bay-backed and Southern Grey Shrikes. It is also a great place for sightings of Striolated, Black-headed and Grey-necked Buntings, the Eurasian Roller, Rain Quail, Painted Francolin, and other birds that inhabit scrubland and bushland.

On the other side of Dive Ghat is Bopdeo Ghat, a grassland habitat. Apart from the larks, pipits and francolins that abound here, birds of prey like hawks and harriers frequent the area. You may also spot foxes and wolves.

Mayureshwar Wildlife Sanctuary This sanctuary is located about 70km from Pune, off the Pune-Solapur highway. The dry deciduous

Coppersmith Barbets

scrub forest is best known for the Chinkara Antelope and occasional sighting of the Striped Hyena and wolves. It is an incredible birding spot. Some of the common birds found here are sandgrouse, shrikes, pipits, buntings, munias, larks and raptors like the Steppe Eagle, Bonelli's Eagle and Short-toed Snake Eagle. In winter, you can see a variety of migrants like wheatears, warblers and some waders. Montagu's Harrier visits the sanctuary during the winter months. There are also records of the Eastern Orphean Warbler from this region.

Access & Accommodation
Pune is the second largest city in Maharashtra and one of the fastest growing cities in India. It is well connected with the rest of the country via air, railways and national highways. Pune Airport is located at the heart of the city. It is an international airport, busy with flights from all around the world. Pune is also well connected by railways, and local as well as express trains run throughout the day. Trains to Pune are available from all major railway stations in India.

Various accommodation options are available in Pune city, from budget hotels to luxurious starred hotels. Private taxis can be booked from your hotel. Ola and Uber taxis are also available in Pune.

Conservation
Pune is one of the fastest growing cities in India, and the ninth most populous. Its rapid growth and urbanization are taking their toll in the form of pollution, deforestation and environmental degradation. The city does not have enough mechanisms for dealing with construction waste. The existing water stock is not sufficient for its citizens, as well as for the surrounding nature and wildlife. The city's green cover is declining with every passing month, and this is one of the most serious conservation issues for the city and its inhabitants.

MANIPUR

A small state in northeastern India, Manipur is bounded by the Indian states of Nagaland to the north, Mizoram to the south and Assam to the west. It also borders the country of Myanmar to the east. The terrain in the state is predominantly hilly, except for a broad central valley. The average altitude is 850m, and the maximum altitude of the hilly region is almost 3,000m. The Manipur (Imphal) and Barak are the main rivers, and there is a large freshwater lake named Loktak to the south of the valley. There is hardly any forest left in the valley; the fertile alluvial plain was cleared for cultivation long ago. The natural forest is mainly found in patches in the hills. Towards the east and south-east, three lakes, Ikop Pat, Kharung Pat and Pumlen Pat, complete the wetland ecosystem of the valley. Imphal, the capital city, lies in an oval-shaped valley of about 1,800km² surrounded by blue mountains, and is at an elevation of 790m above sea level. Hills can be seen from all around the state.

1. Loktak Lake and Keibul Lamjao National Park

Climate

Manipur experiences a mostly tropical and alpine climate. The northeastern region has an amiable climate and is very cold in winter. The temperature in the summer is 32 °C, and in winter it falls to below zero. The rains last from May to mid-October, the average rainfall being about 1,467mm. The climate varies according to the elevations of the landforms in the state. The weather in the plains is similar to that of other Indian states, but the hilly regions are different and enjoy a pleasant climate. The weather in the state is highly influenced by the winds blowing from the Bay of Bengal and is conducive to heavy rainfall in the rainy season. The state has three main seasons, summer, winter and the rainy season. It does not experience extreme climatic conditions, although winter temperatures may drop to below zero degrees. The hills and valleys of the state extend from the Himalayan region and are a part of the Himalayan ranges. The climate therefore supports the greenery here, and the rich flora is a reflection of this.

Access, Transportation & Logistics

Most visitors enter Manipur by air through the beautiful capital city, Imphal, which is connected by air, train and bus. Imphal's Tulihal Airport is about 8km from the heart of the city and is well connected directly to the major cities of India, namely Delhi, Hyderabad, Kolkata and Guwahati, and daily to the major cities of the northeastern states. There is no direct train service to Imphal, but visitors can travel up to Guwahati or Dimapur (the nearest railhead from Imphal), then cover the onward journey by bus or air. Manipur is well connected by the road network from Assam and other northeastern states. There are daily buses from Guwahati via NH 27, NH 29 and NH 2 via Dimapur and Kohima.

KEY BIRDS
Top 5 Birds
1. Manipur Bush Quail
2. Blue Pitta
3. Rufous-throated Hill Partridge
4. Grey Peacock Pheasant
5. Japanese Quail
6. Chinese Francolin
7. Speckled Wood Pigeon
8. Barred Cuckoo Dove
9. Solitary Snipe
10. Ashy Wood Pigeon

Silver-eared Mesia

Health & Safety

Visitors are advised to get a medical check-up before travelling to Manipur and to always keep a personal first-aid box. As the north-east region has an entirely different culture and cuisine from the rest of India, you need to take basic food-safety precautions while travelling here. There are some natural annoyances in rainforests, particularly in the lowlands of the Himalayan region. In addition to terrestrial leeches, ticks, mosquitoes and various species of biting fly occur. The use of long sleeves and insect repellent is recommended. Vaccinations are also recommended for diseases such as hepatitis A and B and typhoid. Other mosquito-borne viruses like dengue fever and malaria are present, and appropriate precautions should be taken.

Crested Goshawk

Birdwatching Highlights

There is one national park (Keibul Lamjao) and five wildlife sanctuaries in Manipur. Loktak Lake, with an area of about 200km², is a wetland of international importance. It has been designated as a Ramsar Site. More than 700 bird species have been identified in Manipur, including some rare and elusive ones like Blyth's Tragopan, Mrs Hume's Pheasant, Green Peafowl (no recent record, presumed extirpated), Dark-rumped Swift and Rufous-necked Hornbill. Records confirm the presence of the White-winged Wood Duck, but the species was always rare and there have been no recent sightings. According to historical records of threatened birds, many birds were recorded in Manipur, such as the Oriental Stork, Greater Adjutant, White-winged Wood Duck, Spot-billed Pelican, Baikal Teal, Baer's Pochard, Pallas's Fish Eagle and Greater Spotted Eagle.

Manipur Fulvetta

Floating islands or Phumdi

Loktak Lake and Keibul Lamjao National Park

Manipur Brow-antlered Deer

KEY FACTS

Nearest Major Towns
Imphal

Habitats
Wetland surrounded by hills

Key Species
Spot-billed Pelican, Greater Spotted Eagle, Lesser Adjutant, Sarus Crane, Hooded Crane, Peregrine Falcon, Chinese Mountain Bamboo Partridge, European White Stork, Green Peafowl, Austen's Brown Hornbill, Rufous-necked Hornbill, Wreathed Hornbill, Oriental Pied Hornbill, Great Hornbill

Other Specialities
Sangai, Indian Hog Deer, Indian Chevrotain, Sambar, Marbled Cat, Hoolock Gibbon, Smooth-coated Otter, Clouded Leopard

Best Time to Visit
October to March

Loktak Lake is the largest natural freshwater lake in north-east India and plays an important role in the ecological and economic security of the region. An oval lake with a maximum length of 26km and a width of 13km, it has an average depth of 2.7m. There are 14 hills varying in size and elevation, appearing as islands, in the southern part of the lake. The lake was designated a wetland of international importance under the Ramsar Convention in 1990.

Keibul Lamjao National Park is home to the highly endangered Manipur Brow-antlered Deer, one of the three subspecies of Thamin Deer. The other two subspecies are found in Myanmar and Indo-China. The park was created to protect this deer, locally known as Sangai. It was reported to be extinct in 1951, but a survey conducted by the IUCN revealed that a few animals still existed in the park. The Sangai is specially adapted to this floating habitat – its characteristic hooves, unlike those of other deer species, help it to walk conveniently over the floating islands. The park forms the southern portion of Loktak Lake, and is a large, continuous mass of swamp with floating mats of vegetation, locally known as *phumdis*, covering much of its surface.

Birdwatching Sites

Loktak Lake provides a refuge for thousands of birds of different species, including many waterfowl. Their numbers easily exceed 20,000. It has records of the Spot-billed Pelican and Greater Spotted Eagle, both globally threatened species. The Lesser Adjutant is regularly seen here. The Critically Endangered Yellow-breasted Bunting, which was considered to be on the

Northern House Martin

Falcated Duck

edge of extinction, was recently spotted in Manipur – in the Thong Jorok River at the mouth of Loktak Lake in the first week of November. Some of the important bird species of the region are the Lesser Whistling Duck, Greylag and Bar-headed Geese, Common and Ruddy Shelducks, Red-crested Pochard, Ferruginous Duck, Japanese Quail, Black-necked Stork and Sarus Crane.

Access & Accommodation

The best place to start a journey to Loktak Lake is Imphal. From here, your first destination is Moirang. This is a nondescript town on the border of the lake, and it serves as a drop-off point for visitors. There is not much to see in Moirang itself. Shared taxis ply the roughly 30km between Imphal and Moirang, and take about an hour. From Moirang, you have to take another shared taxi to Thanga, a cluster of villages that make up a peninsula on Loktak Lake.

For accommodation, Imphal has several mid-range hotels and many homestay options available. There are also a couple of good eco-cottage and homestay options near Thanga.

Conservation

The lake ecosystem changed considerably after the construction of a multipurpose hydroelectric and irrigation project. The natural wetland with a fluctuating water level was converted into a reservoir with a more or less constant water level. Besides bringing about basic hydrological changes, this resulted in severe problems for the lake biota and the communities traditionally dependent on it. The domestic sewage from the floating hutments, which is directly discharged into the lake, is accelerating the process of eutrophication. Populations of both migratory and resident waterfowl and fish have rapidly declined in the past few decades. Shooting, netting, pesticide pollution and hydrological changes, as well as increased human presence, fishing, removal of vegetation and tourism, have all contributed to the decline in birds.

Black-necked Grebe

MIZORAM

Mizoram means 'the land of highlanders'. This state is located in the extreme southern part of northeastern India. It is bounded on the north by Assam and Manipur, on the east and south by Myanmar, and on the west by Bangladesh and Tripura. The terrain is hilly and mostly undulating, with an average altitude of 500–800m, and the maximum reaching 2,157m in the Blue Mountains. The Kolodyna (Chhimtuipui), Tlawng (Dhaleswari) and Tuivai (Tipai) are the major rivers in the state. Mizoram has three major forest types, tropical wet evergreen, tropical moist deciduous and subtropical pine forests.

Climate

Mizoram's climate ranges from moist tropical to moist subtropical. Its temperature varies from region to region with its elevation changes. The overall temperature is quite pleasant and rainy throughout the year. The average annual rainfall ranges from 2,160mm in Aizawl to 3,500mm in Lunglei. During the winter, the temperature varies from 22 °C to 24 °C, and in summer from 18°C to 29°C. The Tropic of Cancer passes through the middle of the state. The summer season continues from the end of March to June, and the winter season is normally from mid-November to February. The monsoon season is long, starting in June and continuing until October. Rain and clouds are very common, and the state receives rainfall both during the monsoon season and other seasons. The average annual rainfall in the state is about 250cm.

Access, Transportation & Logistics

Aizawl is the base for Mizoram. It is well connected with major cities like Guwahati and Kolkata by air, road and rail. It has a domestic airport with good connections to many cities, such as Guwahati, Imphal and Kolkata, through regular daily flights. Mizoram can also be reached from Kolkata via Silchar Airport, which is 200km from Aizawl. A helicopter service

1. Dampa Tiger Reserve

> **KEY BIRDS**
>
> Top 10 Birds
> 1. Green Peafowl
> 2. Mrs Hume's Pheasant
> 3. Hodgson's Hawk Cuckoo
> 4. Eastern Water Rail
> 5. Chinese Pond Heron
> 6. Rufous-bellied Eagle
> 7. Himalayan Buzzard
> 8. Mountain Scops Owl
> 9. Mountain Bamboo Partridge
> 10. Fork-tailed Swift

by Pawan Hans has been started, connecting Aizawl with nearby cities. The closest railway station for Aizawl is Silchar in Assam. Mizoram has a good road network, connecting all the major cities and villages. It is connected to Silchar and Shillong through NH 54. NH 150 connects the state with Seling Mizoram to Imphal Manipur, and NH 40A links the state with Tripura. The distance from Aizawl to other cities is: Guwahati 506km, Imphal 374km, Kohima 479km, Shillong 450km and Agartala 443km.

Health & Safety

Some health and safety precautions have to be taken before coming to this region, as the Himalayan hills and dense forests are pervasive throughout. Travellers, especially those from Western countries, might encounter minor health problems like stomach disorders and fever. It is best to carry some medicine as prescribed by medical practitioners. There are some natural annoyances in rainforests, particularly in the lowlands. In addition to terrestrial leeches, ticks, mosquitoes and various species of biting fly occur. The use of long sleeves and insect repellents is recommended for rainforest birdwatching. Vaccinations are also recommended for diseases such as hepatitis A and B and typhoid. Other mosquito-borne viruses like dengue fever and malaria are present, and appropriate precautions should be taken.

Birdwatching Highlights

Mizoram is very rich in birdlife. More than 450 bird species have been recorded in this region and there is still a lot of potential as this area is yet to be studied well from an ornithology perspective. Wreathed, Rufous-necked, Great, Brown and Oriental Pied Hornbills are found in Mizoram. As in the more famous Jatinga in Assam, where large numbers of birds are killed when they are attracted to artificial lights, in the Lunglei district of Mizoram many birds die in the night after hitting the walls of buildings. This occurs in September and October, when the area receives late rains and the sky is overcast, with fog and mist. Annually, 300–500 birds are killed. Interestingly, most of the birds here are residents such as the Common Moorhen, Grey-fronted Green Pigeon, Asian Emerald Dove, Oriental Dwarf and Ruddy Kingfishers, Hooded

Himalayan Cutia

Pitta, Drongo Cuckoo and several other cuckoos. Some of the other important bird species here are Blyth's Tragopan, Mrs Hume's Pheasant and Dark-rumped Swift.

Mizoram landscape

Dampa Tiger Reserve

Dampa forest

KEY FACTS

Nearest Major Towns
Aizawl

Habitats
Tropical forest and precipitous hills

Key Species
Blue Pitta, Mrs Hume's Pheasant, Grey-backed Shrike, Ruby-cheeked Sunbird, Blue Rock Thrush, Blyth's Tragopan, Green Peafowl, White-cheeked Hill Partridge, Blyth's Kingfisher, Red-breasted Parakeet, Rufous-vented Yuhina, Spot-breasted Laughingthrush, Crested Finchbill, Flavescent Bulbul, Oriental Hobby, Cachar Wedge-billed Wren Babbler, Van Hassell's Sunbird

Other Specialities
Tiger, Hoolock Gibbon, Bamboo Rat, Assamese Macaque, Pigtail Macaque, Stumped-tailed Macaque, Capped Langur, Slow Loris, Golden Cat, Jungle Cat, Marbled Cat, Hog Badger, Large-toothed Ferret Badger

Best Time to Visit
October to March

This reserve is situated in Mizoram on the international border with Bangladesh. It is the largest protected area in Mizoram. The landscape is undulating, with hills running from north to south. Small, perennial rivulets flow all over the reserve. It consists of moist deciduous forests in the lower reaches, and evergreen and semi-evergreen forests with natural grassland at higher altitudes.

Birdwatching Sites

Dampa holds many threatened mammals, including the Tiger, Asian Elephant, Clouded Leopard, Gaur, Goral, Hoolock Gibbon, Stump-tailed Macaque, Binturong and numerous others. However, the area is also justly famous for its birds, with some great rarities to be found only here in India. Mrs Hume's Pheasant (state bird), Blyth's Tragopan, Green Peafowl (not seen for many years and presumed extirpated), White-cheeked

Blue Pitta

Black-eared Shrike-babbler

Black-throated Parrotbill

Mrs Gould's Sunbird

Hill Partridge, Blyth's Kingfisher, Blue Pitta, Moustached, Striped, Rufous-vented, Brown-capped and Spot-breasted Laughingthrushes, Crested Finchbill, Olive and Flavescent Bulbuls, Oriental Hobby, Wedge-billed Wren Babbler and Van Hassell's Sunbird are just some of the species that can be found in the area.

Access & Accommodation

Dampa is easily accessible through Lengpui Airport in Mizoram. This is a domestic airport in Aizawl, connected by flights to Kolkata and Imphal; it is about 32km from Aizawl. There are also flights from Guwahati and Imphal on certain days of the week. Aizawl is connected by road from Silchar, Assam. Dampa Tiger Reserve is about 127km from Aizawl and the journey may take 3–4 hours, depending on the road conditions.

There is forest department accommodation at Teirei, Phuldungsei and Damparengpui. You need to make food arrangements yourself as there are no restaurants inside. For booking contact: The Field Director, Dampa Tiger Reserve, West Phaileng, Tel.: +91-0389-2012298. An inner line permit is required for Indian tourists, and a Restricted Area Permit (RAP) is necessary for foreigners.

Conservation

All the villages surrounding the Tiger Reserve practice slash-and-burn shifting cultivation, and there is always the threat of fire in the reserve during the burning season. The sudden increase in insurgency activity threatens the security of the reserve staff. Funds and staff to manage the reserve are inadequate. There is also some pressure from peripheral villagers for non-timber forest produce and firewood. Poaching of wild animals including birds is another serious threat.

Mrs Hume's Pheasant

MEGHALAYA

1. Balpakram National Park

Meghalaya means 'the abode of the clouds'. It became an autonomous state on 2 April 1970, and was declared a full state of the Indian union on 21 January 1972. Topographically, Meghalaya is a plateau, apart from narrow strips of plains in the northern, western and southern parts. Shillong, the capital, is situated in the centre of a high plateau. The elevation ranges from less than 100m to 1,961m, and the highest peak is Laitkor Peak (1,961m). The state is the homeland of three of India's ancient hill tribes, the Khasis, Jaintias and Garos. Meghalaya is popularly known as the Scotland of India due to its wet and cold climatic conditions. It is a biodiversity hotspot, ranging from lush green coniferous forests to bamboo thickets, from limestone caves to rippling waterfalls, from miles of hilly cropland to forbidding jungles, and its diversity makes it a unique realm. This fog-shrouded land is also home to the wettest place on Earth (Mawsynram).

Climate

The state enjoys a temperate climate. It is directly influenced by the south-west monsoon and the north-east winter wind. The four seasons of Meghalaya are spring, March–April, summer (monsoon), May–September, autumn, October–November, and winter, December–February. The monsoon usually starts by the third week of May and continues to the end of September and sometimes well into the middle of October. The maximum rainfall occurs on the southern slopes of the Khasi Hills, over the Sohra and the Mawsynram platform, which receives the heaviest rainfall in the world. The average rainfall in the state is 12,000mm.

Access, Transportation & Logistics

There is a small airport at Umroi, about 35km from Shillong, the capital city of Meghalaya. At present, Air India operates with ATR42-type aircraft on this route. Buses and taxis are available from Shillong to Umroi and back. The Gopinath Bordoloi Airport in Guwahati (128km from Shillong) is more convenient and better connected to the rest of India, with regular flights. Shared cabs, as well as private taxis, are available outside the airport connecting to Shillong. There is also a helicopter service from Guwahati to Shillong that takes about 20 minutes to reach Shillong from Guwahati. This is a convenient mode of transport if you are running short of time. There are no rail lines in Meghalaya. The nearest railway station is Guwahati. Meghalaya Transport Corporation operates bus services coordinated with train arrivals at Guwahati. You can also get taxis from Guwahati railway station.

Health & Safety

Village and rural tourism have emerged as a new concept in the tourism industry. Meghalaya is now fast evolving as a responsible and sustainable tourism destination with an important social objective through people's participation. There are some natural annoyances in the rainforests, particularly in the lowlands. In addition to terrestrial

KEY BIRDS
Top 10 Birds
1. Dark-rumped Swift
2. White-cheeked Hill Partridge
3. Asian Emerald Cuckoo
4. Jack Snipe
5. Tawny Owl
6. Wreathed Hornbill
7. Darjeeling Pied Woodpecker
8. Golden-throated Barbet
9. Pied Falconet
10. Blue-naped Pitta

leeches, ticks, mosquitoes and various species of biting fly occur here. The use of long sleeves and insect repellents is recommended for rainforest birdwatching. Vaccinations are also recommended for diseases such as hepatitis A and B and typhoid. Other mosquito-borne viruses, like dengue fever and malaria, are present, and appropriate precautions should be taken.

Birdwatching Highlights

Meghalaya has two national parks and five wildlife sanctuaries. Not all of the protected areas fulfill the IBA criteria, and all seven are small. The largest is Balpakram National Park with around 200km², and the smallest is the Baghmara Pitcher Plant Sanctuary.

Despite its relatively small size, Meghalaya is rich in birdlife. Among the threatened species, the following have been reported in the state: the Slender-billed Vulture, Greater Adjutant, White-winged Wood Duck, Pallas's Fish Eagle, Greater Spotted Eagle, Lesser Kestrel, Swamp Francolin, Wood Snipe, Dark-rumped or Khasi Hills Swift, Rufous-necked Hornbill, Grey-sided Thrush, Marsh Babbler, Tawny-breasted Wren Babbler, Slender-billed Babbler and Beautiful Nuthatch. There are historical records of many other species, like the Bengal Florican. Proper systematic surveys of the birdlife are required in the state. There is always the chance of a rare glimpse of the White-winged Wood Duck in the Balpakram area, although recent records are very few. To watch the Dark-rumped Swift (Khasi Hills Swift), try trekking to the cliffs of the Khasi Hills, Cherrapunjee, Lyetkynsew or the Nohkalikai Falls.

Maroon Oriole

Living roots bridge

Balpakram National Park

Balpakram landscape

KEY FACTS

Nearest Major Towns
Baghmara and Tura

Habitats
Subtropical and deciduous forests

Key Species
Kalij Pheasant, Red Junglefowl, Grey Peacock Pheasant, White-cheeked Hill Partridge, Common Hill Partridge, Mountain Bamboo Partridge, Swamp Francolin, Black Francolin, White-winged Wood Duck, Greater Adjutant

Other Specialities
Gaur, Tiger, Marbled Cat, Hoolock Gibbon, Pig-tailed Macaque, Stump-tailed Macaque, Wild Buffalo

Best Time to Visit
October to March

This site includes two protected areas and reserve forest in the South Garo Hills district in western Meghalaya. The protected areas are Balpakram National Park and Siju Wildlife Sanctuary. The Garos believe that this is the land of departed souls. The site, well known for its beautiful scenery, contains expansive tracts of relatively undisturbed, forest-clad hills and gorges. The area supports a large Asian Elephant population. The Balpakram Complex is very rich in avian diversity, with about 250 species identified until now. It lies in the Eastern Himalayas Endemic Bird Area. Some of the important species of the complex are Mountain Bamboo, Common Hill and White-cheeked Hill Partridges, Grey Peacock Pheasant, Grey-headed Parakeet, Striated, White-throated and Black Bulbuls, Blue-throated Barbet, Bay Woodpecker, Short-billed Minivet, Orange-bellied Leafbird, Slaty-backed and Black-backed Forktails, Slaty-bellied Tesia, Streaked Spiderhunter, Maroon Oriole, Grey Treepie, Nepal House Martin, Pale-headed Woodpecker, Lesser Necklaced and Greater Necklaced Laughingthrushes, and Sultan Tit.

Birdwatching Sites

Nokrek National Park This park is located in the Garo Hills, covering parts of the three districts of the East Garo Hills, West Garo Hills and South Garo Hills. Nokrek was declared a national park in 1986, and the final notification was issued in 1997. The park is very small, but it serves as the core area of the Nokrek Biosphere Reserve. More than 150 bird species have been recorded in the park, although more detailed study is required to understand and record its current avian diversity. Some of the important species found in the area are the Bay Woodpecker, Maroon Oriole, Black-winged Cuckooshrike, White-throated Bulbul, Short-billed Minivet and Grey Peacock Pheasant.

Grey-chinned Minivet

Shillong The Upper Shillong Protected Forest is located very close to Shillong, the capital of Meghalaya, in the East Khasi Hills district. These forests have a long history of protection and management (more than 100 years). Tourists and picnickers visit the area especially for the panoramic view of Shillong city. The terrain is an undulating plateau and contains some of the highest reaches of the Meghalaya plateau. Nearly 80 bird species have been recorded in the area so far, but there could be three times as many. The globally threatened Tawny-breasted Wren Babbler has been recorded near Shillong. This poorly known babbler qualifies as Vulnerable because it has a small, declining, severely fragmented population and range due to clearance and degradation of the moist evergreen forest. Many species of the Sino-Himalayan Temperate Forest and Sino-Himalayan Subtropical Forest are found here. A few are: Blyth's Kingfisher, Golden-throated Barbet, Black-winged Cuckooshrike, Black Bulbul, Golden Bush Robin, Aberrant Bush Warbler, Buff-barred Warbler and Ashy-throated Warbler, Rufous-gorgeted Flycatcher and Green-backed Tit.

Cherrapunjee This is one of the highest rainfall areas in the world. Under the Sohra sub-division in the East Khasi Hills, Cherrapunjee is traversed by a number of deep gorges that have cliffs, like Nohkalikai, Mawile, Mawpyrkong, Thankarang and Mawiew. About 100 bird species have been recorded in this IBA, particularly hill birds such as the Mountain Bamboo Partridge, Black Eagle, Black Bulbul, Orange-bellied Leafbird, Ashy Bulbul, Striated and Hill Prinias, and Grey-hooded Leaf Warbler. The site was selected as an IBA based on the presence of the Vulnerable Dark-rumped or Khasi Hills Swift. This bird is specialized to live in crevices on the perpendicular cliffs in this wettest place in India.

Access & Accommodation

Most of the birding sites in Meghalaya are accessed via road. Hiring a vehicle to be at your disposal is recommended for the best birding experiences, as public transport is not frequent or reliable. Many birding places in Meghalaya also demand moderate walking or trekking, so a level of basic fitness is good to have.

There are homestay options available almost everywhere in the state. The local people are friendly and hospitable. Hiring a local birding guide is recommended for exploring the area and getting the maximum result from a birding expedition.

Conservation

As more than 90 per cent of the land belongs to the local communities, community participation in conservation is very important in Meghalaya. The key threats to the birds here are moderate habitat loss (for example due to logging, agriculture and overgrazing) and hunting. Insurgency is a common problem in all northeastern states. In Meghalaya, some parts are still unexplored and may not be safe for tourists. Dam and hydroelectric power projects, tree felling for fuelwood, shifting cultivation, hunting, poaching and coal mining are common, and are some of Meghalaya's very serious conservation issues.

Golden-throated Barbet

Scarlet-backed Flowerpecker

Dark-rumped Swift

NAGALAND

Nagaland is a mystical and beautiful hill state tucked away in the far northeastern corner of India. Its dissected landscape is dominated by crumpled mountain ranges, and it is bounded by Myanmar in the east, the Indian states of Assam in the west, Arunachal Pradesh and part of Assam in the north, and Manipur in the south. The often surreal terrain of towering peaks and deep gorges is interspersed with dense patches of vibrant semi-evergreen rainforests of breathtaking beauty. Nagaland is inhabited by 16 major tribes – Ao, Angami, Chang, Konyak, Lotha, Sumi, Chakhesang, Khiamniungan, Kachari, Phom, Rengma, Sangtam, Yimchungrü, Kuki, Zeliang and Pochury, as well as several sub-tribes. Each tribe is unique, with its own distinct customs, language and dress.

1. Khonoma
2. Benreu
3. Intanki National Park

Climate

Nagaland is part of two biogeographic zones: Mizoram-Manipur-Kachin Rainforests (IM0131) and Northeast India-Myanmar Pine Forests (IM0303). The Mizoram-Manipur-Kachin Rainforests region represents the semi-evergreen submontane rainforests that stretch from the Arakan Yoma and Chin Hills into the Chittagong Hills, the Mizo and Naga Hills, and finally the hills of Myanmar. It is the location at the junction of the Indian, Indo-Chinese, and Indo-Malayan biogeographic regions that results in the presence of biotic elements from all these regions, making it very rich in floral and faunal resources. As a result, the region has the highest bird species richness of all the eco regions in the Indo-Pacific area. Intanki and Puliebadze are protected areas in this region.

The Northeast India-Myanmar Pine Forests region is located in the Burmese-Java Arc. The Patkai, Lushai, Naga, Manipur and Chin Hills are a part of this arc. Pine forests occur in this eco region at 1,500–2,000m. Due to the nature of the forests, the biodiversity of the region is limited. However, they contain some species that are unique to the ecosystem. Fakim is a protected area in this region.

Nagaland has a largely monsoon climate with high humidity levels. Annual rainfall averages around 1,778–2,540mm, concentrated in May–September. The summer months (March–April) and winter months (October–February) are the best times to visit Nagaland.

Access, Transportation & Logistics

Visitors enter Nagaland from Dimapur, which has the only airport in Nagaland. There are direct flights to Dimapur from Guwahati, Kolkata and many other big cities in India. Dimapur also has a railway station on the main line of the Northeast Frontier Railway. It is well connected to Guwahati. NH 39 is the main highway for entry into Nagaland from Assam. It connects Kohima with Dimapur, a distance of 74km. Tarred roads connect much of Nagaland internally.

KEY BIRDS

Top 10 Birds
1. Chestnut-vented Nuthatch
2. Blyth's Tragopan
3. Naga Wren Babbler
4. Mountain Bamboo Partridge
5. Crested Finchbill
6. Striped Laughingthrush
7. Eurasian Woodcock
8. Mountain Scops Owl
9. Yellow-throated Laughingthrush
10. Brown-capped Laughingthrush

Flavescent Bulbul

Amur Falcons

Health & Safety
There are no major health concerns in Nagaland. It is a low- to no-risk area of India when it comes to malaria, but do take precautions especially after the monsoon. Visit a GP or travel clinic 6–8 weeks before departure to make sure you are up to date with any vaccinations required.

Birdwatching Highlights
Nagaland's luxuriant tropical forests amid diverse topographical and climatic conditions favour an abundance of plants and animals. Despite serious hunting pressure and conversion of forests to agriculture, Nagaland's bird list easily exceeds 500 species, and most of them can be found in Nagaland even now. Top of the list of birds is Blyth's Tragopan – the state bird. It is fiercely protected in most places where it occurs. Other key birds include three regional endemics – Striped and Brown-capped Laughingthrushes and the Cachar Wedge-billed Babbler – as well as specialities like the Spot-breasted Parrotbill, (Naga) Long-tailed and Tawny-breasted Wren Babblers, Chestnut-vented Nuthatch, Dark-rumped Swift and Rusty-capped Fulvetta.

Eyebrowed Thrush

Naga Wren Babbler

Khonoma

Nagaland

Khonoma landscape

The picturesque village of Khonoma, which lies 20km west of Kohima, was known for its fighting prowess in the past. It is famous for its forests and a unique form of agriculture, including some of the oldest terraced cultivation in the region. The terrain of the village is hilly, ranging from gentle slopes to steep and rugged hillsides. The hills are covered with lush forest land, rich in various species of flora and fauna. The state bird, Blyth's Tragopan, a pheasant that is now nationally endangered, is reportedly found here.

KEY FACTS

Nearest Major Towns
Kohima

Habitats
Sub tropical and temperate broadleaved forests

Key Species
Blyth's Tragopan, Dark-rumped Swift, Grey Sibia, White-naped Yuhina, Mountain Bamboo Partridge, Striped Laughingthrush, Brown-capped Laughingthrush, Cachar Wedge-billed Babbler, Naga Wren Babbler, Rusty-capped Fulvetta, Chestnut-vented Nuthatch

Other Specialities
Spotted Lingsang, Small-toothed Ferret Badger, Indian Chevrotain, Serow, Leopard Cat, Slow Loris, Orange-bellied Squirrel, Himalayan Musk Deer, Binturong, Palm Civet

Best Time to Visit
October to March

Mountain Bamboo Partridge

Striated Laughingthrush

Birdwatching Sites

Khonoma village's immediate habitat comprises agricultural land, and land under alder, degraded forests and small patches of conifers. The dense and undisturbed tropical forest of considerable diversity is found on the border dividing the Kohima and Peren districts in the Dzulekie area. Some of the key birds found in the Khonoma area are the Mountain Bamboo Partridge, Striped and Brown-capped Laughingthrushes, Manipur Wedge-billed and (Naga) Long-tailed Wren Babblers, Rusty-capped Fulvetta and Chestnut-vented Nuthatch. Crested Finchbills and Grey Sibias are among the common birds found near Khonoma.

Access & Accommodation

Khonoma is connected by road from Dimapur Airport (75km) via NH 29, and the journey takes two and a half hours. Khonoma is 20km from Kohima. There are many homestays available in this region, and local people are warm and welcoming towards travellers.

Conservation

Wildlife hunting used to be a way of life with the Naga tribes, and a large number of birds and other animals were killed every year, including the endangered tragopans. In 1998, the Khonoma village council declared its intention to notify about 20km^2 as the Khonoma Nature Conservation and Tragopan Sanctuary. Its foundation stone was laid in December 1998. It was also decided to ban hunting in the entire village, not only in the sanctuary area.

Mountain Scops Owl

Crested Finchbill

Grey Sibia

Nagaland

Benreu

Benreu landscape

This is a little village perched 1,950m at the foot of Mt Paona. Benreu is famous for housing a unique community where 20 per cent of an animist population dictates the customs and social rules of the majority of Christians. The key to bird diversity in Benreu is the intact natural forests that are characterized by an abundance of broadleaved evergreen and deciduous trees. The area has wild cherry, apple, lemon, banana, walnut, fig, and other trees like Bonsum, Gogra, alder and oak. A rare species of bamboo, believed to be the tallest bamboo in the world, is found in the foothills of Peletkie village close to Benreu.

Birdwatching Sites

Birding in Benreu is along the road, and key birds that can be found in the area are a variety of thrushes (including the Grey-sided Thrush in winter), galliformes (including Blyth's Tragopan), and most of the birds that occur in Khonoma plus the enigmatic Spot-breasted Parrotbill.

KEY FACTS

Nearest Major Towns
Peren and Kohima

Habitats
Dense and hilly forests

Key Species
Grey-crowned Warbler, Yellow-throated Laughingthrush, Striped Laughingthrush, Brown-capped Laughingthrush, Spot-breasted Parrotbill, Long-tailed Wren Babbler, Tawny-breasted Wren Babbler, Chestnut-vented Nuthatch, Dark-rumped Swift, Rusty-capped Fulvetta, Blyth's Tragopan, Grey-sided Thrush

Other Specialities
Spotted Lingsang, Small-toothed Ferret Badger, Indian Chevrotain, Serow, Leopard Cat, Slow Loris, Orange-bellied Squirrel, Himalayan Musk Deer, Binturong, Palm Civet

Best Time to Visit
October to March

Small Niltava

Ashy-headed Laughingthrush

Access & Accommodation

Benreu is about 130km from Khonoma, about six hours away, depending on the road conditions. There are some homestay accommodation options available at Benreu. There is a village resort named Mt Pauna Tourist Village (Tel: (0370) 227 0072), which is an extension of Benreu village (separated from it by about 300m) towards the south along the Pauna mountain range. It has eight double-bedded cottages with attached bathrooms fitted with running hot and cold water.

Conservation

The main threats and conservation issues in Benreu are hunting, habitat destruction and illegal cutting of trees. Some conservation programmes have been started with the joint effort of the local village council and Forest Department.

Rufous-vented Yuhina

Green Cochoa

Spot-breasted Scimitar Babbler

Large Niltava

Intanki National Park

Grey-throated Babbler

KEY FACTS

Nearest Major Towns
Peren and Kohima

Habitats
Tropical semi-evergreen forest, tropical moist deciduous forest

Key Species
White-winged Wood Duck, Rufous-necked Hornbill, Great Hornbill, Austen's Brown Hornbill, Grey Sibia, Golden-throated Barbet, Blue-throated Barbet, Bay Woodpecker, White-throated Bulbul, Maroon Oriole, Grey Treepie, Grey Peacock Pheasant, Pale-headed Woodpecker

Other Specialities
Hoolock Gibbon, Sloth Bear, Monitor Lizard, Malayan Giant Squirrel, Indian Chevrotain

Best Time to Visit
October to March

This park is the centre of all birding activity in the lowlands of Peren district. Intanki is a low-elevation, tropical deciduous forest that abounds in Himalayan foothill species. Some of the key species are the rare Hoolock Gibbon and Golden Langur. There have also been sightings of the Tiger, Palm Civet and Sloth Bear.

Birdwatching Sites

Despite seasonal hunting (daily on Mondays and Thursdays in June–February), Intanki is a birding hotspot with an abundance of lowland birds that include drongos, minivets, bluebirds, thrushes, barbets, woodpeckers, green pigeons, hornbills and bulbuls. Key birds include the Austen's Brown Hornbill, White-winged Wood Duck and Black-headed Bulbul.

Access & Accommodation

The nearest airport and railway station to reach the park is Dimapur, which is about 50km away. Access to Intanki is difficult, with the nearest village at Lilen still 10km short of the park. Infrastructure at Lilen is basic and you have to depend on the hospitality of the villagers to spend any time at this place. The Forest Rest House offers lodging facilities to visiting tourists, who can also use the accommodation facilities at circuit house and Public Works Department Inspection Bungalow, depending on their availability.

Conservation

Intanki was declared a national park in 1993, and in 2005 it was also declared an elephant reserve by the government. This is one of the national parks in India where not much research work has been done, so there is an urgent need to perform detailed scientific research work to understand its rare fauna

Rufous-capped Babbler

and flora. There have been reports of common conservation issues like hunting, poaching, logging and illegal grazing from the park.

Biswapriya Rahut

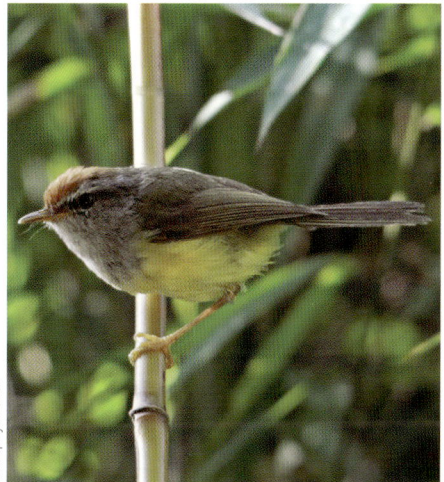

Broad-billed Warbler

Ramki Sreenivasan

Streaked Spiderhunter

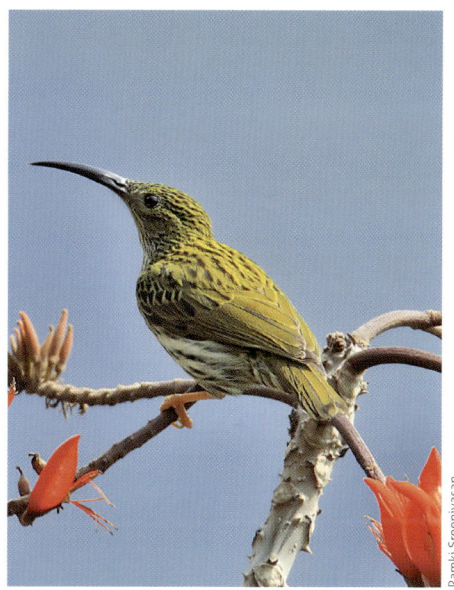

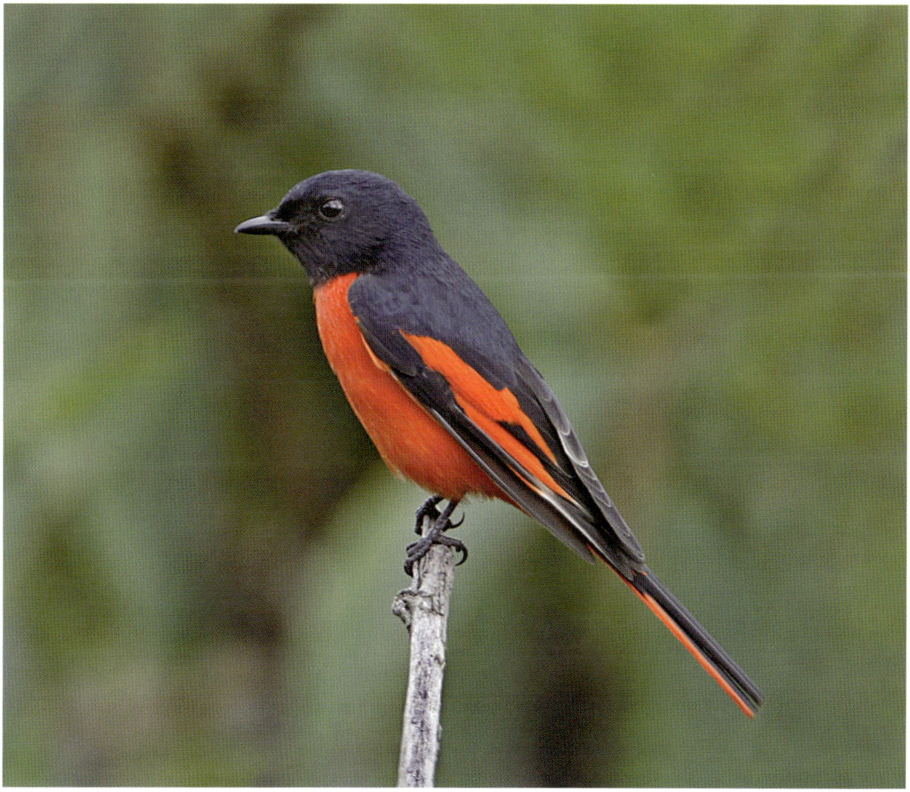
Clement M Francis

Long-tailed Minivet

ODISHA

1. Bhitarkanika Wildlife Sanctuary & National Park
2. Mangalajodi Wetlands
3. Chilika Lake & Nalbana Bird Sanctuary
4. Simlipal National Park

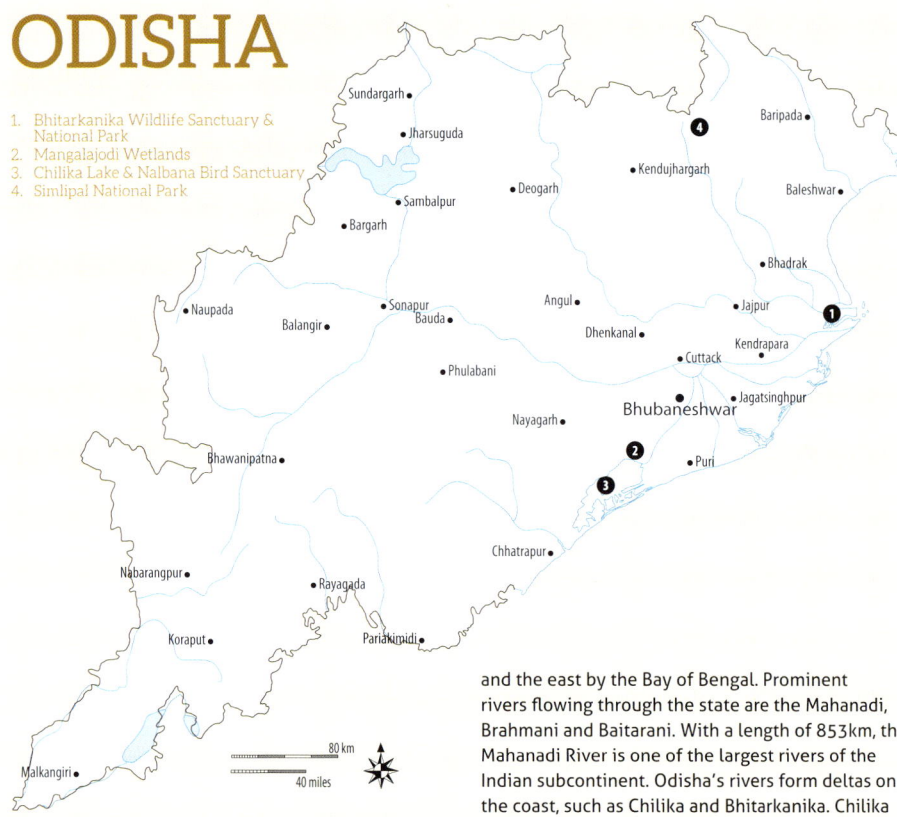

Odisha (formerly Orissa) is an eastern Indian state located on the east coast on the Bay of Bengal. The capital, Bhubaneswar, is home to hundreds of temples, notably the intricately carved Mukteshvara Temple and Lingaraja Temple. Shree Jagannath Temple and Konark Sun Temple of Puri are world famous and important pilgrimage destinations. Physiographically the state can be divided into four distinct regions: Northern Plateau, Eastern Ghats, Central Tableland and Coastal Plains. The north of the state is bounded by Jharkhand, the south-east by Andhra Pradesh, the west by Madhya Pradesh and the east by the Bay of Bengal. Prominent rivers flowing through the state are the Mahanadi, Brahmani and Baitarani. With a length of 853km, the Mahanadi River is one of the largest rivers of the Indian subcontinent. Odisha's rivers form deltas on the coast, such as Chilika and Bhitarkanika. Chilika Lake is a large brackish-water lagoon that sprawls along the east coast of India.

Climate

The climate of Odisha is generally hot and humid. The temperature ranges from 20 °C to 41 °C. Precipitation occurs mainly during the north-east tropical monsoon (September–December), and the mean annual rainfall is 1,200–1,600mm. The coastal area is highly prone to periodical cyclones. The most recent cyclone – Fani (April 2019) – was one of the strongest tropical cyclones to strike India for many years, affecting nearly 15 million people. Four major forest types can be identified in Odisha: tropical semi-evergreen, tropical moist deciduous, tropical dry deciduous and littoral-swamp forests. An agricultural state, more than 76 per cent of its people are dependent on farming for their livelihoods. The major crops are rice, pulses, oil seeds, jute, sugar cane, coconut and turmeric.

Access, Transportation & Logistics

Odisha is well connected to the rest of India via airways, railways and roadways. Biju Patnaik Airport at Bhubaneswar offers the best way to

KEY BIRDS

Top 10 Birds
1. Pale-capped Pigeon
2. Red-billed Tropicbird
3. Savanna Nightjar
4. Lesser Coucal
5. Red-footed Booby
6. Indian Skimmer
7. Black-tailed Godwit
8. Mangrove Pitta
9. Pomarine Skua
10. Saunders's Tern

reach Odisha by air. It is served by regular non-stop flights from Delhi, Bangalore, Kolkata, Mumbai, Visakhapatnam, Guwahati and Chennai. The state has well-established rail connectivity with different parts of India. The major railhead in Odisha is the Bhubaneswar railway station, located in the state capital, and it is served by major train networks that connect it to other parts of the state as well as to other parts of India. An extensive road network also covers the state. It is connected to different parts of the country via national and state highways.

Health & Safety
There are no major health concerns in Odisha. The state is a low- to no-risk area of India when it comes to malaria, but do take precautions especially after the monsoon. There are some common health risks in Odisha, like cholera, dengue fever, dysentery and meningitis. Travellers are advised to take precautionary measures against these. Visit a GP or travel clinic 6–8 weeks before departure to make sure you are up to date with any vaccinations. Avoid roadside food and drink bottled water. During the summer months keep hydrated, and carry a stole, scarf or cap to protect yourself from direct sunlight.

Birdwatching Highlights
The birdlife of Odisha is very rich. More than 500 bird species have been identified in the region, including some very rare ones – White-rumped and Long-billed Vultures are two of the Critically Endangered species recorded. The Mangrove Pitta,

Ruff

Spot-billed Pelican, Lesser Adjutant, Baer's Pochard, Ruddy Kingfisher, Greater Spotted Eagle, Bristled Grassbird and Pale-capped Pigeon are some of the other key birds and are a delight for any birdwatcher. Some of the commonly seen birds are the Oriental Darter, Stork-billed Kingfisher, Indian Paradise-flycatcher, Indian Pitta, and many varieties of duck, geese, heron and raptor.

The Bhitarkanika and Mangalajodi areas of Odisha are quite popular for their unique and rich birdlife, and attract many birdwatchers.

Mangalajodi wetlands

Bhitarkanika Wildlife Sanctuary & National Park

Bhitarkanika landscape

This wildlife sanctuary and national park, located on the eastern coast, together represent one of the finest mangrove forests of India. The area was declared a wildlife sanctuary in 1975 to protect the Estuarine or Saltwater Crocodile, but it was later also found to be a haven for birds. Located in the estuarine region of Brahmani-Baitarani, in the northeastern part of the Kendrapara district of Odisha, the sanctuary covers an area of 672km² of mangrove forests and wetlands.

Birdwatching Sites

Bhitarkanika comprises a rich, lush green, vibrant ecosystem lying in the estuarine region of Brahmani-Baitarani in the northeastern corner of the Kendrapara district of Odisha. The wetland is represented by as many as three protected

KEY FACTS

Nearest Major Towns
Kendrapara

Habitats
Mangroves, tropical moist deciduous forest

Key Species
White-rumped Vulture, Long-billed Vulture, Spot-billed Pelican, Lesser Adjutant, Baer's Pochard, Pallas's Fish Eagle, Greater Spotted Eagle, Lesser Kestrel, Indian Skimmer, Oriental Darter, Painted Stork, Black-necked Stork, Black-headed Ibis, Black-bellied Tern

Other Specialities
King Cobra, Marsh Crocodile, Indian Python, Wild Pig, Spotted Deer, Rhesus Macaque, Water Monitor, Olive Ridley Turtle

Best Time to Visit
October to March

Mangrove Pitta

Brahminy Starling

Black-headed Gulls

areas, namely the Bhitarkanika National Park, Bhitarkanika Wildlife Sanctuary and Gahirmatha Marine Sanctuary. Bhitarkanika is a great place in which to see some of the important birds of Odisha, like different varieties of kingfisher and pitta. This is also a very good location for some of the common birds like woodpeckers, White-bellied Sea Eagles, Ruddy Shelducks, hornbills, waders and Bar-headed Geese.

Access & Accommodation

The nearest airport is Bhubaneswar Airport, about 160km from the park. The airport is well connected to the park by the road network. The nearest railway station is Bhadrak railway station, about 70km from the park, which is also well connected to major cities by the road network. Several government and privately operated vehicles go to the park at frequent intervals.

There are fully furnished tents available for accommodation in the park. The forest rest houses and jungle lodges at Ekakula, Dangmal and Gupti are available for accommodation. For bookings contact Divisional Forest Officer, Mangrove-cum-Wildlife Warden Bhitarkanika National Park, Rajnagar, Kendrapada, Odisha, India, Tel.: 06729-242460, Fax: 06729-242464. For tourism-related issues email dforajnagartourism@gmail.com.

Conservation

Due to conversion of agricultural fields to prawn-culture ponds, Asian Openbills and many other bird species are losing their feeding grounds. This may soon affect the breeding behaviour of these birds and may be detrimental to their survival in the long run. It has been alleged by NGOs working in Odisha that the large numbers of immigrants living and operating around Bhitarkanika Wildlife Sanctuary are adversely affecting the park and could threaten the survival of the Estuarine Crocodiles and Irrawaddy Dolphin. Many mechanized boats reportedly fish illegally in the creeks here – the main home for crocodiles as well as many waterbirds. Large-scale mangrove depletion is also attributed to the large immigrant population.

White-bellied Sea Eagle

Mangalajodi Wetlands

Mangalajodi wetlands

Mangalajodi Wetlands consist of a freshwater swamp at the northeastern edge of Chilika Lake. Mangalajodi hamlet is about 60km south-south-west of Bhubaneswar city off NH 5, and 4km south-east of Tangi town in the Khurda district of Odisha. The area is primarily a freshwater zone connected by channels cut through the reedbeds, with the brackish waters of Chilika lagoon. It is covered with emergent vegetation and reeds consisting mostly of *Typha angustata* and *Phragmites karka*. The main channel runs north to south for about 3km and has a nature trail running parallel to it for about 2km, terminating in a watchtower.

KEY FACTS

Nearest Major Towns
Khurda and Bhubaneshwar

Habitats
Wetlands, reedbeds

Key Species
Spot-billed Pelican, Fulvous Whistling Duck, Bar-headed Goose, Common Shelduck, Ferruginous Duck, Garganey, Falcated Duck, Eurasian Wigeon, Jungle Bush Quail, Grey Francolin, Lesser Flamingo, Little Grebe, Rock Dove, Little Swift, Jacobin Cuckoo, Asian Koel, Eastern Water Rail, Ruddy-breasted Crake, Brown Crake, Grey-headed Swamphen, Common Coot, Asian Openbill, Woolly-necked Stork, Great White Pelican, Yellow Bittern, Cinnamon Bittern, Cattle Egret, Purple Heron, Little Egret, Oriental Darter, Pied Avocet, Grey Plover, Kentish Plover, River Lapwing, Whimbrel, Great Knot, Red Knot, Ruff, Temminck's Stint, Sanderling, Dunlin, Asian Dowitcher, Pintail Snipe, Terek Sandpiper, Marsh Sandpiper, Brown-headed Gull

Other Specialities
Fishing Cat, Jungle Cat, Indian Jackal

Best Time To Visit
October to March

Pied Avocet

Red-crested Pochard

Black-tailed Godwit

Birdwatching Sites
The marshes around Mangalajodi, and the open water between Kalupada Ghat and Teenmuhani, attract a large congregation of waterfowl, especially dabbling ducks such as the Northern Pintail, Northern Shoveler, Garganey and Ruddy Shelduck. Mangalajodi is a waterfowl haven, attracting thousands of winter migrants and playing host to significant and diverse populations of breeding residents. These include the Black-tailed Godwit, Ruff, Tufted Duck, Red-crested Pochard, Fulvous Whistling Duck, Cotton Pygmy-goose, Northern Pintail, Northern Shoveler, Garganey, Eurasian Wigeon, Ruddy Shelduck, Grey-headed Swamphen, Asian Openbill, snipes, sandpipers, pratincoles, crakes, bitterns, egrets, herons and other resident birds. Raptors include the Western Marsh Harrier, Brahminy Kite, Peregrine Falcon and White-bellied Sea Eagle, while the Whiskered Tern is abundant in season. Black-tailed Godwits are the most evident species on the marshes.

Access & Accommodation
Mangalajodi lies at the northeastern tip of Chilika Lake. It is 30km from Khurda Road railway station, and lies off NH 5 connecting Kolkata to Chennai. The nearest railhead is Kalupada Ghat on the SE Railway Howrah-Chennai line, and Tangi on NH 5 is the nearest town. Bhubaneswar city is a two-hour drive from Mangalajodi.

Mangalajodi Ecotourism provides basic accommodation at the site. It can be reached at Tel.: +91 88952 88955 or at its website (www.mangalajodiecotourism.com/contact-us.php). Alternative and better equipped accommodation is available at Balugaon, Barkul (Odisha Tourism Development Corporation, Panthanivas), Kalupada and Khurda Road – all are within driving distance of the site. If you do not have a vehicle, you can arrange one from the place you are staying at. Boat rides at Mangalajodi vary depending on the season and availability. You can carry packed food or ask the guides beforehand to prepare lunch for you. The boatman, Madhu Behera (Tel.: +91 96921 22456), is well aware of the area and surrounding birdlife. The best season to visit Mangalajodi is winter, in November–March.

Conservation
Large-scale poaching has historically been a major threat to the avifauna of Mangalajodi Wetlands. Efforts to curb poaching and provide alternate means of livelihood to families were started in 1996–1997. The organized efforts of Wild Orissa (NGO), along with the Odisha Wildlife Department, have helped to reduce poaching and provided alternative livelihood options to the local villagers. There is a need for more organized efforts to promote ecotourism and the unparalleled biodiversity of this unique area.

Bluethroat

Chilika Lake & Nalbana Bird Sanctuary

Black-bellied Tern

Chilika Lake is situated in Puri, one of the most famous cities of Odisha. The lake is an estuarine lagoon, shallow throughout its massive spread. It is the largest brackish-water wetland in India. The government of India notified Chilika Lake as a Ramsar Site in 1981. The pear-shaped lake is connected to the Bay of Bengal at its north-east end and is subject to minor tidal fluctuations. It receives water from the Daya and Bhargavi Rivers, and several small streams. It is the largest wintering ground for migratory waterfowl in India. Nalbana Bird Sanctuary or Nalbana Island is the core area of the Chilika Lake. It was declared a bird sanctuary under the Wildlife Protection Act in 1972.

KEY FACTS

Nearest Major Towns
Khurda, Puri and Ganjam

Habitats
Brackish wetlands

Key Species
Spot-billed Pelican, Lesser Adjutant, Lesser White-fronted Goose, Baer's Pochard, Pallas's Fish Eagle, Spoon-billed Sandpiper, Indian Skimmer, Caspian Gull, Little Tern, Saunders's Tern, River Tern, Black-winged Kite, White-rumped Vulture, Western Marsh Harrier, Pallid Harrier, Eurasian Sparrowhawk, Brahminy Kite, Spotted Owlet, Common Hoopoe, Brown-headed Barbet, Green Bee-eater, Indian Roller, Common Kingfisher, Red-necked Falcon, Eurasian Hobby, Indian Golden Oriole, Black Drongo

Other Specialities
Irrawaddy Dolphin, Fishing Cat, Marsh Crocodile, Green Sea Turtle, Blackbuck, Limbless Skink

Best Time to Visit
October to March

Cinnamon Bittern

Oriental Darter

Collared Pratincole

Indian Spot-billed Duck

Birdwatching Sites
The shallow lagoon of Chilika Lake can be broadly divided into four ecological sectors based on salinity and depth: southern zone, central zone, northern zone and outer channel. There are several islands in the lagoon, and some of the prominent ones are Krushnaprasad, Nalbana, Kalijai, Somolo and Birds Islands. Chilika Lake in general and the Nalbana area in particular are the most prominent birding sites. This area is among the most important waterfowl habitats in India.

More than 211 species of bird have been recorded in Chilika, including many Vulnerable and Near Threatened species. Large numbers of Northern Pintails, Garganeys, Black-tailed Godwits, Gadwalls, Eurasian Wigeon, Brown-headed Gulls and marine terns congregate in and around the islands at dusk for roosting, and most of them depart in the morning. Thousands of Bar-headed Geese spend winter at Nalbana every year. The rare Asian Dowitcher has also been reported from the area. Other important birds found here are the Spot-billed Pelican, Lesser Adjutant, Lesser White-fronted Goose, Baer's Pochard, Pallas's Fish Eagle, Spoon-billed Sandpiper and Indian Skimmer.

Access & Accommodation
Chilika Lake is 120km from Bhubaneswar, which has an airport and railway station, and very good connectivity with other major cities in India. Regular buses and taxis are available from Bhubaneswar to the Chilika Lake area.

There is an Odisha Tourism Development Corporation rest house in Balugaon, Khurda named Panthnivas (Tel.: 06756-257488/257388, email: otdc@panthanivas.com). There are also many private hotels and resorts near Chilika, where visitors can book accommodation.

Conservation
Chilika Lake has undergone major ecological changes during the last several years, mostly due to salinity changes caused by the choking of the outer channel and closure of the lake's mouth to the sea, and the restoration measures (providing a new mouth) undertaken by the Chilika Development Authority. Conservation measures are urgently needed for avifaunal species like Gull-billed and River Terns to protect them from natural calamities.

Tufted Duck

Simlipal National Park

Simlipal forest

This beautiful park is located in the Mayurbhanj district of Odisha, and is also a Tiger Reserve. It is part of the Similipal-Kuldiha-Hadgarh Elephant Reserve, popularly known as the Mayurbhanj Elephant Reserve, which includes three protected areas, Similipal Tiger Reserve, Hadagarh Wildlife Sanctuary and Kuldiha Wildlife Sanctuary. The highest peak in the Simlipal Hills is Khairi-buru (1,178 m). No locality in the Simlipal Hills suffers from a scarcity of water at any time of the year. Several streams flow through the park and drain into the Bay of Bengal. The major perennial streams are the Budhabalanga, Palpala, Deo, Nekendanacha, Bandan, Kahairi and Khadkei. Simlipal is very popular with tourists, who come to enjoy its scenic beauty and to see the Tiger and the rich birdlife of the area.

KEY FACTS

Nearest Major Towns
Mayurbhanj

Habitats
Tropical semi-evergreen forest, tropical moist deciduous forest, tropical dry deciduous forest

Key Species
White-rumped Vulture, Long-billed Vulture, Greater Spotted Eagle, Pale-capped Pigeon, Green Avadavat, Green-billed Malkoha, Malabar Trogon, Malabar Pied Hornbill, Indian Scimitar Babbler, Malabar Whistling Thrush, Loten's Sunbird, Red-headed Vulture, White-eyed Buzzard, Red-necked Falcon, Rain Quail, Jungle Bush Quail, Painted Bush Quail, Painted Spurfowl, Indian Peafowl, Yellow-wattled Lapwing, Yellow-legged Green Pigeon, Plum-headed Parakeet, White-bellied Drongo, Bank Myna, Brahminy Starling, Chestnut-tailed Starling, Ashy Prinia, Jungle Babbler, Tawny-bellied Babbler, Indian Bushlark

Other Specialities
Asian Elephant, Tiger, Indian Leopard, Gaur

Best Time to Visit
October to March

Pale-capped Pigeons

Western Yellow Wagtail

Chestnut-bellied Nuthatches

Birdwatching Sites

Despite the great importance of Simlipal National Park to the Odisha government and Project Tiger authorities, its birdlife is not well documented. However, more than 250 bird species are found here. Simlipal forest stands as a link between the flora and fauna of southern India and the Himalayas. For instance, the Collared Falconet was sighted in Simlipal, far south of its known range in the Himalayan foothills, Sikkim, Bhutan and Assam. Species at the northernmost extreme of their range are the Malabar Trogon, Malabar Pied Hornbill and Malabar Whistling Thrush. Essentially Himalayan species such as the Green-billed Malkoha and Blue-throated Barbet are near their southern limit in Simlipal. Other important Simlipal-area bird species are Loten's Sunbird, Red-headed Vulture, Red-necked Falcon, White-eyed Buzzard, Rain Quail, Sirkeer Malkoha, Brown-headed Barbet, Small Minivet, Malabar Trogon and Common Woodshrike.

The best time of the year to visit the park is November–May. It is open from 1 October to 15 June.

Access & Accommodation

The nearest airports for the park are Bhubaneswar and Kolkata Airports, about 270 and 240km away, respectively. These airports are well connected to the park by the road network. The nearest railway station is Balasore railway station, about 60km from the park. Simlipal National Park is well connected to major cities and places by the road network. A number of government and privately operated vehicles go to the park at frequent intervals.

Basic forest rest houses are available for accommodation in Simlipal. Other accommodation options are available at Gudgudia, Lulung, Januari, Dhudhruchampa and Maharaja's Log Cabin at Chahala. Aranya Nivas Tourist Complex by the Odisha Tourism Development Corporation is about 10km inside the Baripada entrance (Tel. (official): 06792 252 593, Website: www.similipal.org).

Conservation

Some of the major conservation issues faced by Simlipal are the human-wildlife interface and mitigation of human-wildlife conflict, management of ecotourism, and impacts of incompatible developmental programmes launched by other agencies. Some other conservation issues are related to biodiversity status assessment and monitoring, with particular emphasis on invertebrates and lower plants, and the impacts of habitat alteration.

Jerdon's Baza

PUNJAB

1. Harike Lake Bird Sanctuary

This state in north-west India borders the Pakistani province of Punjab to the west, Jammu and Kashmir to the north, Himachal Pradesh to the north-east, Haryana to the south and south-east, Chandigarh to the south-east and Rajasthan to the south-west. It is broadly divided into three physiographic regions, namely the mountainous Himalayas, the sub-mountainous Himalayas and the alluvial plains. The important rivers draining the state are the Sutlej and Beas. Punjab is considered the wheat bowl of India, a state where India's Green Revolution was started. It is an intensively cultivated state through canal irrigation and/or through underground water. Punjab's capital is Chandigarh, which is administered separately as a union territory since it is also the capital of neighbouring Haryana. Other major cities of Punjab are Ludhiana, Amritsar, Patiala and Jalandhar. The Indus Valley Civilization, one of the world's first and oldest civilizations, was centred around Punjab.

Climate

The Punjab climate is determined by the extreme hot and extreme cold conditions. The region lying near the foothills of the Himalayas receives heavy rainfall, whereas in the region situated at a distance from the hills, the rainfall is scanty and the temperature is high. The climate of the state comprises three main seasons: the summer months that span from mid-April to the end of June, the rainy season from early July to the end of September, and the winter season from early December to the end of February. The transitional seasons are the post-monsoon season and the post-winter season. The average annual rainfall is 400–600mm and the temperature ranges from 2 °C to 45 °C.

Access, Transportation & Logistics

Punjab is considered to have the best infrastructure in India – this includes road, rail, air and river transport links that are extensive throughout the region. The main airports are located in Amritsar and Ludhiana. Regular flights to Delhi can be taken from here. The union territory of Chandigarh, which serves as the capital of both Haryana and Punjab, also has a domestic terminal. Highways in Punjab are very good, with dhabas (roadside restaurants that serve local food) which are very popular. Road transport too is well developed and probably provides the best way to move around in Punjab. There is also good connectivity via railways, and trains can be taken between Amritsar and Delhi, or Chandigarh and Delhi.

Health & Safety

Punjab is one of the well-developed and prosperous states of India, and proper measures have been taken when it comes to the safety of tourists. There are no major health concerns in the state. As a precaution, visitors may take vaccinations for hepatitis A and B and typhoid. Malaria is prevalent in remote areas and prophylaxis should be taken, although this is less of a problem during the dry

KEY BIRDS
Top 10 Birds
1. Yellow-eyed Pigeon
2. Rufous-vented Prinia
3. Slavonian Grebe
4. Red-throated Thrush
5. Long-tailed Duck
6. White-winged Tern
7. Black-bellied Tern
8. Common Raven
9. Spanish Sparrow
10. Common Chaffinch

Eurasian Wigeon

Common Hawk Cuckoo

Bean Goose

season. Dengue fever is also present and appropriate precautions should be taken. Avoid roadside food as far as possible as it may not follow food-safety norms. Visit a GP or travel clinic 6–8 weeks before departure to make sure you are up to date with any vaccinations. In summer adequate precautions need to be taken to avoid the intense heat. Wear light cotton clothing, use a hat or sunshade when going outdoors, and drink plenty of liquids.

Birdwatching Highlights

Punjab is a very thickly populated agricultural state where natural vegetation has disappeared under the plough. It is one of the states that has great potential to support large numbers of wildlife, including avian life, but it lacks systematic study, and not much has been studied and recorded here. Harike has been notified as an IBA and is the best birdwatching site in the state.

Indian Skimmer

Punjab

Harike Lake Bird Sanctuary

Harike wetlands

Harike Lake is a shallow water-storage reservoir created by the construction of a barrage at Harike at the confluence of the Sutlej and Beas Rivers. Locally known as 'Hari-ke-Pattan', it contains a deep-water lake that is a part of the 86km² Harike Bird Sanctuary. The lake is surrounded by agricultural land and is the main source of water for the Indira Gandhi (Rajasthan) Canal. It is a vital source of fish to Punjab. A large waterbody, it attracts large numbers of wintering waterfowl and is home to many specialties. Common waterfowl and other winter migrants include coots, gulls, geese, diving ducks and grebes. They are said to number 200,000–500,000. Harike was designated as a Ramsar Site in March 1990, and has been identified as one of the sites for conservation under the Indian National Wetland Programme.

KEY FACTS

Nearest Major Towns
Amritsar, Kapurthala and Firozpur

Habitats
Freshwater swamp (reservoir)

Key Species
White-rumped Vulture, Slender-billed Vulture, White-headed Duck, Lesser Adjutant, Pallas's Fish Eagle, Greater Spotted Eagle, Eastern Imperial Eagle, Indian Skimmer, Bristled Grassbird, Kashmir Flycatcher, Oriental Darter, Painted Stork, Black-necked Stork, Black-headed Ibis, Rufous-vented Prinia, Pallid Harrier, Red-headed Vulture, Ferruginous Pochard

Other Specialities
Smooth-coated Otter, Wild Pig, Jungle Cat, Indus Dolphin

Best Time to Visit
Throughout the year

Brown-headed Gull

Marbled Duck

Blue-tailed Bee-eater

Birdwatching Sites

Harike Lake was designated as a Ramsar Site on account of its importance as a habitat for a large number and diversity of waterfowl. The lake is a key staging and wintering area for migratory waterfowl. Thousands of ducks have been counted here at the peak of the migratory season – big flocks of Red-crested, Common and Tufted Ducks are pretty common. More than 250 bird species have been recorded at the site. In addition to the common birds, other important species that can be seen here include the Cotton Pygmy-goose, Tufted Duck, Yellow-crowned Woodpecker, Yellow-eyed Pigeon, Watercock, Pallas's, Brown-headed, Black-headed Gulls, Indian Skimmer, White-winged Tern, White-rumped Vulture, Hen Harrier, Eurasian Sparrowhawk, Eurasian Hobby, Horned, Black-necked and Great Crested Grebes, White-browed Fantail, Brown Shrike, Common Woodshrike, White-tailed Stonechat, White-crowned Penduline Tit, Rufous-vented Prinia, Striated Grassbird, Cetti's Warbler and Sulphur-bellied Warbler.

There is only one main entry point into Harike Wetlands. This is from the Nanaksar Gurudwara across the barrage at the southern end on the Ferozepur side of Harike. You can take your vehicle up to the Gurudwara and after that there are birding trails leading into the sanctuary on both sides. On the left is the 22km dirt track of the bund passage. Watchtowers punctuate the passage, and there are two checkpoints along the way where you need to show your permit. Further down there is another checkpoint near Bhootiwala where you can find the rare and Vulnerable Indian Skimmer. There are also some birding trails in the 04km² tract of sarkanda grass to the right of the Gurudwara.

Visitors are permitted to drive their petrol vehicles (speed limit 20km/h). Boating is not possible without special permission. An entry permit is required from the Range Officer (Wildlife), Harike to visit this place. Note that photography of the Harike Barrage is prohibited.

Access & Accommodation

The nearest airport to reach Harike is located in Amritsar, about 55km away, and it takes nearly an hour to reach. The roads are good. There is also a railway station in Amritsar. If you are coming from Delhi take NH 1 and follow the Ambala–Ludhiana–Jalandhar road.

There are no recommended accommodation options in Harike. There are a couple of private hotels near the lake, but visitors need to validate before booking. There are good hotels that should be suitable 40–45km away on the outskirts of Amritsar.

Conservation

The major threats to this important wetland include large-scale utilization of both surface and ground waters for irrigation, expansion of intensive agriculture, resulting in encroachments on the wetland, drainage of agricultural chemicals into the waters, discharge of untreated waste from catchment towns into the rivers that feed the wetland, and deforestation of the lower Shivalik Hills, causing soil erosion and silting. The Indian Army, along with other agencies, launched and completed a major project to clear the Water Hyacinth that was almost choking the lake. Other major problems are siltation and poaching of wildfowl in the lake area.

Rufous-vented Prinia

RAJASTHAN

1. Bikaner/Jorbeer
2. Keoladeo National Park (Bharatpur Bird Sanctuary)
3. Ranthambhore
4. Tal Chappar Wildlife Sanctuary
5. Desert National Park

Rajasthan, which means 'abode of kings', is a vibrant, exotic state where tradition and royal glory meet in a riot of colours against the vast backdrop of sand and desert. The panoramic view of the state is mesmerizing, with the lofty hills of the Aravallis and the golden sand dunes of the Great Indian Desert. The land is endowed with forts, magnificent palaces, *haveli*s (traditional mansions), a rich culture and heritage, beauty and natural resources. This princely state is one of the most exotic locales for tourists the world over. It is so rich in history that every village has its tales of valour and sacrifice, whether for the sake of the nation or for conservation. The state can be divided into four major physiographic regions: the western desert with barren hills, level, rocky plains and sandy plains; the Aravalli Hills, running south-west to north-east starting from Gujarat and

KEY BIRDS
Top 10 Birds
1. Great Indian Bustard
2. Lesser Florican
3. Cream-coloured Courser
4. Green Avadavat
5. Spotted Sandgrouse
6. Black-bellied Sandgrouse
7. Indian Spotted Creeper
8. Yellow-eyed Pigeon
9. Saker Falcon
10. White-bellied Minivet

Lesser Florican

ending in Delhi; the eastern plains with rich alluvial soil, and the southeastern plateau. The state's major rivers are the Mahi, Chambal and Banas.

Climate
The state is located in a unique biological situation in the Indian subcontinent. It is bisected by the Aravalli Range into the western arid region, the Thar Desert, and the eastern semi-arid and sub-humid region. The Aravalli Range is made up of the oldest Archaean granitic rocks and has played an important role as a barrier to desert elements migrating towards the east. In addition to the main range, the Aravallis extend towards south and southeastern Rajasthan, where they merge with the Vindhyan rock system and the Deccan Plateau.

Access, Transportation & Logistics
The majority of visitors enter Rajasthan via the international airport near its capital, Jaipur. There are many other big airports like Jodhpur, Udaipur and Jaisalmer. The state is very well connected with the capital of India, New Delhi. The transport infrastructure is generally good. Hiring a vehicle provides the best way to explore places near Rajasthan.

Health & Safety
There are no major health concerns in Rajasthan. It is a low- to no-risk area of India when it comes to malaria, but do take precautions especially after the monsoon. Visit a GP or travel clinic 6–8 weeks before departure to make sure you are up to date with any

Green Avadavat

vaccinations. Recommended vaccinations for travel to Rajasthan and India generally are hepatitis A, tetanus and typhoid.

Birdwatching Highlights
Rajasthan is an ecologically diversified state, with a wide variety of flora and fauna occurring in different ecological zones. About 627 species of bird have been recorded in the state. It has regular records of Critically Endangered species like White-rumped and Indian Vultures, and the Great Indian Bustard, Endangered species like the Lesser Florican, and several Vulnerable species such as the Sociable Lapwing, Indian Skimmer, Yellow-eyed Pigeon, Indian Spotted Creeper and Stoliczka's Bushchat.

Indian Leopard at Bera

Bikaner/Jorbeer

Carcass dump

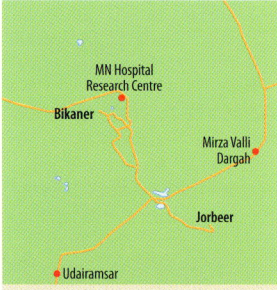

KEY FACTS

Nearest Major Towns
Bikaner

Habitats
Desert

Key Species
Spotted Owlet, Steppe Eagle, Indian Roller, Red-vented Bulbul, Grey-backed Shrike, Griffon Vulture, Cinereous Vulture, Egyptian Vulture, Indian Vulture, House Sparrow, Eastern Imperial Eagle, Cattle Egret, Steppe Eagle, Rosy Starling, Variable Wheatear, Short-toed Snake Eagle, Common Kestrel, Laggar Falcon, Black Drongo, Common Stonechat, Common Myna, Saker Falcon, White-bellied Minivet, White-eared Bulbul, Black-winged Kite

Other Specialities
Blackbuck, Chinkara, Desert Cat, Bengal Fox, Nilgai, Saw-scaled Viper

Best Time to Visit
Throughout the year

Bikaner is a well-known tourist destination in Rajasthan. The city is known for the 16th-century Junagarh Fort, a huge complex of ornate buildings and halls. On the other hand, Bikaner is also very well known among birdwatchers for its large number of hunting birds and endangered vultures.

Birdwatching Sites

Jorbeer An IBA, Jorbeer is situated just 15km from Bikaner. This place is in fact a dumping yard for animal carcasses that eventually attracts thousands of migratory birds every year. The area is mainly famous for the large congregation of migratory and resident vultures, and the seven species reported in the area are Griffon Vulture, Himalayan Vulture, Cinereous, Egyptian, Long-billed, Red-headed and White-rumped Vultures. The area

White-eared Bulbul

Indian Coursers

Black-winged Kite

is also famous for eagles like Steppe, Tawny, Eastern Imperial and Greater Spotted Eagles, and Long-legged Buzzards, Black-eared Kites and Laggar Falcons. Apart from birds, Jorbeer is a good place in which to see and photograph mammals like the Nilgai, Indian Gazelle (Chinkara), Spiny-tailed Lizard and Desert Jird.

Access & Accommodation
Jorbeer is situated 15km from the city of Bikaner, which is a well-known tourist hub. Bikaner is equipped with varied accommodation options, from luxury resorts to budget lodges.

Conservation
The area of the protected forest block of Jorbeer has recently been declared a Conservation Reserve. Heavy biotic pressure, train tracks passing very close to dump sites, feral dogs, poultry and fish waste dumps, and poisoned carcasses are the key threats to vultures and other scavenging species.

Cinereous Vultures

Himalayan Vultures

Common Buzzard

Tawny Eagle

Keoladeo National Park (Bharatpur Bird Sanctuary)

Bharatpur landscape

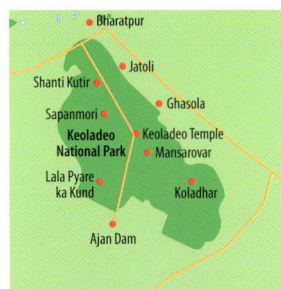

Formerly known as the Bharatpur Bird Sanctuary, Keoladeo National Park is situated in Bharatpur, in the state of Rajasthan. The park is a man-made and man-managed wetland. As it lies on the Central Asian Flyway of the Asia Pacific Global Migratory Flyway, it is a wintering as well as a breeding ground for a large number of migratory waterfowl that breeds in the Palaearctic region.

Birdwatching Sites

The park is an area of just 29km² situated on the extreme western edge of the Gangetic Basin, but this whole area is very rich in avifaunal life forms. About 400 birds have been recorded in the area, including pelicans, herons, storks, ibises, many ducks and endangered Sarus Cranes. Inside Bharatpur, visitors can walk or rent a bicycle, rickshaw or tonga (horse carriage). The paved path in the park is about 11km long. For most tourists and bird photographers with heavy lenses, a cycle-rickshaw is probably the best choice. Inside the park there are various spots with colonies of Painted Storks, many ducks, and sometimes Sarus Cranes.

Access & Accommodation

New Delhi and Jaipur are the two nearest airports for Bharatpur, and it is well connected by road from Delhi and Jaipur. It is roughly three hours by road from Jaipur and 4–5 hours from Delhi, depending on the traffic. The nearest railway station to Bharatpur is Bharatpur Junction, which is well connected with New Delhi and Jaipur.

KEY FACTS

Nearest Major Towns
Bharatpur

Habitats
Wetland, woodland

Key Species
White-rumped Vulture, Long-billed Vulture, Common Crane, Spot-billed Pelican, Lesser Adjutant, Lesser White-fronted Goose, Baer's Pochard, Pallas's Fish Eagle, Greater Spotted Eagle, Eastern Imperial Eagle, Sarus Crane, Sociable Lapwing, Indian Skimmer, Stoliczka's Bushchat, Dalmatian Pelican, Oriental Darter, Painted Stork, Black-necked Stork, Black-headed Ibis, Lesser Flamingo, Ferruginous Pochard, White-tailed Sea Eagle, Grey-headed Fish Eagle, Cinereous Vulture, Red-headed Vulture, Pallid Harrier, Black-bellied Tern

Other Specialities
Nilgai, Rhesus Macaque, Spotted Deer, Indian Python, Indian Porcupine, Indian Jackal, Jungle Cat, Grey Mongoose

Best Time to Visit
Throughout the year

Lesser Whistling Duck

There are plenty of accommodation options near the park, depending on budget and requirements. Iora Guest House, Hotel Sunbird and Birder's Inn are among the known places among birders.

Conservation
Keoladeo National Park was established as a national park in March 1982 and since then grazing has been banned inside the park. However, grazing livestock and poaching are still the biggest threats to the park.

Indian Nightjar

Indian Spotted Eagle

Common Teal

Common Pochard

Ranthambhore

Tiger

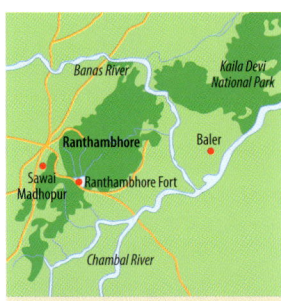

KEY FACTS

Nearest Major Towns
Sawai Madhopur

Habitats
Tropical dry deciduous forest

Key Species
White-rumped Vulture, Long-billed Vulture, Lesser Adjutant, Eastern Imperial Eagle, Sarus Crane, Stoliczka's Bushchat, Oriental Darter, Painted Stork, Black-necked Stork, Black-headed Ibis, Ferruginous Pochard, Cinereous Vulture, Red-headed Vulture, Black-bellied Tern, Painted Francolin, Rain Quail, Jungle Bush Quail, Painted Bush Quail, Painted Spurfowl, Indian Peafowl, Yellow-wattled Lapwing, Yellow-legged Green Pigeon, Sirkeer Malkoha, Indian Nightjar, Indian Grey Hornbill, Yellow-crowned Woodpecker, Indian Bushlark, Ashy-crowned Sparrow Lark, Black-headed Cuckooshrike, Bank Myna, Brahminy Starling, White-browed Fantail, Jungle Prinia, Ashy Prinia, Rufous-fronted Prinia, Large Grey Babbler, Jungle Babbler, Brown Rock Chat, Indian Robin, Marshall's Iora, Common Woodshrike

Other Specialities
Tiger, Indian Leopard, Sloth Bear, Lynx

Best Time to Visit
October to April

The Ranthambhore National Park is situated in the Sawai Madhopur district of eastern Rajasthan. It is part of the much larger Ranthambhore Tiger Reserve. Ranthambhore is primarily known for the Indian or Bengal Tiger, and this is the single largest area in Rajasthan where you can find a good number of Tigers. Apart from Tigers and other big mammals, Ranthambhore is home to many bird species due to its varied terrain and abundance of waterbodies. About 300 bird species have been documented in the area. Greylag and Bar-headed Geese, Indian Coursers, Indian Skimmers, Ruddy Shelduck, Painted Spurfowl, Indian Peafowl and Rufous Treepies are among the important species found in the park.

Birdwatching Sites

Mansarovar Dam Mansarovar, one of the largest waterbodies in the area, is about 30km from the park. It is a good place for waterbirds like geese, Northern Pintails and pelicans in large numbers during the winter season.

Soorwal Dam This is home to different species of migratory bird in winter. Painted Storks, Lesser Flamingos, Eurasian Spoonbills, Sarus and Demoiselle Cranes, and Greylag Geese can be seen in large numbers in November–March.

Ranthambore Fort The base of the Ranthambhore Fort is the best place in which to see the Painted Spurfowl. The area is also good for parakeets, doves, flycatchers and the Indian Peafowl.

White-naped Woodpecker

Access & Accommodation

The nearest airport for Ranthambhore is Jaipur, which is about 180km from the park. Sawai Madhopur is the nearest railway station. Both Jaipur and Sawai Madhopur are well connected to all major cities of India.

Ranthambore National Park is situated in the Sawai Madhopur district, which is well equipped with many private accommodation options. The Ranthambhore Bagh (http://ranthambhore.com) is well known among wildlife enthusiasts. There is also a Rajasthan Tourism Development Corporation facility available in Sawai Madhopur.

Conservation

Ranthambhore was designated as a national park in 1980. Like many other national parks in India, it faces a serious threat of habitat loss and fragmentation due to development projects and encroachment. Other serious conservation issues are human-wildlife conflict and poaching. However, poaching is not as widespread or frequent as it was in the past.

White-bellied Drongo

Ranthambhore landscape

Tal Chappar Wildlife Sanctuary

Demoiselle Cranes

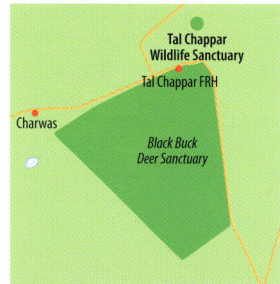

KEY FACTS

Nearest Major Towns
Churu

Habitats
Tropical arid zone

Key Species
White-rumped Vulture, Long-billed Vulture, Eastern Imperial Eagle, Cinereous Vulture, Red-headed Vulture, Desert Wheatear, Black Redstart, Large Grey Babbler, White-eared Bulbul, Laggar Falcon, Indian Roller, Long-legged Buzzard, Common Sandpiper, Demoiselle Crane, Greater Coucal, Chestnut-bellied Sandgrouse, Eurasian Collared Dove, Bar-headed Goose, Indian Silverbill, Water Pipit, Tawny Pipit, Long-billed Pipit, Barn Swallow, Bimaculated Lark, Large-billed Crow, House Crow, Rufous Treepie, White-browed Fantail, Rose-ringed Parakeet

Other Specialities
Blackbuck, Indian Jackal, Desert Jird, Striped Hyena, Spiny-tailed Lizard

Best Time to Visit
Throughout the year

Located 220km from Jaipur, this is a small but rich wildlife sanctuary in the Churu district of Rajasthan. The sanctuary is known for its enchanting natural beauty and unique and fragile ecosystem. Situated in the path of the migratory passage of exotic birds, the sanctuary is renowned for its rich population of Blackbucks and thousands of beautiful migratory birds, especially the harriers that pass over the sanctuary in September. During this season, the Eurasian Sparrowhawk, Hen and Montagu's Harriers, Eastern Imperial Eagle, Western Marsh Harrier and Short-toed Snake Eagle can easily be seen. The terrain of the sanctuary is almost flat and has areas with open grassland interspersed with *Acacia* and *Prosopis* trees, giving the sanctuary a look of the savannahs.

Birdwatching Sites

Tal Chappar Wildlife sanctuary has been identified as an IBA. The entire Tal Chappar area is quite popular among birdwatchers, especially those who have an interest in raptors. There is abundant wildlife even outside the park. You can plan a trip to Gaushala (a community grazing ground outside the park). This area has a large population of Spiny-tailed Lizards that attracts a number of raptors, foxes, cats and other reptiles. The neighbouring villages have several waterbodies that are visited by a large number of waders. Tal Chappar is about five hours' drive from Khichan, a small village but a great place in which to see thousands of Demoiselle Cranes. Every year a great number of the cranes visit this place after the monsoon season. You can combine this destination with your Tal Chappar trip.

White-bellied Minivet

Indian Scops Owls

Indian Spotted Creeper

Access & Accommodation
The nearest airport for Tal Chappar is Jaipur, about 230km away, and the nearest railway station is Sujangarh, which has frequent connectivity with New Delhi. While travelling inside the park, a vehicle with high ground clearance is necessary. Jeeps can be rented from the forest guest-house premises.

There is a Forest Department guest house just outside the park, which is comfortable but has very limited rooms; getting a reservation in winter is challenging. The nearest town is Sujangarh, with a few private hotels.

Conservation
Tal Chappar was declared a sanctuary in 1962. In recent years, due to blockage of water channels in the Gopalpura Hills and illegal establishment of saltworks in the peripheral belt of the sanctuary, not only has the water quantity decreased, but also human intervention has caused much disturbance in the region, coupled with many other problems threatening the very existence of the protected area. Livestock grazing and intrusion of domestic dogs are among the serious issues for the faunal elements of the sanctuary.

Bimaculated Lark

Yellow-eyed Pigeon

Desert National Park

Sand dunes

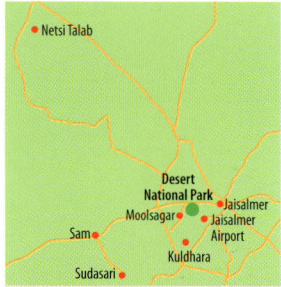

KEY FACTS

Nearest Major Towns
Jaisalmer and Barmer

Habitats
Desert, thorny scrub, grassland

Key Species
White-rumped Vulture, Long-billed Vulture, Great Indian Bustard, Greater Spotted Eagle, Stoliczka's Bushchat, Cinereous Vulture, Red-headed Vulture, MacQueen's Bustard, Spotted Sandgrouse, Sykes's Nightjar, Desert Lark, Greater Hoopoe Lark, White-eared Bulbul, Trumpeter Finch

Other Specialities
Chinkara, Blackbuck, Bengal Fox, Indian Fox, Desert Cat, Indian Wolf, Spiny-tailed Lizard, Monitor Lizard, Desert Hedgehog, Saw-scaled Viper, Russell's Viper, Common Krait

Best Time to Visit
Throughout the year

This park near the town of Jaisalmer is on the easternmost edge of the Sahara-Arabian Desert Biome. It comprises a large area of about 3,162km² in the Jaisalmer and Barmer districts of Rajasthan. It is an important area for the state bird of Rajasthan, the Great Indian Bustard. Despite the park's fragile ecosystem, the birdlife and wildlife are abundant. The region is a haven for migratory and resident birds of the desert. Many eagles, harriers, falcons, buzzards, kestrels and vultures are spotted here. Short-toed Snake, Tawny and Greater Spotted Eagles, Laggar Falcons and kestrels are the most common among these.

Birdwatching Sites

Sudasari This is one of the prime spots in which to look for the Great Indian Bustard. There are large enclosures where livestock is not permitted. These have good grass and serve as a refuge for the species. Carrying binoculars is recommended as this is a shy bird that prefers to keep a safe distance from the roadside.

Kuldhara The abandoned village of Kuldhara contains ruins and a rocky area in its vicinity as you approach the area. Most of the birding is done on the roadside, so entering the ruins is optional and depends on your interest. Some of the target species in this area are the Desert Lark, Rufous-tailed Wheatear and Striolated Bunting.

Netsi Talab This lake is situated just near Netsi village. The driving time from Sam to Netsi is about an hour. Netsi Talab is a good place in which to look for waterfowl and hundreds of

Cream-coloured Courser

Bengal Foxes

White-eared Bulbul

sandgrouse (Chestnut-bellied, Black-bellied and Spotted). The Trumpeter Finch is also found near the lake.

Access & Accommodation
Desert National Park is easily accessible from the district of Jaisalmer, which is about 20km from the park. Jaisalmer Junction is the nearest railway station. Jodhpur is the nearest airport, approximately 300km from the park.

Sam and Khuri are the two places nearby where various accommodation options, to suit different budget, are available. During the winter season (November–February) many tourists visit Jaisalmer, so advance booking is recommended.

Conservation
Desert National Park is facing considerable pressure of different types, like overgrazing of livestock, encroachment of land, poaching and windmills, which is expected to reduce the numbers of Great Indian Bustards to a critical level. It is evident that the number of these birds has reduced very sharply throughout India, and only this park and a few other areas are still fortunate enough to have it.

Black-bellied Sandgrouse

Great Indian Bustard

SIKKIM

A small yet very beautiful, rugged montane state, Sikkim is located in the north-east part of India. It is a paradise for birdwatchers and ornithologists. Wedged in between the Himalayan kingdom of Nepal in the west and Bhutan in the east, Sikkim is bounded by the Darjeeling district of West Bengal in the south, and a stretch of the Tibetan Plateau in the north. There are four districts in Sikkim, namely North Sikkim, West Sikkim, South Sikkim and East Sikkim. The Teesta River is the lifeline and main river of Sikkim, running north to south across the state.

Climate
Sikkim has a varied climate, topography and altitudinal ranges within a small area. The climate varies between the tropical heat of valleys and the alpine cold of Tibetan plateaus and snowy ranges. The best season for birding is November–February, when rainfall is low and the weather is normally clear. Sikkim's vegetation can be categorized into three major types, tropical, temperate and alpine. The tropical zone contains deep valleys and well-drained slopes covered with evergreen jungle. Above it is the temperate zone comprising dense, tall evergreen forests of oaks and rhododendrons, and towards its upper limits predominantly silver firs, oaks, magnolias and Blue Pines. The alpine zone is composed of small crooked trees and shrubs. Sikkim is well known for its varied rhododendrons.

Access, Transportation & Logistics
The majority of visitors enter Sikkim via the Bagdogra Airport, which is linked with New Delhi, Kolkata and Guwahati. Sikkim is very well connected by road to the rest of India, and the transport infrastructure is generally good. Visitors usually enter Sikkim by driving up the Sevoke road from Siliguri. This national highway follows the course of the Teesta River. The driving time to Gangtok (the capital of Sikkim) is four hours (120km). Buses, shared taxis and private taxis are freely available

1. West Sikkim
2. Maenam Wildlife Sanctuary
3. Pangolakha Wildlife Sanctuary, Zuluk
4. North Sikkim: Lachen, Thangu & Gurudongmar

for transport in Sikkim. On the way to North Sikkim, Mangan is a good place in which to stay overnight, then spend the morning of the next day on birding. It is good for lowland birding.

Health & Safety
There are no major health concerns in Sikkim. It is a low- to no-risk area of India when it comes to malaria, but do take precautions especially after the monsoon. Visit a GP or travel clinic 6–8 weeks before departure to make sure you are up to date with any vaccinations. If you suffer from altitude sickness, consult a doctor and carry your medicine with you.

Birdwatching Highlights
Sikkim is a well-known destination for eastern Himalayan birds. About 11 IBAs have been selected in this small state, and more than 550 bird species have been recorded here. The western and southern parts of Sikkim have some of the well-known birding destinations, like Pelling, Yuksom, Barsey Rhododendron Sanctuary, Ravangla and Kanchenjunga National Park. The eastern part has birding hotspots like Fambong Lho Wildlife Sanctuary and Changu Lake. Northern Sikkim offers a varied ecology, with many birding hotspots like Lachen, Thangu, Lachung, Yumthang and the Gurudongmar area.

KEY BIRDS
Top 10 Birds
1. Himalayan Snowcock
2. Tibetan Snowcock
3. Satyr Tragopan
4. Blood Pheasant
5. Snow Pigeon
6. Silver-backed Needletail
7. Himalayan Cuckoo
8. Ibisbill
9. Golden Eagle
10. Golden-breasted Fulvetta

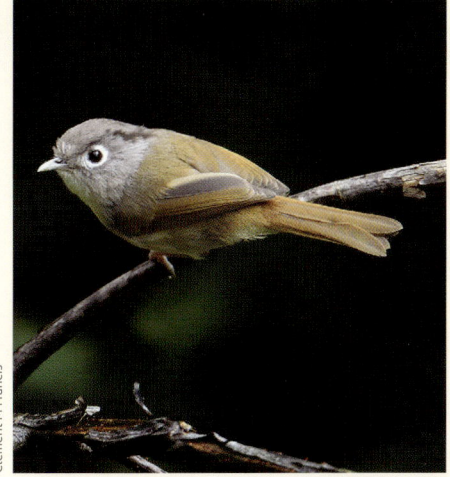

Clement M Francis

Nepal Fulvetta

Bhanu Singh

White-capped Redstart

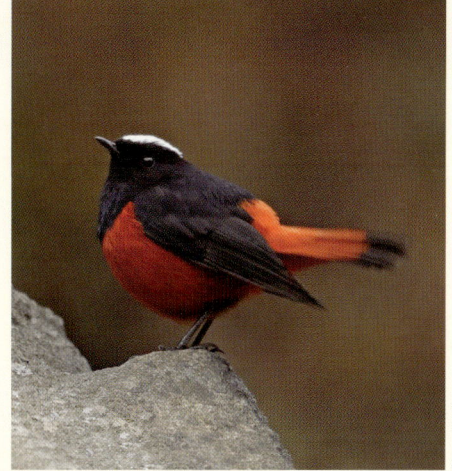

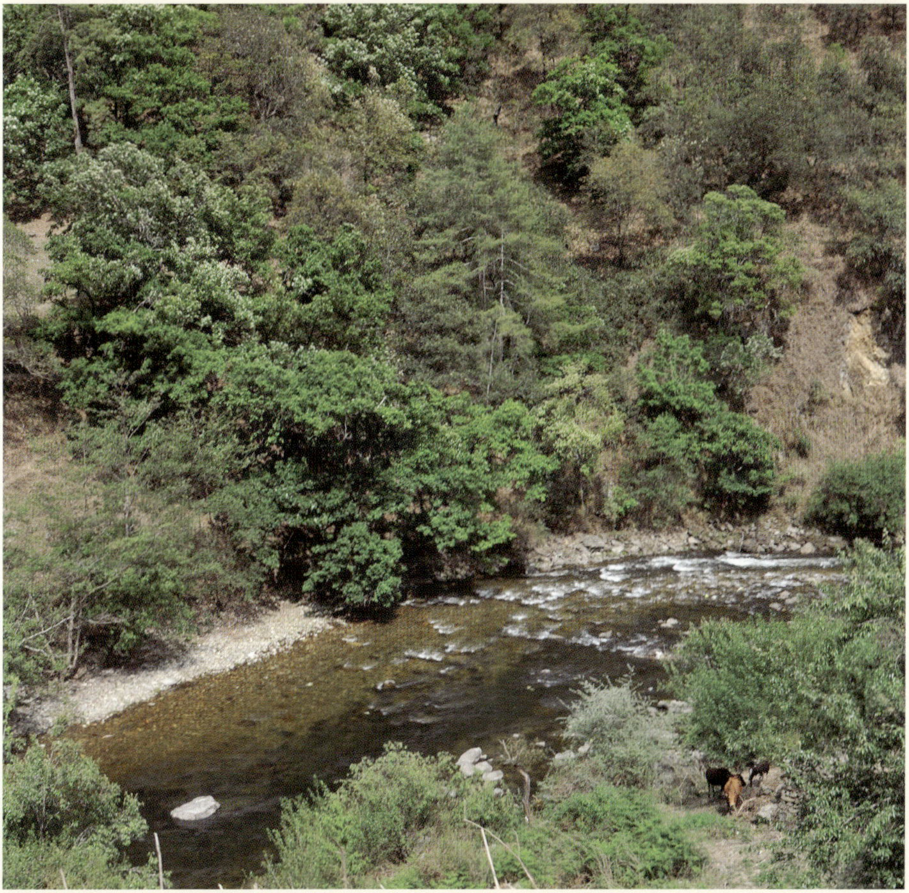

Bhanu Singh

Sikkim forest

West Sikkim

Sikkim landscape

The West Sikkim district is known among trekkers and birdwatchers for its high elevations. Pelling and Yuksom are among the well known destinations for birdwatching.

Birdwatching Sites

Pelling This popular birding destination in western Sikkim is 50km from Ravangla and 25km from Kewzing. Some of the birds that can be spotted here are the Indian Cuckoo, Grey Headed and Bay Woodpeckers, Striated Bulbul, Mountain Hawk Eagle, Himalayan Vulture, Grey-sided and Greater Necklaced Laughingthrushes, Lesser Yellow-naped Woodpecker, Hoary-throated and Rusty-fronted Barwings, White-throated Needletail and Black Eagle.

KEY FACTS

Nearest Major Towns
Gangtok

Habitats
High-altitude coniferous forests

Key Species
Wedge-tailed Green Pigeon, Himalayan Swiftlet, Scarlet Minivet, Grey-headed Canary-flycatcher, Green-backed Tit, Black Bulbul, Red-billed Leiothrix, Small Niltava, Verditer Flycatcher, Blue Whistling Thrush, Grey-winged Blackbird, Fire-breasted Flowerpecker, Scarlet Finch, Yellow-breasted Greenfinch, House Crow, Red-vented Bulbul, Common Myna, House Sparrow, Russet Sparrow, Spotted Dove, Barred Cuckoo Dove, Large Hawk Cuckoo, Alpine Swift, Mountain Hawk Eagle, Great Barbet, Blue-throated Barbet, Long-tailed Shrike, Black Drongo, Hair-crested Drongo, White-throated Fantail

Other Specialities
Red Panda, Snow Leopard, Himalayan Musk Deer, Bharal, Himalayan Tahr, Northern Plains Grey Langur, Leopard Cat, Himalayan Black Bear, Blue Sheep

Best Time to Visit
Throughout the year

Fire-breasted Flowerpecker

Chestnut-crowned Warbler

Crimson-breasted Pied Woodpecker

Striated Laughingthrush

Yuksom Yuksom, the first capital of Sikkim, is a historical town in the Geyzing subdivision. some of the birds here are the Yellow-rumped Honeyguide, Striated Bulbul, Maroon Oriole, Mountain Hawk Eagle, Short-billed and Grey-chinned Minivets, and Crimson-breasted Pied Woodpecker.

Khecheopalri Lake The scenic Khecheopalri Lake in Khecheopalri village is 34km from Pelling town. The lake is known for birds like the Black-necked Grebe, Speckled Wood Pigeon, many ducks and laughingthrushes. You can also find the elusive Baer's Pochard in this area.

Barsey Rhododendron Sanctuary This sanctuary is famous for rosefinches, and in September and October there are huge gatherings of species like Common, Dark-breasted, Himalayan White-browed and even Himalayan Beautiful Rosefinches. Other birds to look out for here are the Spotted Laughingthrush, parrotbills, Fire-capped Tit, White-browed and Lesser Shortwings, Kalij Pheasant and Satyr Tragopan. The state animal of Sikkim, the Red Panda, is also found in the sanctuary.

Access & Accommodation
West Sikkim is connected by road from Darjeeling (75km) and Gangtok (125km). Road conditions are average and depend on the season, as this is a hilly terrain and landslide-prone area. There are many homestays available in the region, and local people are warm and welcoming towards travellers. There are also a good number of budget through to luxury hotels, depending upon individual needs.

Conservation
Even though about two-thirds of Sikkim is under forest cover and the human population is relatively low, the state suffers from some conservation issues such as the need for fuelwood and fodder, increase in human-wildlife conflicts, and increasing populations of stray and feral dogs, as well as feral cats, in urban areas. Hunting is not a major issue in any IBA.

Grey-crested Tit

Maenam Wildlife Sanctuary

Maenam forest

This sanctuary is situated in the southern part of Sikkim near Kewzing, and covers an area of about 35km². The distance from the state capital, Gangtok, to the sanctuary is about 65km. The sanctuary is home to more than 200 bird species and is well known for the Satyr Tragopan, one of the rarest and quite elusive pheasant species found in the eastern Himalayas. The sanctuary is also known as a treasure house of medicinal plants and herbs.

KEY FACTS

Nearest Major Towns
Gangtok

Habitats
Subtropical dry evergreen forest, subtropical broadleaved hill forest

Key Species
Greater Spotted Eagle, Chestnut-breasted Hill Partridge, Blyth's Tragopan, Rufous-necked Hornbill, Rusty-bellied Shortwing, Rufous-throated Wren Babbler, Sikkim Wedge-billed Babbler, Hoary-throated Barwing, Broad-billed Warbler, White-naped Yuhina, Common Hill Partridge, Satyr Tragopan, Bay Woodpecker, Tickell's Thrush, Gould's Shortwing, White-collared Blackbird, Striated Laughingthrush, Scaly Laughingthrush, Greater Scaly-breasted Wren Babbler, Bar-throated Minla, Red-tailed Minla, Black-faced Laughingthrush

Other Specialities
Golden Cat, Himalayan Musk Deer, Chinese Pangolin, Marbled Cat

Best Time to Visit
Throughout the year

Pink-browed Rosefinch

Hodgson's Redstart

Birdwatching Sites

The sanctuary starts at 2,000m, with subtropical forests gradually being taken over by moist temperate forests at about 2,500–3,000m, then by subalpine conifer forests at an altitude higher than 3,000m. Important birds in this region include the Satyr Tragopan, Kalij Pheasant, Rusty-bellied Shortwing, minivets, yuhinas and flycatchers. You can also find another star bird in this region, the Fire-tailed Myzornis.

Access & Accommodation

The main access to the sanctuary is from Ravangla Town. It takes approximately four hours to trek from Ravangla to Maenam Peak. Various accommodation options are available at Namchi and Ravangla. Hiring a guide is recommended before going on the trail.

Conservation

The main threats and conservation issues of Maenam Wildlife Sanctuary are uncontrolled recreation and tourism, illegal cutting of trees, erosion and occasional poaching. Increasing populations of stray and feral dogs are becoming an issue in general.

Rahul & Khushboo Sharma

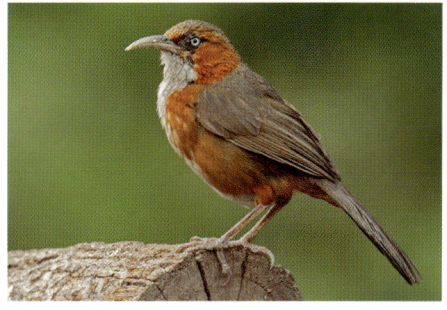

Rusty-cheeked Scimitar Babbler

Rahul & Khushboo Sharma

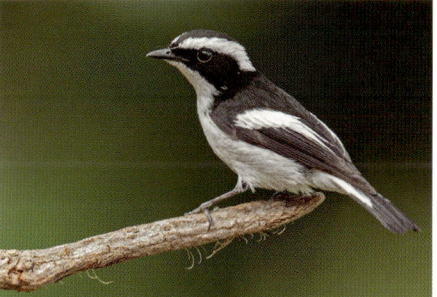

Little Pied Flycather

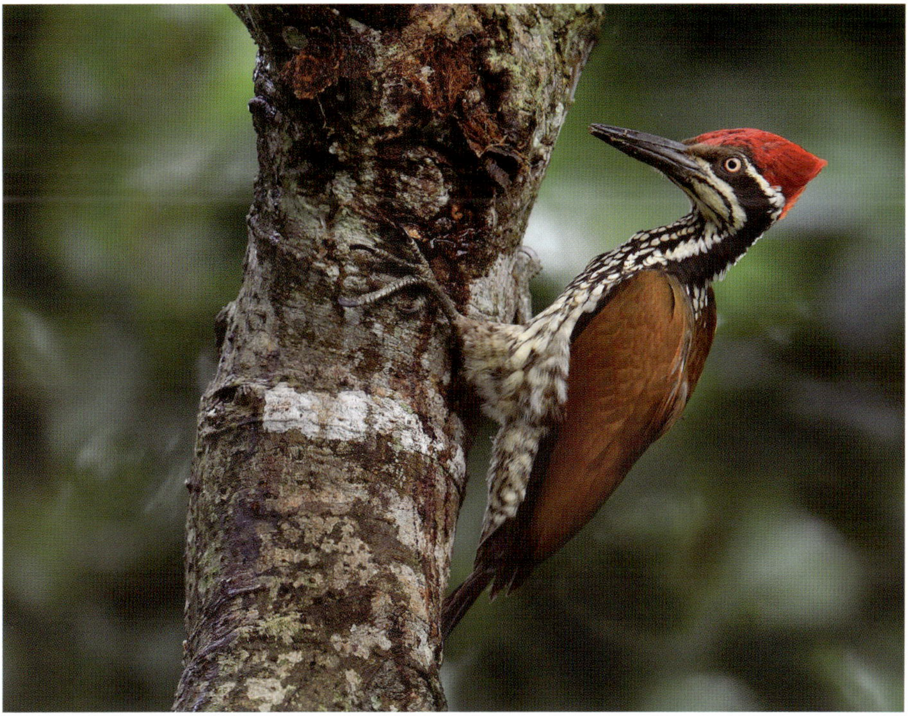

Clement M Francis

Greater Golden-backed Woodpecker

Pangolakha Wildlife Sanctuary, Zuluk

Pangolakha forests

This sanctuary is about 30km from the Rangpo district in east Sikkim. The Pangolakha Range extends below the Chola Range and separates Sikkim from Bhutan, so this range offers some great birdwatching trails.

Birdwatching Sites

The Pangolakha area is located between Bedang Lake and the Nathu La complex of north-east Sikkim. The mountain passes of Nathu La and Jelep La (*la* means pass) form the routes for migratory waterbirds, many of which stop over at the various wetlands in the area, especially Bedang Tso Lake. Among them

KEY FACTS

Nearest Major Towns
Gangtok

Habitats
Subtropical pine forest, subtropical broadleaved hill forest, alpine moist pastures

Key Species
Pallas's Fish Eagle, Greater Spotted Eagle, Chestnut-breasted Hill Partridge, Wood Snipe, Rufous-necked Hornbill, Slender-billed Babbler, Black-breasted Parrotbill, Grey-crowned Prinia, Ward's Trogon, Hoary-throated Barwing, Broad-billed Warbler, Himalayan Vulture, Snow Partridge, Solitary Snipe, Plain-backed Snowfinch, Rufous-necked Snowfinch, Red-fronted Rosefinch, Plain Mountain Finch, Grandala, Altai Accentor, Alpine Accentor, Rosy Pipit, Common Hill Partridge, Rufous-breasted Accentor, Gould's Shortwing, Himalayan Rubythroat, Fire-tailed Myzornis, Great Parrotbill

Other Specialities
Clouded Leopard, Fishing Cat, Golden Cat, Himalayan Tahr, Red Panda, Indian Wolf

Best Time to Visit
Throughout the year

Crimson Sunbird

Black-throated Accentor

Himalayan Snowcock

are the Eurasian Woodcock and the Wood Snipe, a globally threatened species. Hill Pigeons are commonly seen near local houses. The Snow Pigeon, Snow Partridge, Himalayan Monal and Gold-naped Finch are among the other birds that are found on the alpine slopes.

Access & Accommodation

There are many homestay options at Lingtam, and some at Nimachen, Padamchen and Zuluk. Zuluk has an army cantonment, and in some of the areas photography is restricted. All these places are well connected through road transport. Hiring a private vehicle is recommended for ease of travel and to maximize the birdwatching experience.

Conservation

There are no major conservation issues. However, as the area is at high altitude bordering Tibet, it is manned by the Indian Army, which occupies it in short shifts of about six months to a year. Any biodiversity-sensitization programme is hence short-lived. Most camps are around or near waterbodies, with resultant pollution, especially of non-biodegradable rubbish, and the spread of stray dogs around these settlements. In addition to preying on wildlife such as the Ruddy Shelduck, there have been reports of human casualties due to these dogs. The State Forest Department has set up eco-development committees around all wildlife protected areas to address the conservation issues.

Blood Pheasant

North Sikkim: Lachen, Thangu & Gurudongmar

North Sikkim

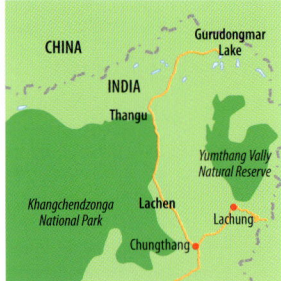

KEY FACTS

Nearest Major Towns
Gangtok

Habitats
High-altitude coniferous forest

Key Species
Snow Pigeon, Blood Pheasant, Himalayan Monal, White-throated Dipper, Spot-winged Grosbeak, White-winged Grosbeak, Bearded Vulture, Red-headed Vulture, Himalayan Vulture, White-rumped Vulture, Mountain Hawk Eagle, Rufous-bellied Eagle, Himalayan Buzzard, Upland Buzzard, Little Owl, Indian Eagle Owl, Altai Accentor, Alpine Accentor, Rufous-breasted Accentor, Robin Accentor

Other Specialities
Clouded Leopard, Red Panda, Himalayan Black Bear, Himalayan Musk Deer, Blue Sheep

Best Time to Visit
Throughout the year

Situated at a height of 2,750m, Lachen is one of the most beautiful towns in North Sikkim. The gateway to the Chopta Valley and the scenic Gurudongmar Lake, this Sikkimese town is known for its pictorial beauty and rustic cultures. Thangu and Gurudongmar Lakes are two of the closest destinations to Lachen.

Birdwatching Sites

Lachen This is a small town at an elevation of 2,700m in North Sikkim. Generally, it is the base for most of the birdwatching expeditions in North Sikkim. It is well known for many bird species, like Common Hill, Rufous-throated Hill and Chestnut-breasted Hill Partridges, Grey Peacock Pheasant, Himalayan and Tibetan Snowcocks, Chukar Partridge, Snow and Tibetan Partridges, Black and Swamp Francolins, Red Junglefowl, Himalayan Monal, Satyr Tragopan, and Kalij and Blood Pheasants.

Thangu and Gurudongmar These are high-altitude regions of North Sikkim. Gurudongmar Lake is situated at an astounding 5,425m altitude, and at such a high altitude the birdlife is unique and special. Some of the birds in the region are the Blood Pheasant, Grandala, Golden Eagle, White-rumped Snowfinch, Hume's Ground Tit, Little Owl and Upland Buzzard. The Terek Sandpiper has also been recorded here.

Access & Accommodation

Lachen is well connected via road from Gangtok and Mangan. Most roads are in good condition, with some occasional rough patches.

Chestnut Thrush

Yellow-rumped Honeyguide

White-rumped Snowfinch

Hiring a private vehicle is recommended because public transport is not available beyond Lachen.

Due to the proximity of international borders staying is not allowed near Gurudongmar. Hence Lachen, with many homestays and hotels, is a good place in which to set up a base. Carrying appropriate clothing is recommended, as there may be severe weather conditions in North Sikkim.

Conservation

There are no major conservation threats in this region, but there are some common issues like intense grazing pressure on the land by both domestic and wild herbivores. One of the most important threats are the stray and feral dogs around military settlements.

Horned Lark

White-browed Fulvetta

Little Owl

TAMIL NADU

Formerly Madras State, Tamil Nadu is situated on the southeastern side of the Indian peninsula. Chennai, the capital city, is one of the metropolitan areas of India with many historical monuments, temples and industries. Tamil Nadu is bordered by the union territory of Puducherry, and the South Indian states of Kerala, Karnataka and Andhra Pradesh. It is bounded by the Eastern Ghats in the north, the Nilgiri Mountains, Anaimalai Hills and Kerala in the west, the Bay of Bengal in the east, the Gulf of Mannar and Palk Strait in the south-east, and the Indian Ocean in the south. It has the country's third-longest coastline. The state can be divided into three physiographic regions: the eastern coastal region, the western hilly region and the plains. The northern and western parts of the state are mainly hilly areas of the Western Ghats with an average elevation of 1,220m, going up to 2,440m, which is the highest point. The major rivers flowing through the state are the Palar, Cheyyar, Ponnaiyar, Cauvery, Moyar, Bhavani, Amravati, Vaigai, Chittar and Tamaraparni. At 805km in length, the Cauvery is one of the largest rivers in the Indian subcontinent.

1. Point Calimere Wildlife Sanctuary
2. Mudumalai National Park
3. Anamalai Tiger Reserve
4. Udhagamandalam (Ooty)

Climate

Tamil Nadu's climate is generally tropical, ranging from dry sub-humid to semi-arid, and features fairly hot temperatures over the year except during the monsoon season. The state has two distinct periods of rainfall: the south-west monsoon in June–September, with strong south-west winds; and the north-east monsoon in October–December, with dominant north-east winds. The temperature ranges from as low as 0 °C in the higher reaches of the Western Ghats to as high as 42 °C in the hot plains. There are three seasons: pre-monsoon (July–September), monsoon (October–December) and post-monsoon (January–June). The summer runs throughout March–May and is characterized by intense heat and scant rainfall across the state. Cold weather starts in early December and ends in mid-March; the climate at this time is pleasant.

KEY BIRDS

Top 10 Birds
1. White-tailed Tropicbird
2. Cape Petrel
3. Great Frigatebird
4. Red Knot
5. Broad-billed Sandpiper
6. Spoon-billed Sandpiper
7. Great Snipe
8. South Polar Skua
9. Lesser Noddy
10. Bridled Tern

Access, Transportation & Logistics

Tamil Nadu is very well connected to the rest of India and many international destinations via airways. The Anna International Airport, 16km south of Chennai, has flights connecting most major Indian cities as well as many international ones. Other international airports connecting Tamil Nadu with the rest of the world are Coimbatore, Tiruchirapalli and Madurai.

Tamil Nadu is conveniently connected by railways, and both inbound and outbound commutes are made simpler through a series of direct trains operating from the state's major cities. The Tamil Nadu railway network is part of the Southern Railways and offers connectivity to almost all major cities of India. There are also frequent buses plying daily from Bangalore, Thiruvananthapuram, Ernakulam, Hyderabad, Tirupathi and other prime destinations to Chennai and other major cities in Tamil Nadu. App-based cabs such as Uber and Ola provide the most convenient way of getting around. Auto-rickshaws are plentiful, but fares are relatively expensive and rarely charged as per the meter.

Health & Safety

Tamil Nadu is a safe state for visitors to India, with no major health concerns. It is a low- to no-risk area of India when it comes to malaria, but do take precautions especially after the monsoon. Visit a GP or travel clinic 6–8 weeks before departure to make sure you are up to date with any vaccinations. Avoid roadside food and drink bottled water. During the summer months keep hydrated, and carry a stole, scarf or cap to protect youself from direct sunlight. Tamil Nadu experiences high humidity throughout the year, so pack breathable cotton clothes.

Birdwatching Highlights

Tamil Nadu has many important protected areas, such as Anamalai, Mudumalai, Point Calimere and Vedanthangal. All of them are identified as IBAs. More than 555 bird species have been recorded from the state. Among Critically Endangered species, White-rumped and Long-billed Vultures are present; among Endangered species, the Nilgiri Laughingthrush and the Spotted Greenshank have been recorded. The Lesser Adjutant, Wood Snipe and Lesser Kestrel are occasionally seen.

The Western Ghats region of Tamil Nadu contains high diversity and a great number of endemic rainforest plants and animal taxa. Some of the key species in this region are the Nilgiri Wood Pigeon, Malabar Parakeet, Malabar Grey Hornbill, Nilgiri Pipit, Grey-headed Bulbul, Black-and-Orange, Nilgiri Flycatchers, Broad-tailed Grassbird, Crimson-backed Sunbird, Palani and Nilgiri Laughingthrushes, Rufous Babbler, White-bellied Blue Flycatcher, White-bellied Shortwing and White-bellied Treepie.

Clement M Francis

Black-winged Stilts

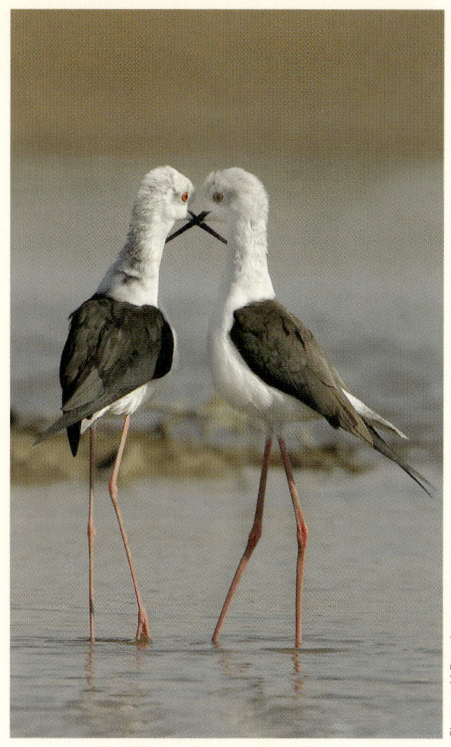

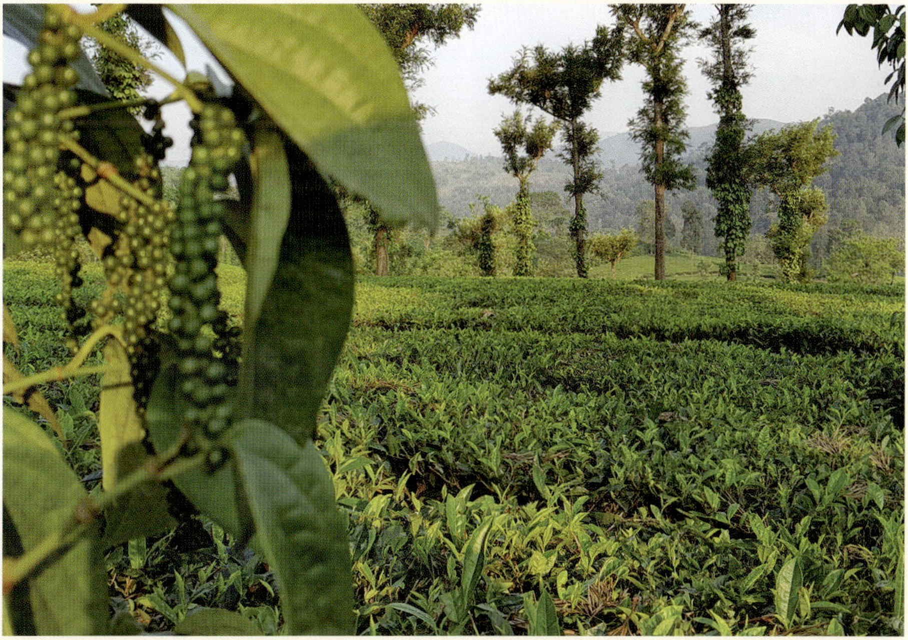

Clement M Francis

Plantations

Point Calimere Wildlife Sanctuary

Point Calimere

KEY FACTS

Nearest Major Towns
Nagapattinam

Habitats
Tropical dry evergreen forest, littoral forest

Key Species
Spotted Greenshank, Spot-billed Pelican, Spoon-billed Sandpiper, Broad-tailed Grassbird, Oriental Darter, Painted Stork, Black-necked Stork, Black-headed Ibis, Lesser Flamingo, Pallid Harrier

Other Specialities
Blackbuck, Spotted Deer, Indian Jackal, Bonnet Macaque, Monitor Lizard, Short-nosed Fruit Bat, Small Indian Civet, Star Tortoise, Indian Grey Mongoose, Black-naped Hare, feral horses

Best Time to Visit
Throughout the year

This compact sanctuary comprises just 21.47km² of the island formed by the Bay of Bengal, the Palk Strait and swampy backwaters at the southeastern tip of the Nagapattinam district in Tamil Nadu. The sanctuary was created in 1967 for the conservation of the Near Threatened Blackbuck antelope, an endemic mammal of India. The sanctuary is well known for large congregations of waterbirds, especially Lesser Flamingos.

Birdwatching Sites

The sanctuary is an area of high biodiversity, with many unique animal species, including birds. The site has recorded the second largest congregation of migratory waterbirds in India, with a peak population of more than 100,000 individuals, representing 103 species. In October these waterbirds arrive from the Rann of Kutch, eastern Siberia, northern Russia, Central Asia and parts of Europe for their feeding season, and start returning to those breeding places in January. They include threatened species such as the Spot-billed Pelican, Nordmann's Greenshank, Spoon-billed Sandpiper and Black-necked Stork. Near Threatened species include the Black-headed Ibis, Asian Dowitcher, Lesser Flamingo, Eurasian Spoonbill, Oriental Darter and Painted Stork.

Access & Accommodation

The sanctuary entrance is 5km south of Vedaranyam, 55km south of Nagapattinam and 380km south of Chennai. A 6km paved road leads from the checkpoint to the villages of Kodaikadu and Kodaikarai. Vehicles are prohibited in the core of the sanctuary. Forest Department guides are available, with

Grey Heron

an advance request to the Wildlife Warden to escort visitors for bird and other wildlife watching in the core of the sanctuary. Several watchtowers accessible to the public provide panoramic views of the sanctuary. It is open all year from 6 a.m. to 5 p.m. The best weather is in November–December, when the area is cooled by the north-east monsoon and the grassland is the most luxuriant. The best time for birdwatching is in October–January, and the best time for mammal viewing is in March–August. The nearest railway station is Nagapattinam (60km) and the nearest airport is Tiruchirapalli (150km).

Lodging and restaurants are available at Vedaranyam. Lodging near the sanctuary is available at the Forest Department Rest House named Flamingo House (Poonarai illam), in Kodiyakarai, with the prior approval of the Wildlife Warden, District Forest Office, District Collectorate Campus, Nagapattinam, Tamil Nadu 611003, Tel.: 04365-253092, Email: wlwngp@gmail.com.

Conservation

Major threats to the natural biodiversity and ecological balance of the sanctuary are loss of habitat for waterbirds, soil and water salinization by adjacent salt pans, the spread of the invasive *Prosopis juliflora*, cattle grazing and scarcity of fresh water. Sanctuary staff conduct programmes to alleviate all these issues.

Painted Stork

Lesser Flamingo

Mudumalai National Park

Tamil Nadu

Spotted Deer

KEY FACTS

Nearest Major Towns
Bengaluru, Masnagudi, Ooty and Coimbatore

Habitats
Tropical moist deciduous, tropical dry deciduous and southern tropical thorn forests

Key Species
White-rumped Vulture, Long-billed Vulture, Lesser Adjutant, Nilgiri Wood Pigeon, Yellow-throated Bulbul, Broad-tailed Grassbird, Black-and-orange Flycatcher, Rufous Babbler, White-bellied Blue Flycatcher, Crimson-backed Sunbird, White-bellied Treepie, Green-billed Malkoha, Edible-nest Swiftlet, Malabar Trogon, Malabar Pied Hornbill, White-cheeked Barbet, Malabar Barbet, Yellow-browed Bulbul, Malabar Whistling Thrush, Bonelli's Eagle, Changeable Hawk Eagle

Other Specialities
Asian Elephant, Tiger, Indian Leopard, Gaur, Sloth Bear, Black-naped Hare

Best Time to Visit
October to March

This park is located in the Nilgiri district of Tamil Nadu, in the Western Ghats. It is mainly known for its large mammals, but also harbours a rich avian diversity. This declared Tiger Reserve is about 150km north-west of Coimbatore city in Tamil Nadu. It shares its boundaries with the states of Karnataka and Kerala. The sanctuary is divided into five ranges – Masinagudi, Thepakadu, Mudumalai, Kargudi and Nellakota. Mudumalai is endowed with a diversity of habitats, which support a rich variety of flora and fauna. There are three main types of forest: tropical moist deciduous, tropical dry deciduous and southern tropical thorn. In certain places, mixed vegetation types are also present. The protected area is home to several Endangered and Vulnerable species, including the Asian Elephant, Bengal Tiger, Indian Gaur and Indian Leopard.

Birdwatching Sites

More than 270 bird species have been recorded in the sanctuary, including the Critically Endangered White-rumped and Long-billed Vultures. Regional endemics include the Malabar Trogon, Black-and-orange Flycatcher, Broad-tailed Grassbird and Malabar Grey Hornbill. Some rare birds of prey, like the Rufous-bellied Eagle, can occasionally be seen in the sanctuary. Other predatory birds include Crested and Changeable Hawk Eagles, Crested Serpent Eagle, Black and Bonelli's Eagles, Oriental Honey Buzzard, Jerdon's Baza, Crested Goshawk, Besra, Mottled Wood Owl, Brown Hawk Owl, and several minivets.

Crested Serpent Eagle

Mottled Wood Owl

Black Eagle

Access & Accommodation

Mudumalai National Park is about 240km from Bangalore, 90km from Mysore, 68km from Udhagamandalam (Ooty) and 124km from Kozhikode by road. The reserve straddles the Ooty-Mysore interstate national highway, roughly midway between the two cities, and thus may be approached with equal facility either way. From Ooty, another approach is through the very steep ghat road via Kalhatti – there are 36 hairpin bends on the narrow road, but the distance is truncated to about 40km. The nearest railway stations to Mudumalai are Mysore, about 100km, and Udhagamandalam, about 68km by hill track.

The nearest airports are Coimbatore (160km), Bangalore (240km) and Calicut (124km). The nearest towns for medical facilities, petrol bunks and telephone networks are Gudalur (15km) and Masinagudi (7km).

There are plenty of accommodation options near the park and visitors can check availability and book rooms online (www.mudumalaitigerreserve.com).

Conservation

The Nilgiri area has undergone dramatic changes to the landscape, with the replacement of forests and grassland by monoculture plantations and agriculture. Other developmental processes, such as the construction of dams, reservoirs, canals and tunnels for hydroelectric projects, have impacted the ecology of the area. Tourism, especially in the Sigur/Masinagudi area, is claimed by some to pose a threat to the region, but this is strongly repudiated by those who live and work in the area. Extensive growth of invasive plants such as *Lantana*, which hinder the natural regeneration process of the forests, has occurred as a result of excessive cattle grazing. Construction activities on the proposed India-based Neutrino Observatory at Singara, Masinagudi, are likely to have significant impacts on the local wildlife. The five-year work plan, the high volume of debris and waste disposal, blasting activities, extensive vehicular activity and a large number of outside workers and their support infrastructure all threaten to disrupt the wildlife corridor on the Sigur Plateau, including Mudumalai Sanctuary, connecting the Western Ghats and the Eastern Ghats.

Asian Fairy-bluebird

Anamalai Tiger Reserve

Tamil Nadu

Anamalai landscape

Earlier known as the Indira Gandhi Wildlife Sanctuary and National Park, Anamalai Tiger Reserve is a protected area in the Anaimalai Hills of the Pollachi and Valparai talukas in the Coimbatore district, and the Udumalaipettai taluka in the Tiruppur district, Tamil Nadu. The reserve is carved out of the Tamil Nadu portion of the Anamalais Falls within the Western Ghats mountain chain of south-west India, a region designated as one of the 25 Global Biodiversity Hotspots. The reserve possesses diverse fauna and flora, well representative of the region. The Kozhikamudhi Elephant Camp inside the reserve is a public attraction – earlier this camp was used to train wild elephants. The reserve is also home to one of the world's rarest primates, the Lion-tailed Macaque.

KEY FACTS

Nearest Major Towns
Coimbatore

Habitats
Tropical dry evergreen forest, tropical moist scrub, tropical grassland, tropical secondary scrub

Key Species
White-bellied Shortwing, Broad-tailed Grassbird, Oriental Darter, Pallid Harrier, Great Hornbill, Black-and-orange Flycatcher, Nilgiri Flycatcher, Nilgiri Wood Pigeon, Malabar Parakeet, Malabar Grey Hornbill, Grey-headed Bulbul, Nilgiri Pipit, Wayanad Laughingthrush, Palani Laughingthrush, Rufous Babbler, White-bellied Blue Flycatcher, Crimson-backed Sunbird, Heart-spotted Woodpecker, Great Eared Nightjar, Yellow-browed Bulbul, Brown-backed Needletail, Indian Swiftlet, Red Spurfowl, Mountain Imperial Pigeon

Other Specialities
Tiger, Asian Elephant, Gaur, Lion-tailed Macaque, Nilgiri Tahr, Indian Leopard

Best Time to Visit
October to March

Wayanad Laughingthrush — *Crimson-backed Sunbird*

238

White-bellied Shortwing

Birdwatching Sites

The reserve is home to nearly 300 bird species, including a good number of South Indian endemic birds. Some important birds of the region are the Wayanad Laughingthrush, Sri Lanka Frogmouth, White-bellied Treepie, Malabar Trogon, White-bellied Blue Flycatcher and Black-throated Munia. Other key birds include a large breeding population of Great Hornbills, Malabar Grey Hornbill, Red Spurfowl, Grey Junglefowl, Malabar Parakeet, Heart-spotted Woodpecker, Sri Lanka Bay Owl, Mountain Imperial Pigeon, Indian Swiftlet, Brown-backed Needletail, Rufous Babbler, Yellow-browed Bulbul, Crimson-backed Sunbird, Greater Racket-tailed Drongo, Great Eared Nightjar and Large-billed Leaf Warbler. Many of these birds can be seen on the 1km track from the edge of Karian Shola to the watchtower overlooking a reservoir.

Access & Accommodation

Anamalai is best approached from Pollachi, which is a fair-sized town south of Coimbatore. The usual route is by road from Coimbatore to Pollachi (40km) to the Wildlife Warden's Office, then by road to Top Slip (35km), the entry point to the park. There are regular buses from Coimbatore and Palani to Pollachi, and two buses a day from Pollachi to Top Slip. Hired cars/taxis are also available at Pollachi. The nearest airport for Anamalai is Coimbatore, about 75km away.

There are Forest Rest Houses at Topslip Varagaliar, Amaravathi and Sethumadai. The Ambuli Illam Guesthouse, at Topslip, 2km from the reception centre and equipped with a restaurant, is a good place to stay. Alternatively, it is possible to stay at Pollachi or Coimbatore, where there are several accommodation options, and do day trips to the park. Visitors can book hotels online (www.atrpollachi.com/where-to-stay).

Conservation

Illegal cutting of trees, smuggling of wood, illegal grazing of livestock and poaching of wild animals, including birds, are some of the common and serious conservation challenges faced by Anamalai. The reserve has some areas on the western side adjoining Kerala where entry is relatively easy due to the existence of private estates nearby. People from across the state border often enter the reserve to carry out illegal activities. Since the area is very remote from the Tamil Nadu side, frequent patrolling becomes difficult.

Yellow-browed Bulbul

Udhagamandalam (Ooty)

Udhagamandalam tea gardens

KEY FACTS

Nearest Major Towns
Udhagamandalam (Ooty)

Habitats
Tropical forests, wetland, hills

Key Species
Grey Junglefowl, Common Pigeon, Spotted Dove, Common Coot, Indian Pond Heron, Common Hoopoe, House Crow, Velvet-fronted Nuthatch, Red-whiskered Bulbul, Tickell's Leaf Warbler, Blyth's Reed Warbler, Kashmir Flycatcher, Oriental Magpie Robin, Pied Bushchat, Indian Blackbird, Common Rosefinch, House Sparrow, Oriental Honey Buzzard, Black-winged Kite, Short-toed Snake Eagle, Crested Serpent Eagle, Common Kestrel, Indian Pitta, Heart-spotted Woodpecker, Brown-backed Needletail

Other Specialities
Wild Pig, Nilgiri Tahr, Malabar Giant Squirrel, Grizzled Giant Squirrel, Bonnet Macaque, Indian Chevrotain

Best Time to Visit
Throughout the year

Udhagamandalam, popularly known as Ooty by visitors, is the queen of hill stations. Of many hill stations in South India the town of Ooty, at an altitude of 2,200m, is the largest and most important. It was developed as a summer retreat during the British Raj. Coffee and tea plantations, and trees like conifers, eucalyptus, pine and wattle, dot the hillside in Ooty and its environs. The summer temperature is a maximum of 25 °C and a minimum of 10 °C. During the winter it is a minimum of 5 °C and a maximum of 21 °C, when it is slightly warmer.

Birdwatching Sites

Ooty is a great place in which to see and photograph Western Ghats endemic bird species. You can visit at any time of the year as the main species are resident, though migrants such as the Kashmir Flycatcher and Tickell's Leaf Warbler are only present in winter. The specialties of the region are the Painted Bush Quail, Nilgiri Wood Pigeon, Nilgiri Laughingthrush, Black-and-Orange and Nilgiri Flycatchers, and White-bellied Shortwing. There are several sites and trails for birdwatching. Most of the birding is done on foot, although having a vehicle at your disposal is an added advantage, as it is then possible to cover more of the area and spend more productive time in birding.

Access & Accommodation

Ooty is a major tourist attraction of South India. Coimbatore, about 88km away, is the nearest domestic airport to Ooty; it is well connected to most Indian cities. There are flights from Bangalore (295km from Ooty). Both government and luxury buses from Bangalore, Mysore and Chennai go to Ooty – they are overnight buses and the journey is quite comfortable. A special government bus leaves daily at 5 p.m. and 7 p.m. from

Black-and-orange Flycatcher

Painted Bush Quail

Kozhikode and takes just six hours to reach Ooty. The nearest railhead is Mettupalyam, 40km from Ooty. There are trains from Chennai, Coimbatore, Mysore and Bangalore connecting to Mettupalayam. From here, there is the heritage train known as the Nilgiri Mountain train, which goes uphill on an exciting journey. However, if you wish to travel faster, take a taxi from the railway junction.

There is a wide range of accommodation options in Ooty. If visiting in the summer season, advance booking is a must as Ooty attracts a lot of visitors during this time. Visitors can find hotels online using the popular travel apps Oyo and Airbnb. Shared rooms are also available in Ooty.

Conservation

The site has experienced significant habitat degradation and loss over the past few decades. Some tea estates and many resorts/hotels located in and around Ooty are major sources of pollution in Nilgiris, dumping untreated sewage into the Ooty lake. Industrial and chemical waste is also increasing day by day. Being one of the main travel destinations in South India, Ooty attracts tourists not only from various parts of India but also from abroad. This brings pressure on the pristine beauty of the hill station. There is an immediate need to enforce strict regulations on tourism. Transportation of construction and building materials should be regulated and people should be encouraged to build homes with locally available wood products.

Nilgiri Flycatcher

UTTAR PRADESH

1. Dudhwa National Park
2. Nawabganj Bird Sanctuary
3. National Chambal Wildlife Sanctuary

Uttar Pradesh, a vibrant northern state of India, is a very popular tourist destination. It shares an international border with Nepal to the north. Indian states that share borders with Uttar Pradesh are Uttarakhand, Haryana and Delhi to the north and north-west, Rajasthan in the west, Madhya Pradesh in the south, Chhattisgarh and Jharkhand in the south-east, and Bihar in the east. With over 200 million inhabitants, this is the most populous state in India. It is very diverse, with the Himalayan foothills in the extreme north and the Gangetic Plains in the centre. The state attracts visitors from all over the world to India's most visited sites, the Taj Mahal, and Hinduism's holiest city, Varanasi.

KEY BIRDS

Top 10 Birds
1. Great Slaty Woodpecker
2. Greater Scaup
3. Grey Nightjar
4. Sarus Crane
5. Black Stork
6. White-rumped Vulture
7. Lesser Fish Eagle
8. Finn's Weaver
9. Long-billed Pipit
10. Red-headed Bunting

Climate

Uttar Pradesh comprises three geographic regions: the submontane region lying between the Himalayas and the plains, the vast alluvial Gangetic Plains, and the southern hills and plateau. The state has a diverse range of habitats for birds and biodiversity, including forest areas in the sub-Himalayan terai and the dry deciduous forests of the Bundelkhand regions bordering Madhya Pradesh and Chhattisgarh.

It has a tropical climate with a wide variance in temperatures. There are three main seasons: summer in March–mid-June, the rainy season in mid-June–September, and winter in October–February.

Access, Transportation & Logistics

Uttar Pradesh state is well connected to the rest of India via airways, railways and roadways. The state has the largest railway network in the country, and a large, multi-modal transport system with the largest road network in India. It well connected to its nine neighbouring states and almost all other parts of India through the national highways. Chaudhary Charan Singh International Airport at Lucknow is the biggest and busiest airport. There is another international airport in Varanasi named Lal Bahadur Shastri International Airport. These airports connect Uttar Pradesh with the rest of India through regular flights.

Health & Safety

There are no major health concerns in Uttar Pradesh. As a precaution visitors may take vaccinations for hepatitis A and B and typhoid. Malaria is prevalent in remote areas and prophylaxis should be taken, although this is less of a problem during the dry season. Dengue fever is also present and appropriate precautions should be taken. Leeches, ticks and a variety of biting insects occur in the terai region. Avoid roadside food as it may not follow food-safety norms. Visit a GP or travel clinic 6–8 weeks before departure to make sure you are up to date with any vaccinations.

Birdwatching Highlights

The birdlife of Uttar Pradesh is rich and varied. More than 800 species are found here, including some extremely rare ones. Among the Critically

Baya Weaver

Endangered species, White-rumped, Long-billed and Slender-billed Vultures occur in the state. Bengal and Lesser Floricans, White-headed Duck, Black-necked Stork and Indian Grassbird are some other important birds recorded in this region. Another species that need special mention is Hodgson's Bushchat, which is a winter visitor in the terai region of Uttar Pradesh and some other northeastern states of India.

Swamp Francolins in Dudhwa National Park

Dudhwa National Park

Asian Elephants in Dudhwa National Park

This park is situated in the terai region of the Lakhimpur Kheri district in Uttar Pradesh. The forest region of Lakhimpur Kheri was earlier known as Dudhwa Sanctuary and was a safe place for Swamp Deer conservation. This area and some of its surroundings was declared as Dudhwa National Park in 1977. In 1987, the park was brought under Project Tiger, with the addition of Kishanpur Wildlife Sanctuary. Dudhwa National Park falls under the Terai-Bhabar biogeographic subdivision of the Upper Gangetic Plains. Apart from being home to some of the most endangered animals like the Tiger, Indian Rhinoceros and Hispid Hare, the park is also an excellent home to rare avian species like the Bengal Florican.

KEY FACTS

Nearest Major Towns
Lakhimpur Kheri

Habitats
Moist deciduous forest, dry deciduous forest, tropical grassland

Key Species
White-rumped Vulture, Long-billed Vulture, Slender-billed Vulture, Bengal Florican, Lesser Florican, Spot-billed Pelican, Lesser Adjutant, Marbled Duck, Pallas's Fish Eagle, Greater Spotted Eagle, Sarus Crane, Wood Snipe, Grey-crowned Prinia, Swamp Francolin, White-tailed Stonechat, Striated Babbler, Indian Grassbird, Black-breasted Weaver

Other Specialities
Asian Elephant, Indian Rhinoceros, Tiger, Indian Leopard, Hispid Hare, Swamp Deer, Sloth Bear, Jungle Cat, Fishing Cat, Leopard Cat

Best Time to Visit
October to March

White-tailed Stonechat

Slender-billed Vultures

Great Slaty Woodpeckers

Birdwatching Sites
Dudhwa National Park is rich in avifauna, and more than 400 bird species have been recorded here. Of these, about 40 per cent are resident breeding birds, including important ones such as the Vulnerable Bengal Florican and Swamp Francolin. Dudhwa is also home to many migrant birds. Every year hundreds of waterfowl and several species of leaf-warbler visit the park. The Critically Endangered White-rumped Vulture used to breed in and around the park in large numbers, but its population has crashed since the mid-1990s. Pallas's Fish Eagle is regularly seen in the park. The Spot-billed Pelican and Marbled Duck occur in small numbers.

Access & Accommodation
Dudhwa can be reached from Lucknow via NH 30. Lucknow has an international airport that has regular flights from all major cities of India. Lucknow is about 240km from Dudhwa, roughly five and a half hours by road. There are regular buses from Lucknow to Lakhimpur Kheri, which is about 100km from Dudhwa.

There are many accommodation options available near Dudhwa. Uttar Pradesh ecotourism also has a few cottages and a dormitory in Dudhwa. Reservations can be made online or from the office of the Regional Manager (Ecotourism) Uttar Pradesh Forest Corporation, 21/475 Indira Nagar Lucknow, Tel.: 0522-2358950.

Conservation
Some of the key conservation issues in the Dudhwa area are poaching of wild animals including birds, excessive grazing, illegal cutting of trees and forest fires. Every year in Dudhwa controlled burning is practised to check the growth of herbs and woody species. Burning has two effects in the park. In some situations it is a useful exercise, but in others it is detrimental, altering the habitat to the extent that it becomes less attractive to many endangered species.

Savanna Nightjar

Nawabganj Bird Sanctuary

Nawabganj scene

The Nawabganj Bird Sanctuary, renamed the Shahid Chandra Shekhar Azad Bird Sanctuary, is located on the Kanpur-Lucknow highway, 45km east of Lucknow, near the village of Nawabganj in the Unnao district. The sanctuary consists of a lake and the surrounding environment. It is one of the many wetlands of northern India. In 1974, the Forest Department declared Nawabganj as a sanctuary and planted thousands of trees and built mounds, like in Keoladeo National Park in Bharatpur (Rajasthan). The sanctuary used to attract a good number of waterfowl, but recently its population has decreased dramatically.

KEY FACTS

Nearest Major Towns
Unnao

Habitats
Freshwater swamp

Key Species
Greater Spotted Eagle, Sarus Crane, Painted Stork, Black-necked Stork, Black-headed Ibis, Indian Peafowl, Pied Cuckoo, White-breasted Waterhen, Bronze-winged Jacana, Asian Openbill, Oriental Darter, Little Cormorant, Purple Heron, Cattle Egret, Common Hoopoe, Indian Grey Hornbill, White-throated Kingfisher, Coppersmith Barbet, Brown-headed Barbet, Brown-capped Pygmy Woodpecker, Yellow-crowned Woodpecker, Black-hooded Oriole, Oriental Magpie Robin, Oriental White-eye, Common Tailorbird, Common Chiffchaff, Indian Nuthatch

Other Specialities
Indian Chevrotain, Spotted Deer, Rat Snake, Indian Krait, Checkered Keelback

Best Time to Visit
Throughout the year

Northern Shovelers

Greater Spotted Eagle

Eurasian Spoonbill

Black Stork

Birdwatching Sites
The lake in Nawabganj Bird Sanctuary is important for resident and migratory waterfowl, and more than 200 species have been identified in the region. Some of the common birds seen here are the Oriental Darter, Black-crowned Night Heron, many species of egret and cormorant, and Eurasian Spoonbill. Other resident species include the Grey-headed Swamphen or Swamphen, and Pheasant-tailed and Bronze-winged Jacanas. The lake is also an important area for the Common Coot. Nawabganj additionally has many raptors such as Pallas's Fish Eagle, Greater Spotted Eagle and Western Marsh Harrier. A couple of pairs of Sarus Cranes have also been reported from the sanctuary.

Access & Accommodation
Nawabganj Bird Sanctuary is just 30km from Lucknow International Airport, which has good connectivity with all major cities in India. There are regular buses and other local transport options available from Lucknow.

There are a couple of resorts and hotels around the sanctuary. The Uttar Pradesh Forest Department has also developed a guest house inside the sanctuary, and bookings can be made online (www.upforest.gov.in/StaticPages/nawabganjbird_Accomodation.aspx). Lucknow has various accommodation options, from the most luxurious hotels to less expensive facilities.

Conservation
This wetland has been listed as a high priority, with high ecological and socio-economic potential, but with poor data availability. Pollution from adjoining industries' drains is the biggest conservation challenge, resulting in heavy weed infestation in the lake. Other conservation challenges are the introduction of unsuitable plant species that have adversely affected the whole area, including the birdlife.

Common Coot

National Chambal Wildlife Sanctuary

Uttar Pradesh

Marsh Crocodile and Indian Skimmers

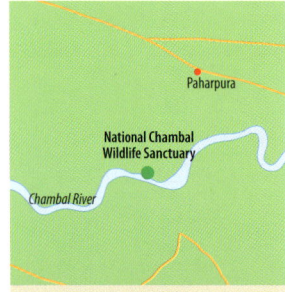

KEY FACTS

Nearest Major Towns
Agra and Etawah

Habitats
Riverine vegetation, wetland

Key Species
Pallas's Fish Eagle, Greater Spotted Eagle, Sarus Crane, Indian Skimmer, Bar-headed Goose, Greylag Goose, Goosander, Common Shelduck, Common Pochard, Ferruginous Duck, Gadwall, Eurasian Wigeon, Indian Spot-billed Duck, Northern Pintail, Common Teal, Comb Duck, Indian Peafowl, Common Quail, Rain Quail, Jungle Bush Quail, Black Francolin, Lesser Flamingo, Little Grebe, Rock Dove, Spotted Dove, Laughing Dove, Chestnut-bellied Sandgrouse, Little Swift, Lesser Coucal, Jacobin Cuckoo, Asian Koel, Common Moorhen, Common Coot

Other Specialities
Garial, Smooth-coated Otter, Wild Pig, Marsh Crocodile, Common Palm Civet, Nilgai, Indian Jackal, Black-naped Hare, Jungle Cat

Best Time to Visit
Throughout the year

This sanctuary lies in three states, Rajasthan, Madhya Pradesh and Uttar Pradesh, stretching from Kota in Rajasthan to the confluence of the Chambal River with the Yamuna River in Uttar Pradesh. In Uttar Pradesh, the sanctuary lies in the Agra and Etawah districts. The sanctuary was created mainly to protect the Endangered Garial and Gangetic Dolphin.

Birdwatching Sites

The Chambal area is very important for both resident and migratory waterfowl, especially the Common Teal, Northern Pintail, Bar-headed Goose, Ruddy Shelduck, Red-crested Pochard and Indian Skimmer. About 300 migratory and resident bird species have been recorded in the area. During the winter

Black Drongo

Indian Thick-knee

Grey-headed Swamphens

Brown Hawk Owl

season the Black-necked Stork, Common and Sarus Cranes, and Black-bellied Tern are found along the river. The sanctuary is one of the most important bird areas in India, being the breeding site of the Indian Skimmer.

Access & Accommodation
The nearest airport for the sanctuary is Agra, which is almost 60km away from Chambal. You can hire a taxi from the airport. There is also a train station in Agra that has good connectivity from all major cities in India.

There are a couple of private resorts and hotels available near the Chambal area. Chambal Safari Lodge (www.chambalsafari.com) is one of the premium resorts; it arranges birdwatching tours.

Conservation
The sanctuary is protected under India's Wildlife Protection Act of 1972. The major problem in this riverine sanctuary is illegal mining of sand. Although the Forest Department is trying to promote tourism, it seems that this needs some more targeted efforts. Another main conservation issue for birdlife is the cultivation of summer vegetables and fruits in the region, which disturbs the nesting islands of Indian Skimmers. Illegal fishing, illegal cutting of trees, poaching and feral dogs are some of the common conservation challenges for the birdlife of the Chambal area. The increasing demand for drinking and irrigation water from nearby towns is a long-term threat.

Indian Skimmer

UTTARAKHAND

1. Chopta, Tungnath & Kedarnath Wildlife Sanctuary
2. Corbett National Park
3. Sattal
4. Pangot

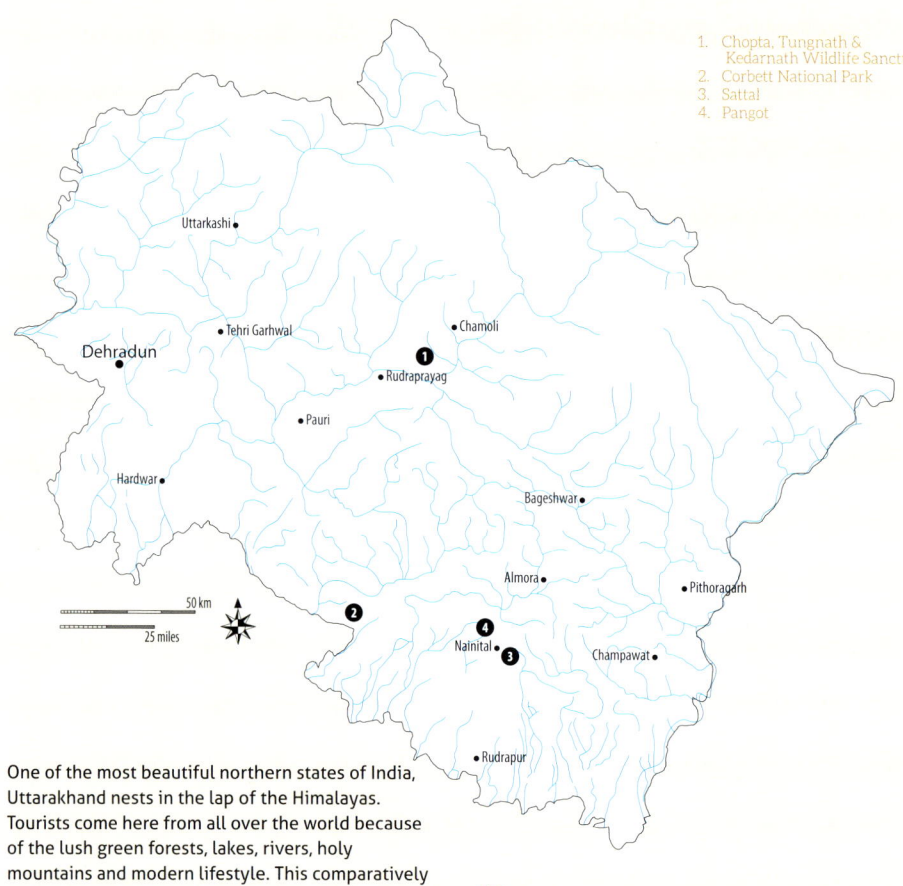

One of the most beautiful northern states of India, Uttarakhand nests in the lap of the Himalayas. Tourists come here from all over the world because of the lush green forests, lakes, rivers, holy mountains and modern lifestyle. This comparatively young state, which is also known as Dev Bhoomi or the land of the gods, was carved out of Uttar Pradesh in November 2000. The state has a total area of 53,483km², of which about 86 per cent is mountainous and 70 per cent is covered by forests. Nanda Devi, the highest peak in India, lies in the Kumaon Himalayas in Uttarakhand.

KEY BIRDS

Top 10 Birds
1. Grey-crowned Prinia
2. Common Hill Partridge
3. Cheer Pheasant
4. Koklass Pheasant
5. Himalayan Monal
6. Black-and-yellow Grosbeak
7. Plain Mountain Finch
8. Red-fronted Serin
9. Yellow-billed Blue Magpie
10. Scarlet Finch

Climate

Uttarakhand has a varied climate, topography and altitudinal range within a small area. The habitat varies from the terai region in the plains, the Bhabhars (the outermost foothills of the Himalaya), the Shivaliks (sub-Himalayan range) and the Lower Himalayas, to the high-altitude Greater Himalayas and Trans-Himalayas. The amazing diversity in habitat from the plains up to the numerous mountain peaks results in marvellous avifaunal diversity. November–May is the season for birding. January–February are the peak winter months, and visitors should expect medium to heavy snow at higher elevation ranges. This time is best for winter migrants, whereas April–May is best for seeing breeding birds.

Access, Transportation & Logistics

Uttarakhand state is well connected to the rest of India via airways, railways and roadways. The Jolly

Spot-bellied Eagle Owl

has rail connectivity with Delhi, Dehradun and Howrah (Kolkata). Trains to Dehradun railway station come from Delhi, Maharashtra, Gujarat, Uttar Pradesh and Punjab. Jan Shatabdi Express, Dehradun Express, Doon Express and Nanda Devi Express are major trains serving Uttarakhand. The state is also well connected by NH 34 and NH 7 with the capital city, New Delhi.

Health & Safety
There are no major health concerns in Uttarakhand. It is a low- to no-risk area of India when it comes to malaria, but do take precautions, especially after the monsoon. Visit a GP or travel clinic 6–8 weeks before departure to make sure you are up to date with any vaccinations. If you are prone to altitude sickness, consult a doctor and carry your medicine with you.

Birdwatching Highlights
The state of Uttarakhand is a paradise for birdwatchers. About 14 IBAs have been selected in this state, and more than 700 bird species have been recorded in these areas. Of these, Baer's Pochard, Slender-billed, White-rumped and Red-headed Vultures, and Himalayan Quail are listed as Critically Endangered. Nine species are endemic to the Western Himalaya, the Western Tragopan, Cheer Pheasant, Himalayan Quail, White-throated and White-cheeked Tits, West Himalayan Bush Warbler, Tytler's Leaf Warbler, Spectacled Finch and Orange Bullfinch.

Grant Airport (DED) in Dehradun is the biggest airport. It is served by regular non-stop flights from Delhi, Bangalore, Mumbai and Chennai. Travelling to Uttarakhand by rail is the cheapest and perhaps most convenient way to visit. The train station in Kathgodam, 35km from Nainital, is the last terminus of the North East railways. Kathgodam train station

Himalayan sunset

Chopta, Tungnath & Kedarnath Wildlife Sanctuary

View from Chopta

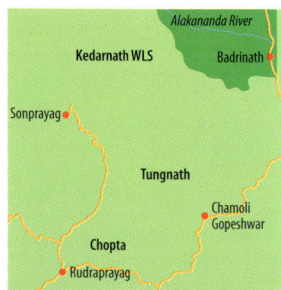

KEY FACTS

Nearest Major Towns
Rudraprayag

Habitats
Himalayan evergreen and coniferous forests

Key Species
Himalayan Swiftlet, Crested Goshawk, Snow Partridge, Himalayan Vulture, Mountain Hawk Eagle, Common Goldeneye, Common Hill Partridge, Koklass Pheasant, Himalayan Monal, Maroon Oriole, Cheer Pheasant, Chestnut-crowned Warbler, Scarlet Finch, Spot-winged Rosefinch, Long-billed Thrush, Pied Thrush

Other Specialities
Himalayan Musk Deer, Indian Leopard, Indian Jackal, Himalayan Langur, Himalayan Black Bear, Jungle Cat, Goral, Bharal

Best Time to Visit
Thoughout the year, though winters are exceedingly cold

Chopta is a small hamlet in the Rudraprayag district of Uttarakhand. This less travelled place offers an imposing view of the Himalayan range, including Trishul, Nanda Devi and Chaukhamba. The altitude of the terrain varies from 800 to 4,000m as you travel north from the Sub-Himalayan ranges to the Lesser Himalayan peaks of the Garhwal Himalayas. The area is drained by the Mandakini River and its tributaries. The land is covered by wet temperate mixed forests, subtropical forests and subtropical mixed forests interspersed with alpine meadows (bugyal) at higher altitudes. Tungnath is the highest Shiva temple in the world at 3,680m, and lies just below the Chandrashila peak. Tungnath is part of the Kedarnath Wildlife Sanctuary (Kedarnath Musk Deer Sanctuary), and is the home of the much sought-after Himalayan Monal – the species voted the most beautiful bird in India.

Birdwatching Sites

Chopta Chopta lies between Kedarnath and Badrinath, at the edge of Kedarnath Wildlife Sanctuary. It takes about eight hours to reach Chopta from Dehradun Airport. It is the best place to see and photograph the Himalayan Monal and Koklass Pheasant, along with many rare and elusive birds like the Kalij Pheasant, Golden Eagle, Bearded Vulture, Upland Buzzard, Himalayan Wood Owl, Spot-winged Rosefinch, Rufous-bellied Woodpecker, European Goldfinch, Ashy Wood Pigeon, Fire-capped Tit, Spotted and Variegated Laughingthrushes, Nepal Wren Babbler, Pied and Long-billed Thrushes, Scarlet Finch, Spot-winged Grosbeak, Golden and White-browed Bush Robins, Coal Tit, Red-fronted Serin, Spot-winged Starling, Himalayan Beautiful

Rusty-flanked Treecreeper

Scarlet Finch

Rosefinch and Yellow-rumped Honeyguide. Many wild mammals are also found here, like the Himalayan Black Bear, Indian Leopard, Wild Pig, Indian Jackal, Indian Fox, Jungle and Marbled Cats, Himalayan Langur, Yellow-throated Marten, Pika, Himalayan Tahr, Goral and Indian Chevrotain.

Tungnath Tungnath is a 3km trek from Chopta. The approximately 1km trek from Tungnath to Chandrashila peak is quite steep. From the Chandrashila peak there is a 360-degree view of the Himalayan ranges. The Himalayan Monal and Snow Partridge can be seen here at a relatively close distance.

Mandal Mandal is a small settlement about 100km from Chopta at the confluence of the Balkhilya Ganga and Atri Ganga Rivers. This is a good place for low-elevation birding. There are many species of finch, particularly the Scarlet Finch and Red-headed Bullfinch. Some of the common birds found in the area are the Yellow-billed Blue Magpie, Grey Treepie, Speckled Piculet, Himalayan Bulbul, Kalij Pheasant and Black Francolin.

Access & Accommodation

Chopta can be reached from Rudraprayag by driving along the Mandakini River via Ukhimath. There is also a route to Chamoli along the bank of the Alaknanda River, which involves turning left towards Gopeshwar, then driving to Chopta via the Mandal forest. There are many options for setting up a base for birding in the area.

Sparrow House Birders' Camp is a homestay in Makku village (Makkumath). Not only is it strategically located, but it comes with one of the best bird guides of the region–Yashpal Negi (Tel.: +91-9412909399 and +91-9720709499).

Duggalbitta (2,360m) is on the way to Chopta and is a good location in which to set up a base, especially in winter. There are some hotels/homestays in Duggalbitta.

Chopta has a wide choice of accommodation, and most birders choose to stay here. Chopta Green View Resort at Bhulkan, about 2km from Chopta, is strategically located for Himalayan Monal spotting.

Mandal (1,720m) is 24km south-east of Chopta and is an alternative site for birders visiting the area.

Conservation

The Chopta and Tungnath area comes under Kedarnath Wildlife Sanctuary. It was declared a protected area in 1972 specially for the conservation of the rare and endangered Himalayan Musk Deer. Since then many measures have been taken for the conservation of flora and fauna of the area. There have been reports of some conservation issues such as encroachment for fuelwood and fodder, and an increase in human-wildlife conflicts. Hunting is not a major issue in the area.

Himalayan Monal

Corbett National Park

Asian Elephants at Corbett National Park

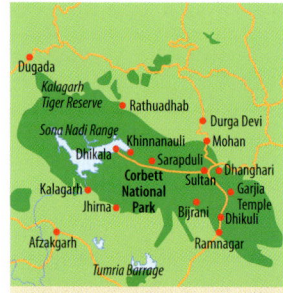

KEY FACTS

Nearest Major Towns
Nainital and Ramnagar

Habitats
Subalpine forest, tropical dry deciduous forest, tropical moist deciduous forest, tropical grassland, reservoir

Key Species
White-rumped Vulture, Slender-billed Vulture, Spot-billed Pelican, Lesser Adjutant, Pallas's Fish Eagle, Greater Spotted Eagle, Eastern Imperial Eagle, Sarus Crane, Sociable Lapwing, Wood Snipe, Grey-crowned Prinia, Oriental Darter, Painted Stork, Black-necked Stork, Black-headed Ibis, Ferruginous Pochard, White-tailed Sea Eagle, Lesser Fish Eagle, Cinereous Vulture

Other Specialities
Asian Elephant, Tiger, Indian Leopard, Spotted Deer, Hog Deer, Sambar, Indian Jackal, Gharial, Marsh Crocodile, Yellow-throated Marten

Best Time to Visit
Thoughout the year, though winters are exceedingly cold

India's first and oldest Tiger Reserve, Corbett National Park and Tiger Reserve is known as the land of roars, trumpets and songs. It is situated 280km north-east of India's capital city, New Delhi, in the Nainital and Pauri Garhwal districts of Uttarakhand. The park was named after the well-known hunter turned naturalist Jim Corbett, who immortalized the area through his famous books. The park comprises an impressive 1,318km² area, which is bounded between the Western Himalayas and the adjacent terai region. The area encompasses a wide range of habitats 400–1,200m above sea level. The Ramganga River and its tributaries are the lifeline of the region and support a hotspot of biodiversity.

Birdwatching Sites

Corbett is justly famous for its healthy population of large mammals, including the Indian Tiger and Asian Elephant; it is also famous for its birds, with more than 600 species recorded in the area. The Cinereous Vulture, Pallas's and Lesser Fish Eagles, Tawny Fish and Spot-bellied Eagle Owls, Great Slaty Woodpecker, Ibisbill, Wallcreeper, Hodgson's Bushchat, Bright-headed Cisticola, Rosy Minivet and Himalayan Rubythroat are just some of the species that are regularly seen in and around the park.

Dhikuli, Garjia and Dhangari Dhikuli and Garjia villages are situated in Ramnagar quite near to Dhangari, which is the entrance to Corbett National Park. Garjia Devi is one of the famous temples of Ramnagar – it is a huge rock placed in the middle of the Kosi River, a perennial river that flows along the long stretches of Jim Corbett National Park. The river forms the eastern boundary of the park from Mohan through Dhikuli to Ramnagar. It is inhabited by a kind of fish called the Mahseer, which attracts many migratory birds.

Lesser Fish Eagle

Dhikala The Dhikala zone, one of the most sought-after zones in Corbett, has some of the most picturesque landscapes. There are a couple of forest rest houses in Dhikala where an overnight stay is permitted. To reach either of them, entry into the forest is through the Dhangarhi gate, where an accommodation permit and identity card have to be shown to the forest of officials. The prime attractions in Dhikala are the Royal Bengal Tiger and the Asian Elephant. The Dhikala zone is named after the Dhikala grassland, which is the largest grassland in the Corbett reserve landscape. Apart from the Tigers and Asian Elephants, the zone is home to a large number of varied wild fauna, including Indian Leopards, Spotted, Sambar and Hog, Terai Langur, Wild Pig, Marsh Crocodile, and many other mammals and reptiles. Dhikala is also famous for its abundant birdlife. The rare winter migrant Hodgson's Bushchat can be seen in the grassland. Some of the other birds that can be seen here include Bright-headed Cisticolas, many larks, pipits, stonechats, Small Pratincoles, Black-necked Storks, Eurasian Spoonbills and Blue-tailed Bee-eaters.

Sambar Road Sambar Road is a small side road that connects Dhikala to Khinnanauli and runs parallel to a stream of the Ramganga River. This road is a good place for birding, and some of the birds that can be seen here are the Brown Fish Owl, Rusty-cheeked Scimitar Babbler, Black-naped Monarch, Indian Paradise-flycatcher and Orange-headed Thrush.

Bijrani Among the six safari zones in Corbett, the Bijrani zone is considered the best zone after the Dhikala zone in which to spot the Tiger. The picturesque landscape, which consists of large *chaurs* (grass fields) and dense sal forest, plenty of waterholes and river streams, is quite enchanting. The zone is quite good for birding. Visitors can easily see Oriental Pied Hornbills, darters, herons, storks, Common Kingfishers and Spangled Drongos.

Durga Devi Zone This zone is situated in the northeastern corner of Corbett. The hilly terrain, with its lush green forest and intense tranquility, defines the magnificent beauty of the Durga Devi tourism zone. This zone is a paradise for birdwatchers, and some of the commonly seen birds here are the Orange-headed Thrush, Pied Cuckoo, Common Hawk Cuckoo and Great Slaty Woodpecker. With some luck you can also locate the Long-tailed Broadbill.

Access & Accommodation

The main access to Corbett National Park is from Ramnagar town, 280km from New Delhi. There are regular buses from New Delhi to Ramnagar, and the road journey takes about 6–8 hours. Ramnagar also has a train station with an overnight train service from New Delhi.

There is a tourist complex at Dhikala and forest rest houses at Bijrani, Malani, Kanda, Gairal, Khinanauli, Sarapduli and Sultan. Visitors are required to make advance bookings before arrival. For online bookings visit http://corbettonline.uk.gov.in. There are also many resorts and hotels outside the park to suit different needs and budgets.

Conservation

Many conservation projects are running in Corbett National Park, like Project Tiger, the Crocodile Conservation Project and Project Elephant, aimed at safeguarding key species. Apart from these important conservation programmes, many initiatives have been started with the help and coordination of the Forest Department and local NGOs. Poaching of wild animals predominantly for their skins and body parts has been an ongoing issue and challenge for the conservation programmes. Uncontrolled recreation and tourism, and illegal cutting of trees, are some of the other issues.

Tawny Fish Owl

Ashy Bulbul

Spotted Forktail

Sattal

Sattal Lake

Sattal is situated in the Nainital district of Uttarakhand. In the local language *sat* means seven and *tal* means lake, and Sattal derives its name from seven interconnected lakes. Sattal lies in the Lower Himalayan Range and is surrounded by pine and oak forests – this whole area is unparalleled for diverse avifauna. The region is home to more than 500 resident and migratory bird species.

Birdwatching Sites

Sattal is one of the best places in which to see middle-elevation western Himalayan bird species. Recently, local birding experts have developed birdwatching hides in Sattal, and these offer good views of Himalayan birds at a close distance without the need for tiring climbing and trekking up the mountains. Many birding trails can also be explored for the birds that do not visit hides, or by those interested in seeing and finding birds in their natural habitats. Some of the common birds found

KEY FACTS

Nearest Major Towns
Nainital

Habitats
Himalayan evergreen and coniferous forests

Key Species
Oriental Turtle Dove, Spotted Dove, Great Barbet, Black Drongo, Ashy Drongo, Black-headed Jay, Red-rumped Swallow, Himalayan Bulbul, Black Bulbul, Streaked Laughingthrush, Russet Sparrow, Red-billed Blue Magpie, Grey-headed Canary-flycatcher, Green-backed Tit, Rufous Sibia, Tickell's Thrush, Brown Wood Owl, Crested Kingfisher, Himalayan Pied Woodpecker, Blue-throated Barbet, Kalij Pheasant, Eurasian Jay

Other Specialities
Yellow-throated Marten, Himalayan Langur, Indian Leopard, Sloth Bear, Wild Pig, Goral

Best Time to Visit
Throughout the year, though winters are exceedingly cold

Rufous-chinned Laughingthrush

Long-tailed Broadbill

White-crested Laughingthrush

Rufous-throated Hill Partridge

in the area are Greater and Lesser Yellow-naped Woodpeckers, Himalayan and Pygmy Woodpeckers, Great and Blue-throated Barbets, many species of laughingthrush, Kalij Pheasant, Brown Wood Owl and Crested Kingfisher.

Access & Accommodation

The best mode of transport to Sattal is via railways. The nearest railway station is Kathgodam, which is about 27km from Sattal. The nearest major airports are Dehradun (about 5–6 hours) and New Delhi (7–8 hours). There is a small airport at Pantnagar which has very limited connectivity. Sattal is a seven-hour road journey from New Delhi via Haldwani.

There are various accommodation options available in Sattal. Some of the best known among birdwatching groups are Sattal Forest Resort (sattalforestresort@gmail.com), Sattal Birding Lodge (www.sattalbirdinglodge.com) and Lama Birding Trails (http://lamabirdingtrails.com). The best season to visit Sattal is October–March.

Conservation

There are no major conservation issues in the Sattal area, but there are some environmental challenges. Sattal is an ecologically fragile group of lakes and is under the impact of heavy environmental degradation. Extensive deforestation, dumping of non-biodegradable waste, and uncontrolled urbanization of the catchments and the nearby forest is harming the ecology of the area. This is resulting in the decrease of many plant and animal species, including birds. There are occasional incidents of poaching in the area.

Spot-winged Starling

Pangot

Pangot landscape

Located just 15km from Nainital, Pangot is a popular tourist destination. The road from Sattal to Pangot passes through the Cheena Peaks and is also the road to Vinayak past Kilbury. The area around Pangot is mixed forests dominated by thick oak, pine and rhododendrons. A large part of the landscape is characterized by dense vegetation. Numerous perennial creeks and streams crisscross the area. It and its surrounding hills are hot places for birding.

Birdwatching Sites

The Pangot area is well known for high-altitude west-central Himalayan birds, and it boasts about 250 bird species. It is a popular destination for the elusive and endangered Cheer

KEY FACTS

Nearest Major Towns
Nainital

Habitat
Himalayan evergreen and coniferous forests

Key Species
Black Francolin, Oriental Turtle Dove, Spotted Dove, Wedge-tailed Green Pigeon, Common Cuckoo, Grey Nightjar, Black Eagle, Great Barbet, Blue-throated Barbet, Brown-fronted Pied Woodpecker, Rufous-bellied Woodpecker, Himalayan Pied Woodpecker, Streak-throated Woodpecker, Slaty-headed Parakeet, Long-tailed Minivet, Long-tailed Shrike, Hume's Leaf Warbler, Greenish Leaf Warbler, Black-throated Tit, Whiskered Yuhina, Oriental White-eye, Hodgson's Treecreeper

Other Specialities
Yellow-throated Marten, Himalayan Langur, Indian Leopard, Sloth Bear, Wild Pig, Goral

Best Time to Visit
Throughout the year, though winters are exceedingly cold

Chestnut-crowned Laughingthrush

Grey Treepie

Rufous-bellied Woodpecker

Common Green Magpie

Pheasant and Koklass Pheasants. Key birds include the Himalayan Vulture, Bearded Vulture, Blue-winged Siva, Rufous-bellied Niltava, Rufous-bellied Woodpecker, Black-headed Jay, many species of laughingthrush, babblers, tits, flycatchers and sunbirds, and a great variety of woodpeckers.

Access & Accommodation
Pangot is a 7–8 hour journey from New Delhi by road and a one-hour (17km) drive from Nainital. Many trains connect Nainital with New Delhi either through Kathgodam or Ramnagar.

There are varied accommodation options in Pangot. Some of the best-known places among birdwatchers are Jungle Lore Birding Lodge (www.pangot.com) and Kafal House (www.kafalhouse.com).

Conservation
There are no major conservation threats in this region, but there are some common issues like deforestation, illegal cutting of trees, and intense grazing pressure by both domestic and wild herbivores on the land. There are occasional cases of poaching, and one of the most key threats is the existence of stray and feral dogs.

Grey-crowned Prinia

Scaly-breasted Wren Babbler

WEST BENGAL

This eastern state of India is wedged between the Himalayas and the Bay of Bengal. The state is bounded in the north by Sikkim and Bhutan, in the east by Assam and Bangladesh, in the south by the Bay of Bengal, and in the west by Odisha, Bihar and Nepal. The state can be divided into two distinct regions, the Himalayas and Sub-Himalayas and their associated forest types in the northern parts, and the thickly populated Gangetic Plains, merging with the sea in the form of vast, tangled mangroves. Kolkata (formerly Calcutta) is the capital city, and one of the oldest and culturally rich cities of India. In the past, Kolkata was a trading post and capital of the British Raj. The Victoria Memorial, dedicated to Britain's queen, is one of the iconic landmarks of Kolkata. There are five national parks and 16 wildlife sanctuaries in the state, and many of them are IBAs.

1. Buxa Tiger Reserve
2. Lava-Neora Valley National Park
3. Mahananda Wildlife Sanctuary
4. Singalila National Park
5. Sundarbans Biosphere Reserve

Climate

The climate of West Bengal is generally humid tropical monsoon. It varies from moist-tropical in the south-east to dry tropical in the south-west, and from subtropical to temperate in the mountains of the north. The temperature ranges from 0 °C to 45 °C. The mean annual rainfall ranges from 900mm in the south-west to 6,000mm in parts of the north. There are eight forest types in the state: tropical semi-evergreen, tropical moist deciduous, tropical dry deciduous, subtropical broadleaved hill, subtropical pine, Himalayan moist temperate, montane wet temperate, littoral and swamp forests. The varied climatic conditions of the state support a great variety and abundance of resident birdlife, and make the area arguably one of the best birding hotspots of its size anywhere in the world.

Access, Transportation & Logistics

West Bengal state is well connected with the rest of India via airways, railways and roadways. The Netaji Subhash Chandra Bose International Airport in Kolkata is the biggest airport, and is served by regular non-stop flights from Delhi, Bangalore, Mumbai, Guwahati and Chennai. Bagdogra is the second international airport in the state, located near Siliguri. This airport is more suited for visitors to the hilly regions of North Bengal. It is also served by regular flights from all major cities of India. The state is well known for its good rail connectivity with different parts of India. The main railheads are Howrah railway station and Sealdah railway station. Both the rail junctions are located in Kolkata. Then there is New Jalpaiguri railway station, near Siliguri. The Darjeeling Toy Train, a UNESCO World Heritage Site, offers a scenic rail journey to the hill station of Darjeeling. An extensive road network also covers the state, and it is connected to different parts of the country via national and state highways.

Health & Safety

There are no major health concerns in West Bengal. The state is a low- to no-risk area of India when it comes to malaria, but do take precautions, especially after the monsoon. Visit a GP or travel clinic 6–8 weeks before departure to make sure you are up to date with any vaccinations. If you are prone to altitude sickness, consult a doctor and carry your medicine with you. Avoid roadside food and drink bottled water. During the summer months keep

---

KEY BIRDS

Top 10 Birds
1. Ruddy Kingfisher
2. Masked Finfoot
3. Golliath Heron
4. Mangrove Pitta
5. Mangrove Whistler
6. Striated Bulbul
7. White-troated Redstart
8. Brown Parrotbill
9. Black-browed Reed Warbler
10. Tibetan Siskin

hydrated, and carry a stole, scarf or cap to protect yourself from direct sunlight. Carry appropriate woollen clothes if heading towards northern hilly regions.

Birdwatching Highlights

The birdlife of West Bengal is very rich, and more than 950 species have been identified in the region, including some very rare ones. White-rumped, Long-billed and Slender-billed Vultures are some of the Critically Endangered species recorded here. Other key birds found in the state include the White-bellied Heron, Greater Adjutant, Baer's Pochard, Ruddy Kingfisher, Buffy Fish Owl, Eastern Imperial Eagle, Masked Finfoot, Black-necked Stork and Bengal Florican. Some of the commonly seen birds are the Stork-billed Kingfisher, Lineated and Blue-throated Barbets, Fulvous-breasted Pied Woodpecker, Black-hooded Oriole, Bronze-winged Jacana, Chestnut-tailed Starling, Richard's Pipit, Bengal Bushlark and Brown Shrike.

Blood Pheasants

Sundarbans mangroves

Buxa Tiger Reserve

Buxa forest

KEY FACTS

Nearest Major Towns
Jalpaiguri

Habitats
Riverine vegetation, tropical dry deciduous forest, tropical dry evergreen forest, tropical grassland

Key Species
White-rumped Vulture, Slender-billed Vulture, Lesser Adjutant, Swamp Francolin, Chestnut-breasted Hill Partridge, Black-necked Crane, Wood Snipe, Indian Skimmer, Rufous-necked Hornbill, Jerdon's Babbler, Black-breasted Parrotbill, Grey-crowned Prinia, Bristled Grassbird, Beautiful Nuthatch, Finn's Weaver, Oriental Darter, Common Teal, Black Baza, Pied Harrier, Eurasian Wigeon, Cinnamon Bittern, Indian Grey Hornbill

Other Specialities
Asian Elephant, Indian Leopard, Gaur, Sambar, Clouded Leopard, Malayan Giant Squirrel, Chinese Pangolin, Hispid Hare

Best Time to Visit
October to March

This reserve is located in the northeastern corner of the Jalpaiguri district. Buxa was declared a Tiger Reserve in 1983 and received the status of national park in 1992. It is about 180km from the city of Jalpaiguri. Its northern boundary lies along the international border with Bhutan, its eastern side forms the interstate boundary with Assam, demarcated by the Sankosh River, and its western and southern sides are bounded by tea gardens and agricultural fields. The reserve is located at the meeting ground of three major biogeographical provinces, namely the Lower Gangetic Plains, Central Himalayas and Brahmaputra Valley. The park holds innumerable flora and fauna. It is a safe ground for the maximum number of elephants, Gaur, Indian Leopard, many species of deer and different bird species. Because of the inaccessible terrain, some parts of the Buxa Hills in the Sinchula Range are still unexplored.

Birdwatching Sites

The reserve is known for its rich avifauna diversity. More than 530 bird species have been recorded here by various renowned organizations and eminent ornithologists, among which about 50 per cent are residents, with some long-distance migrants and some local migrants. The highest species richness was recorded in undisturbed semi-evergreen forests and hill forests. The campsite is in Buxa duar, which is situated at an altitude of about 700m and is approachable on foot from Santrabari (about 3.5km) or by vehicle up to a point, but invariably involving an uphill trek of about 2km. There are many birding trails from Buxa duar, including Buxa duar > Lal Bangla > Tashigaon, Buxa duar > Sadar Bazar and Zero point.

Some of the key birds recorded on the trails are Scarlet and Long-tailed Minivets, Rufous-vented, Stripe-throated and

Plain-backed Thrush

Small Niltava

Rufous Woodpecker

Whiskered Yuhinas, Small Niltava, Greater Rufous-headed and Black-throated Parrotbills, Chestnut-tailed Minla, Red-tailed Minla, Himalayan Cutia, Coral-billed Scimitar Babbler, Little and Slaty-backed Forktails, Great, Oriental Pied, Wreathed and Rufous-necked Hornbills, several warblers, laughingthrushes and tits. Around the waterbody, there are winter migratory birds like the Chinese Spot-billed Duck, Ferruginous Duck and Common Pochards, Common Teal, Eurasian Wigeon, Lesser Whistling Duck, Gadwall, Northern Pintail, Northern Shoveler and Common Coot. The forests of the reserve are also famous for various raptors like the Crested Serpent Eagle, Changeable and Mountain Hawk Eagles, Black Eagle, Jerdon's and Black Bazas, Common Kestrel, Collared Falconet, Eurasian Hobby, Osprey, Oriental Honey Buzzard and Northern Goshawk.

Access & Accommodation

Most tourists visiting Buxa come through either New Jalpaiguri or Alipurduar rail stations. The distance from New Jalpaiguri to Rajabhatkhawa is about 160km and the journey takes close to four hours. Alipurduar is much nearer and the junction rail station is located just 10km from the forest check post at Rajabhatkhawa. Several small and large vehicles to Rajabhatkhawa are available, including shared vehicles at Alipurduar station.

Buxa Jungle Lodge (www.wbfdc.org/rbk-buxa-jungle-lodge) is a good place to stay near the park. It is run by the West Bengal Forest Development Corporation and is situated in Rajabhatkhawa, very close to the forest entry gate. A forest rest house and a couple of homestays are also available near the park. Buxa Valley Homestay (www.buxavalley.com) is one of them.

Conservation

Some of the common conservation issues faced by the park are grazing, poaching of wild animals including birds, illegal cutting of trees and firewood collection. A World Bank-supported eco-development project has been implemented in the reserve to reduce the dependency of local people on its natural resources. There have been many incidents of human-animal conflict, especially due to a good population of wild elephants in the vicinity. Crop damage by elephants is one of the major problems. The Forest Department, under Project Elephant, has enclosed many settlements and villages within electric fences, but their maintenance is an ongoing challenge.

Lava-Neora Valley National Park

Neora Valley forest

This park is situated in the Kalimpong district of West Bengal, and is named after the Neora River, which flows through it. The park is one of three protected areas in the Darjeeling district, and probably the most undisturbed patch of forest in West Bengal. It is unique and ecologically important as it includes a relatively inaccessible patch of late-successional forest with rich diversity and a wide range of environmental gradients. The park has four main habitat types: subtropical mixed broadleaved forests, lower temperate evergreen forests, upper temperate mixed broadleaved forests and rhododendron forests.

KEY FACTS

Nearest Major Towns
Darjeeling and Jalpaiguri

Habitats
Subtropical broadleaved hill forest, montane wet temperate forest, subtropical pine forest, subtropical secondary scrub

Key Species
Eastern Imperial Eagle, Lesser Kestrel, Pale-capped Pigeon, Rufous-necked Hornbill, Rusty-bellied Shortwing, Black-breasted Parrotbill, Beautiful Nuthatch, Satyr Tragopan, Ward's Trogon, Speckled Wood Pigeon, Darjeeling Pied Woodpecker, Long-billed Thrush, White-collared Blackbird, Chestnut Thrush, Gould's Shortwing, Indian Blue Robin

Other Specialities
Red Panda, Indian Leopard, Tiger, Himalayan Black Bear, Sloth Bear, Golden Cat, Leopard Cat, Goral, Serow, Himalayan Flying Squirrel

Best Time to Visit
October to March

Grey-winged Blackbird

White-tailed Robin

Birdwatching Sites

Neora Valley and Lava are very popular among birdwatchers, and more than 345 bird species have been recorded here. There are no forest safaris or forest trail treks inside the park, so most birding is done on foot along the roadside. Hiring a local bird guide is recommended for the best birding experience.

Some of the key species of this area are the Rufous-throated Hill Partridge, Satyr Tragopan, Crimson-breasted, Darjeeling and Bay Woodpeckers, Golden-throated Barbet, Hodgson's Hawk Cuckoo, Lesser Cuckoo, Brown Wood Owl, Ashy Wood Pigeon, Mountain Imperial Pigeon, Jerdon's Baza, Black Eagle, Mountain Hawk Eagle, Dark-throated Thrush, White-browed Bush Robin, White-tailed Robin, Yellow-browed Tit, Striated Bulbul, Chestnut-crowned and Black-faced Warblers, Streak-breasted Scimitar Babbler, Scaly-breasted, Pygmy Wren Babblers, Black-headed Shrike-babbler, Fire-breasted Flowerpecker, Fire-tailed Sunbird and Maroon-backed Accentor.

Access & Accommodation

Bagdogra is the nearest airport, about 100km from the park and well connected to it by the road network. The nearest railway station is Darjeeling railway station, about 30km from the park. The park has two entry points: one is from Lava, the other from Samsing. The trek route through Lava starts from Zero Point, which is about 25km from Lava.

Neora Valley Jungle Camps is a nice resort in Kolakham, run by Help Tourism, offering good accommodation. There are five fully furnished wooden cottages. For bookings contact Neora Valley Jungle Camps, Kolakham Village, Lingesykha Gram Panchayat, P.O. Lava, Kalimpong, Tel.: (Help Tourism, Siliguri): +91-353-2433683/2535896, Tel.: (Help Tourism, Kolkata): +91-33-24550917/24549719. Red Panda Eco Huts is a budget accommodation located in Kolakham. There are four wooden cottages with twin or double beds. For bookings contact Red Panda Eco Huts, Kolakham-Neora Valley National Park (Near Lava), Tel.: (+91) 98 30077180/8420394747/8420434545, Email: redpanda.ecotourism@gmail.com.

Conservation

Like any other protected area in India, Neora-Valley National Park incurs illegal cattle grazing, firewood collection, encroachment on the fringes and poaching. However, due to its inaccessibility and difficult terrain, the biotic pressures are not very acute. The park has some intact forest patches. There is a need for more organized efforts to promote ecotourism and the park's unparalleled biodiversity.

Golden-breasted Fulvetta

White-browed Bush Robin

Streak-breasted Scimitar Babbler

Mahananda Wildlife Sanctuary

West Bengal

Mahananda scene

This sanctuary is located on the west bank of the Teesta River in the southern part of the Darjeeling district of West Bengal. It is named after the Mahananda River. Its terrain extends across hills as well as plains, and because of this the forest types are quite varied and support many forms of rare and elusive flora and fauna.

Birdwatching Sites

The birdlife here is very rich as the sanctuary is at the crossroads of two biomes, the Sino-Himalayan Subtropical Forest and the Indo-Chinese Tropical Moist Forests, with a small portion of the Indo-Gangetic Plain. More than 300 bird species have been recorded in the region. There are regular day safaris in the sanctuary from the Sukna gate both in the morning and in the afternoon. However, no vehicles are provided, and visitors must come in their own vehicles. The total distance of the usual safari route is 8km, of which 2km is unpaved or dirt road.

KEY FACTS

Nearest Major Towns
Darjeeling

Habitats
Tropical semi-evergreen forest, tropical dry deciduous forest, montane wet temperate forest

Key Species
White-rumped Vulture, Slender-billed Vulture, Bengal Florican, Lesser Adjutant, Rufous-necked Hornbill, Black-breasted Parrotbill, Jerdon's Babbler, Oriental Darter, Black-headed Ibis, Red-headed Vulture, Pallid Harrier, Great Hornbill, Rusty-bellied Shortwing, Asian Emerald Cuckoo, Bay Woodpecker, Black-winged Cuckooshrike, Himalayan Bulbul, White-throated Bulbul, Long-billed Thrush, Purple Cochoa, Green Cochoa, Rufous-chinned Laughingthrush

Other Specialities
Asian Elephant, Tiger, Indian Leopard, Himalayan Serow, Leopard Cat

Best Time to Visit
October to March

Asian Emerald Dove

Tickell's Blue Flycatcher

Some of the key species found here are the Rufous-necked Hornbill, Chestnut-winged Cuckoo, Red-headed Trogon, Dollarbird and Sultan Tit. Other birds that can be seen are the Lesser Adjutant, Black-breasted Parrotbill, Jerdon's Babbler, Pallid Harrier, Great Hornbill and Rusty-bellied Shortwing.

Access & Accommodation
The main entry point to the sanctuary is via Sukna, a small village 12km from Siliguri on the Siliguri-Darjeeling road. NH 31 passes through the southern part of the sanctuary. It is perhaps the largest compact block of forested habitats situated at the western end of the migratory route of the Asian Elephant.

There is a Forest Lodge run by the Forest Department in the Sukna area a few metres inside the jungle. For bookings contact Divisional Forest Officer, Wildlife Division – I, Darjeeling, Tel.: (0354) 2257314; or Range Officer, Sukna Range, Mahananda Wildlife Sanctuary, Sukna, Darjeeling, Tel.: (0353) 2573323. Latpanchar, which is part of the sanctuary about 44km from Siliguri, is an alternative for accommodation. There are a couple of good homestay options in Latpanchar, which also offers some good birding opportunities.

Conservation
Poaching and illegal cutting of the forest trees are some of the major conservation issues of Mahananda Wildlife Sanctuary, as it is easily accessible via road. Some other common issues are grazing of livestock, forest fires and a road that cuts through the sanctuary.

Green-backed Tit

Chestnut-winged Cuckoo

Swinhoe's Minivet

Singalila National Park

Red Panda

KEY FACTS

Nearest Major Towns
Darjeeling

Habitats
Montane wet temperate forest, subtropical broadleaved hill, subalpine dry scrub, subalpine forest

Key Species
Greater Spotted Eagle, Chestnut-breasted Hill Partridge, Wood Snipe, Rusty-bellied Shortwing, Beautiful Nuthatch, Chestnut-breasted Hill Partridge, Grey Sibia, Rosy Pipit, Olive-backed Pipit, Grey Wagtail, Eurasian Tree Sparrow, Fire-tailed Sunbird, Black-throated Parrotbill, Himalayan Beautiful Rosefinch, Little Bunting, White-winged Grosbeak, Brown Bullfinch, Maroon-backed Accentor, White-bellied Erpornis

Other Specialities
Red Panda, Indian Leopard, Tiger, Himalayan Black Bear, Sloth Bear, Golden Cat, Leopard Cat, Goral, Serow, Himalayan Flying Squirrel

Best Time to Visit
October to March

This beautiful park is located at the northwestern border of the Darjeeling district, and was established in 1986. The park is part of the Eastern Himalayas and is situated at a high elevation of 1,900–3,630m. It is on the border with the neighbouring country of Nepal to the west, and with Sikkim state to the north. The park is home to the exotic Red Panda and Himalayan Black Bear. It was declared an IBA because it harbours many endemic and threatened bird species. Thick bamboo, oak and rhododendron forest at 2,000–3,640m covers the Singalila Ridge, which runs roughly north to south.

Birdwatching Sites

The park is divided into two ranges, North Range (Rimbick) and South Range (Maneybhanjyang). Within these two ranges, there are four beats, namely, Gairibas, Sandakphu, Rammam and Gorkhey. The popular Sandakphu trek passes through the park. Treks begin at Manebhanjan, which is 30km from Darjeeling. The popular birding trek is from Manebhanjan (2,100m) to Sandakphu (3,640m) and back, with halts at Gairibas (2,620m) or Tumling/Tonglu (3,000m).

More than 300 bird species have been recorded in these areas, and some of the key ones are Blood and Kalij Pheasants, Satyr Tragopan, Himalayan Monal, Common Hill Partridge, Himalayan, White-rumped, Indian and Slender-billed Vultures, Black and Golden Eagles, Mountain Hawk Eagle, Hen Harrier, Brown and Fulvous Parrotbills, Rufous-fronted Tit, Fire-tailed Myzornis and Golden-breasted Fulvetta.

White-throated Redstart

Satyr Tragopan

Access & Accommodation

The park is easily accessible from Darjeeling or Siliguri, West Bengal. While it is possible to drive up to Sandakphu on very rough gravel roads, treks from Manebhanjan with halts at Tumling, Gairibas and Sandakphu are the preferred alternative for the best birding experience. The whole trek should take 5-6 days if the visitor wants to spend some quality time here. There is also a four-wheel-drive motorable road to Sandakphu, so a cab can be hired down to Manebhanjan and Darjeeling via Tomling and Tonglu if needed.

Tourist lodges and/or trekkers huts are located at all strategic halts, as this is a very popular trekking route. Some of the contacts for lodges are:
Tumling (Nepal): Shikhar Lodge, Tumling. Contact Keshav Gurung, Tel.: +91-9232695120/ 9564797551.
Gairibas: Gairibas Trekkers' Hut (very basic). Bookings can be made through DDT, GTA, Silver Fir, Bhanu Sarani, Darjeeling, Tel.: +91-9832375065.
Sandakphu Peak: Sherpa Chalet Lodge (Nepal). Tel.: +91 9332599161/9933488159.

Conservation

Human settlements in the vicinity of Singalila are completely dependent on forest resources for their sustenance and livelihood. They are dependent on the park for firewood, fodder and to some extent timber, which they use for construction. Edible plants, bamboo shoots, various herbs, and other medicinal plants, as well as mushrooms, are also seasonally harvested by the communities for subsistence. Communities around Singalila are already motivated for participatory conservation. The Singalila Environment Protection Committee, comprising community members of Nepal and India, is working actively to reduce the threats of the unsustainable harvesting of forest resources.

Common Hill Partridge

Sundarbans Biosphere Reserve

Tiger at Sundarbans

The Sundarbans is the largest delta covered with mangrove forests and vast saline mudflats in the world. It is named after the mangrove plant locally known as Sundari. It contains the largest mangrove forest with perhaps the largest Tiger population in the world. It is a World Heritage Site and Biosphere Reserve, both in India and in Bangladesh, and is known for its rich biodiversity, especially with regard to fish, crustaceans, reptiles and birds. Along with the Tiger, there are many more mammals in the park, including Jungle and Fishing Cats, Spotted Deer, macaques, Wild Pigs, flying foxes, pangolins and Indian Grey Mongooses. It is well-known for rare birds like the Mangrove Pitta, Buffy Fish Owl, Brown-winged Kingfisher and Goliath Heron.

KEY FACTS

Nearest Major Towns
Kolkata

Habitats
Tropical moist deciduous forest, littoral forest, wetland

Key Species
Greater Adjutant, Lesser Adjutant, Baer's Pochard, Pallas's Fish Eagle, Greater Spotted Eagle, Swamp Francolin, Masked Finfoot, Oriental Darter, Black-necked Stork, Ferruginous Pochard, Black-bellied Tern

Other Specialities
Tiger, Wild Pig, Fishing Cat, Jungle Cat, Spotted Deer, Irrawaddy Dolphin, Estuarine Crocodile, Eurasian Otter, River Terrapin, Russell's Viper

Best Time to Visit
October to March

Mangroves

Buffy Fish Owl

Birdwatching Sites

The park area is divided into two ranges, and each range is further subdivided into beats. The park also has floating watch stations and camps to protect the property from poachers. It has a landscape dominated by great tidal creeks and waterways, and the only way to access and enjoy the area is in motorized boats that come in various sizes and shapes. There are also many watchtowers for bird as well as wild mammal observations. The fringes of the Sundarbans play host to many local endemics, and visitors are advised to spend time on land outside the core area before venturing into the park.

More than 370 bird species have been recorded in this region, and some of the rare and elusive ones are the Masked Finfoot, Spoon-billed Sandpiper, Buffy Fish Owl and Greater Adjutant. Of the 12 kingfisher species found in India, six occur in this IBA, Common, Brown-winged, Stork-billed, White-throated, Black-capped and Collared Kingfishers. Sunderbans is one of two sites where the Mangrove Whistler is found. Commonly seen birds include the Ferruginous Pochard, Lesser Adjutant, Black-necked Stork, Oriental Darter, Black-bellied Tern, Pallas's Fish Eagle and Greater Spotted Eagle.

Access & Accommodation

The Sundarbans is accessed from Kolkata city by travelling towards either the south-east or the south-west. The south-west route goes through Diamond Harbour to Kakdwip and Namkhana. The south-east route is more popular. There is a 97km drive through wetlands and agricultural land to reach Gadkhali jetty. The tourist facilities are a two-hour boat ride from Gadkhali.

Forest lodges and rest houses are available at Sajnekhali, Bakkhali and Piyali for accommodation. Lodging facilities are also available at Sundarbans Tiger Camp on Dayapur Island, a resort overlooking the national park, and at Sundarbans Jungle Camp on Bali Island run by the Help Tourism Group in collaboration with local communities and members of the Bali Nature and Wildlife Conservation Society. For bookings contact Tourism Centre, West Bengal Tourism Development Corporation Limited, 3/2-BBD Bag (Near Great Eastern Hotel), Kolkata, Tel.: +91 33 2210 3199/+91 33 2248 8271.

Conservation

Despite its status as a World Heritage Site, Biosphere Reserve, Tiger Reserve, National Park and Wildlife Sanctuary, the Sundarbans suffers from many anthropogenic problems. Illegal fishing, cutting of mangroves, poaching and encroachment are the biggest threats. With the help of the eco-development committees and forest departments some eco-conservation, eco-development, training, education and research programmes have been started to increase alternate income and reduce human-animal conflicts. The Mangrove Interpretation Centre has been established at Sajnekhali to make local people and visitors aware of the importance of nature conservation in general and mangrove ecosystems particularly.

Brown-winged Kingfisher

Goliath Heron

Recommended Reading

Ali, S. & Ripley, D. (1964–74) *Handbook of the Birds of India & Pakistan* (Vols 1–10). Bombay: OUP.
Ali, S. & Ripley, D. (1983) *A Pictorial Guide to the Birds of the Indian Subcontinent*. Bombay: OUP.
Ali, S. (1985) *Fall of a Sparrow*. Bombay: OUP.
Ali, S. (1996) *The Book of Indian Birds*. (12th ed.) New Delhi: BNHS & OUP.
Daniels, R. (1992) *Birds of Urban South India*. Bangalore: *Indian Institute of Science*.
Daniels, R. (1996) *Field Guide to the Birds of Southwest India*. New Delhi: OUP.
Grewal, B. (1995) *Birds of the Indian Subcontinent*. Hong Kong: *The Guidebook Company Limited*.
Grewal, B. (1995, 2008) *Birds of India & Nepal*. London: New Holland.
Grewal, B., Sen S. & Sreenivasan (2012) *Birds of Nagaland*. Kohima: Nagaland Govt.
Grewal, B., Bhatia, G. (2014) *A Naturalist's Guide to the Birds of India*. Oxford: John Beaufoy Publisher.
Grimmet, R. Inskipp, T., & Inskipp, C. (1998) *Birds of the Indian Subcontinent*. UK: A&C Black.
Kazmierczak, K. & van Perlo, B. (2000) *A Field Guide to the Birds of the Indian Subcontinent*. UK: Pica Press.
Lainer, H. (2004) *Birds of Goa*. Goa: The Other India Bookstore.
Naoroji, R. (2006) *Birds of Prey of the Indian Subcontinent*. London: Christopher Helm.
Neelakantan, K.K. Sashikumar & Venugopalan (1993) *A Book of Kerala Birds*. Trivandrum: WWF.
Rasmussen, P. & Anderton, J. (2005) *Birds of South Asia*: *The Ripley Guide*. Barcelona: Lynx Editions.
Woodcock, M. (1980) *Collins Handguide to the Birds of the Indian Subcontinent*. London: Collins.

Acknowledgements

Bikram Grewal and Bhanu Singh would like to thank the following people for their help in producing this book: Aidan Fonseca, Alka Vaidya, Alpa Seth, Amala Shah, Amano Samarpan, Amit Sharma, Anand Arya, Arka Sarkar, Arpit Bansal, Arpit Deomurari, Arshiya Sethi, Avinash Khemka, Barnita Newar, Biswapriya Rahut, Bittu Sahgal, Chinmay Agnihotri, Clement M Francis, D B Newar, Deboshree Gogoi, Dhritiman Chatterjee, Dhairya Jhaveri, Dushyant Prashar, Falguna Shah, Ganesh Adhikari, Garima Bhatia, Goutam Mohapatra, Gopinath Kollur, Gururaj Moorching, Ingo Waschaies, James Eaton, Kalyan Verma, Keya Khare, Khushboo Sharma, Kintoo Dhawan, Koshy Koshy, Kunan Naik, Lila Newar, Lima Rosalind, M V Shreeram, Maheep Singh, Manjula Mathur, Mohit Mishra, Mukund Thakkar, Natalia Parkina, Nikhat Grewal, Nikhil Devasar, Niranjan Sant, Nitin Bhardwaj, Nitin Srinivasamurthy, P B Biju, P S Anand, Panchami Manoo Ukil, Parth Satvalekar, Prasanna Parab, Pia Sethi, Pratap Chahal, Purushottam Lad, Rahul Sharma, Ramki Sreenivasan, Raju Kasambe, Rathika Ramasamy, Samiha Grewal Mishra, SarwanDeep Singh, Satinder Sharma, Satya Singh, Savio Fonseca, Shahnawaz Khan, Shashank Dalvi, Shyam Ghate, Sonu Anand, Sugata Goswami, Subhoranjan Sen, Sujan Chatterjee, Sumit K Sen, Supriyo Samanta, Swati Kulkarni, Tapas Misra, Tejas Naik, Tripta Sood, Uma & Ganesh, Urmi Nath, Vaibhav Deshmukh, Vaidehi Gunjal, Vivek Sinha. We would especially like to thank Alpana Khare and her team of Neeraj Aggarwal, Raghuvir Khare, Diya Kapur, Samiha Grewal Mishra, Ajmal Nayab Siddiqui, Raj Kishore Beck and Harshit Bist.

The authors would like to especially thank Rajyashree Dutt for all her help with the book.

Conservation & Birdwatching Organizations

India Region
Sanctuary Asia
   www.sanctuaryasia.com
Bombay Natural History Society (BNHS)
   www.bnhs.org
Conservation India
   www.conservationindia.org
Centre for Environment Education
   www.ceeindia.org
Indian Bird Conservation Network
   http://ibcn.in
National Centre for Biological Sciences
   www.ncbs.res.in
Nature Conservation Foundation
   http://ncf-india.org
Nature Science Initiative
   www.naturescienceinitiative.org
Salim Ali Centre for Ornithology and Natural History (SACON)
   www.sacon.in
Wildlife Institute of India
   www.wii.gov.in

North India
Delhi Bird Watching Society
   www.delhibird.com
Wildlife Conservation and Birds Club of Ladakh
   www.wcbcl.org
Kalpavriksha
   https://kalpavriksh.org

North-east India
Aaranyak
   www.aaranyak.org
Sundarvan Nature Discovery Centre
   http://sundarvan.org
Balipara Foundation
   https://baliparafoundation.com

West India
Maharashtra Pakshimitra
   www.pakshimitra.org
Goa Bird Conservation Network
   www.birdsofgoa.org
Ela Foundation
   www.elafoundation.org
India BirdRaces
   www.indiabirdraces.in

South India
Nilgiri Natural History Society
   http://nnhs.in
Prakriti Wildlife Club, IIT-Madras
   https://home.iitm.ac.in/prakriti/prakriti
Mysore Nature
   www.mysorenature.org
Salem Ornithological Foundation
   https://salembirds.wordpress.com
Madras Naturalists' Society
   www.blackbuck.org.in
Deccan Birders (formerly Bird Watcher Society of Andhra Pradesh)
   www.bsap.in
The Nature Trust
   www.thenaturetrust.org

Facebook Groups
Bengalbird
Birds Planet – Strabo Pixel Club
Biodiversity of Paschim Bardhaman
Bird Katta
Birds of Maharashtra
Birds of Mumbai
Birds of Orissa
Birds of Pune
Birds of Thane & Raigad District
Birds of Vidarbha
Birds of Goa
Birdwatchers of Kerala
Coastal Karnataka Birdwatchers Network
CLaW-Conservation Lenses & Wildlife
Hyderabad Birding Pals
Indian Birds
kolkatabirds.com
Manipal Birders Club
The Bhubaneswar Bird Walks

Other Bird Watching Resources
Bird Sounds
Xeno-canto (xeno-canto.org)
AVoCet (avocet.zoology.msu.edu)
Macaulay Library of Natural Sounds
   (macaulaylibrary.org)

Bird Image Repositories
The Oriental Bird Images (Database of Bird Pictures–http://orientalbirdimages.org)

Bird Records Submission
Ebird (ebird.org)

Web Portals
Kolkatabirds (www.kolkatabirds.com)
India Nature Watch
   (www.indianaturewatch.net)

Checklist of Birds of India

Taxonomy follows Praveen J., Jayapal, R., & Pittie, A., 2016. A checklist of the birds of India. Indian BIRDS, 11:113-170.

IUCN Red List Status
LC	Least Concern	CR	Critically Endangered
VU	Vulnerable	NT	Near Threatened
EN	Endangered	DD	Data Deficient

English Name	Scientific Name	IUCN Category
Anseriformes		
Anatidae (Ducks, Geese and Swans)		
Fulvous Whistling Duck	Dendrocygna bicolor	LC
Lesser Whistling Duck	Dendrocygna javanica	LC
White-headed Duck	Oxyura leucocephala	EN
Mute Swan	Cygnus olor	LC
Tundra Swan	Cygnus columbianus	LC
Whooper Swan	Cygnus cygnus	LC
Red-breasted Goose	Branta ruficollis	VU
Bar-headed Goose	Anser indicus	LC
Greylag Goose	Anser anser	LC
Bean Goose	Anser fabalis	LC
Greater White-fronted Goose	Anser albifrons	LC
Lesser White-fronted Goose	Anser erythropus	VU
Long-tailed Duck	Clangula hyemalis	VU
Common Goldeneye	Bucephala clangula	LC
Smew	Mergellus albellus	LC
Common Merganser	Mergus merganser	LC
Red-breasted Merganser	Mergus serrator	LC
Common Shelduck	Tadorna tadorna	LC
Ruddy Shelduck	Tadorna ferruginea	LC
Marbled Teal	Marmaronetta angustirostris	VU
White-winged Wood Duck	Asarcornis scutulata	EN
Red-crested Pochard	Netta rufina	LC
Common Pochard	Aythya ferina	VU
Baer's Pochard	Aythya baeri	CR
Ferruginous Duck	Aythya nyroca	NT
Tufted Duck	Aythya fuligula	LC
Greater Scaup	Aythya marila	LC
Pink-headed Duck	Rhodonessa caryophyllacea	CR
Garganey	Spatula querquedula	LC
Northern Shoveler	Spatula clypeata	LC
Baikal Teal	Sibirionetta formosa	LC
Falcated Duck	Mareca falcata	NT
Gadwall	Mareca strepera	LC
Eurasian Wigeon	Mareca penelope	LC
Chinese Spot-billed Duck	Anas zonorhyncha	LC
Indian Spot-billed Duck	Anas poecilorhyncha	LC
Mallard	Anas platyrhynchos	LC
Andaman Teal	Anas albogularis	VU
Northern Pintail	Anas acuta	LC
Common Teal	Anas crecca	LC
Comb Duck	Sarkidiornis melanotos	LC
Mandarin Duck	Aix galericulata	LC
Cotton Teal	Nettapus coromandelianus	LC
Galliformes		
Megapodiidae (Megapode)		
Nicobar Megapode	Megapodius nicobariensis	VU
Phasianidae (Partridges and Pheasants)		
Common Hill Partridge	Arborophila torqueola	LC
Rufous-throated Hill Partridge	Arborophila rufogularis	LC
White-cheeked Hill Partridge	Arborophila atrogularis	NT
Chestnut-breasted Hill Partridge	Arborophila mandellii	VU
Indian Peafowl	Pavo cristatus	LC
Green Peafowl	Pavo muticus	EN
Grey Peacock Pheasant	Polyplectron bicalcaratum	LC
Common Quail	Coturnix coturnix	LC
Japanese Quail	Coturnix japonica	NT
Rain Quail	Coturnix coromandelica	LC

English Name	Scientific Name	IUCN Category
Blue-breasted Quail	Synoicus chinensis	LC
Himalayan Snowcock	Tetraogallus himalayensis	LC
Tibetan Snowcock	Tetraogallus tibetanus	LC
Chukar Partridge	Alectoris chukar	LC
Snow Partridge	Lerwa lerwa	LC
Jungle Bush Quail	Perdicula asiatica	LC
Rock Bush Quail	Perdicula argoondah	LC
Painted Bush Quail	Perdicula erythrorhyncha	LC
Manipur Bush Quail	Perdicula manipurensis	EN
Himalayan Quail	Ophrysia superciliosa	CR
Black Francolin	Francolinus francolinus	LC
Painted Francolin	Francolinus pictus	LC
Chinese Francolin	Francolinus pintadeanus	LC
Grey Francolin	Francolinus pondicerianus	LC
Swamp Francolin	Francolinus gularis	VU
Mountain Bamboo Partridge	Bambusicola fytchii	LC
Red Junglefowl	Gallus gallus	LC
Grey Junglefowl	Gallus sonneratii	LC
Himalayan Monal	Lophophorus impejanus	LC
Sclater's Monal	Lophophorus sclateri	VU
Western Tragopan	Tragopan melanocephalus	VU
Satyr Tragopan	Tragopan satyra	NT
Blyth's Tragopan	Tragopan blythii	VU
Temminck's Tragopan	Tragopan temminckii	LC
Mrs Hume's Pheasant	Syrmaticus humiae	NT
Cheer Pheasant	Catreus wallichii	VU
Kalij Pheasant	Lophura leucomelanos	LC
Tibetan Partridge	Perdix hodgsoniae	LC
Red Spurfowl	Galloperdix spadicea	LC
Painted Spurfowl	Galloperdix lunulata	LC
Koklass Pheasant	Pucrasia macrolopha	LC
Blood Pheasant	Ithaginis cruentus	LC
Phoenicopteriformes		
Phoenicopteridae (Flamingos)		
Greater Flamingo	Phoenicopterus roseus	LC
Lesser Flamingo	Phoeniconaias minor	NT
Podicipediformes		
Podicipedidae (Grebes)		
Little Grebe	Tachybaptus ruficollis	LC
Red-necked Grebe	Podiceps grisegena	LC
Great Crested Grebe	Podiceps cristatus	LC
Horned Grebe	Podiceps auritus	VU
Black-necked Grebe	Podiceps nigricollis	LC
Columbiformes		
Columbidae (Pigeons and Doves)		
Rock Pigeon	Columba livia	LC
Hill Pigeon	Columba rupestris	LC
Snow Pigeon	Columba leuconota	LC
Yellow-eyed Pigeon	Columba eversmanni	VU
Common Wood Pigeon	Columba palumbus	LC
Speckled Wood Pigeon	Columba hodgsonii	LC
Ashy Wood Pigeon	Columba pulchricollis	LC
Nilgiri Wood Pigeon	Columba elphinstonii	VU
Pale-capped Pigeon	Columba punicea	VU
Andaman Wood Pigeon	Columba palumboides	NT
European Turtle Dove	Streptopelia turtur	VU
Oriental Turtle Dove	Streptopelia orientalis	LC
Eurasian Collared Dove	Streptopelia decaocto	LC
Red Collared Dove	Streptopelia tranquebarica	LC
Spotted Dove	Streptopelia chinensis	LC
Laughing Dove	Streptopelia senegalensis	LC
Barred Cuckoo Dove	Macropygia unchall	LC
Andaman Cuckoo Dove	Macropygia rufipennis	LC
Orange-breasted Green Pigeon	Treron bicinctus	LC
Ashy-headed Green Pigeon	Treron phayrei	NT
Grey-fronted Green Pigeon	Treron affinis	LC
Andaman Green Pigeon	Treron chloropterus	NT
Thick-billed Green Pigeon	Treron curvirostra	LC
Yellow-legged Green Pigeon	Treron phoenicopterus	LC
Pin-tailed Green Pigeon	Treron apicauda	LC
Wedge-tailed Green Pigeon	Treron sphenurus	LC

English Name	Scientific Name	IUCN Category
Nicobar Pigeon	*Caloenas nicobarica*	NT
Asian Emerald Dove	*Chalcophaps indica*	LC
Namaqua Dove	*Oena capensis*	LC
Green Imperial Pigeon	*Ducula aenea*	LC
Mountain Imperial Pigeon	*Ducula badia*	LC
Pied Imperial Pigeon	*Ducula bicolor*	LC
Pterocliformes		
Pteroclidae (Sandgrouse)		
Tibetan Sandgrouse	*Syrrhaptes tibetanus*	LC
Pallas's Sandgrouse	*Syrrhaptes paradoxus*	LC
Pin-tailed Sandgrouse	*Pterocles alchata*	LC
Chestnut-bellied Sandgrouse	*Pterocles exustus*	LC
Spotted Sandgrouse	*Pterocles senegallus*	LC
Black-bellied Sandgrouse	*Pterocles orientalis*	LC
Painted Sandgrouse	*Pterocles indicus*	LC
Phaethontiformes		
Phaethontidae (Tropicbirds)		
Red-billed Tropicbird	*Phaethon aethereus*	LC
Red-tailed Tropicbird	*Phaethon rubricauda*	LC
White-tailed Tropicbird	*Phaethon lepturus*	LC
Caprimulgiformes		
Podargidae (Frogmouths)		
Sri Lanka Frogmouth	*Batrachostomus moniliger*	LC
Hodgson's Frogmouth	*Batrachostomus hodgsoni*	LC
Caprimulgidae (Nightjars)		
Great Eared Nightjar	*Lyncornis macrotis*	LC
Grey Nightjar	*Caprimulgus jotaka*	LC
Jungle Nightjar	*Caprimulgus indicus*	LC
European Nightjar	*Caprimulgus europaeus*	LC
Sykes's Nightjar	*Caprimulgus mahrattensis*	LC
Jerdon's Nightjar	*Caprimulgus atripennis*	LC
Large-tailed Nightjar	*Caprimulgus macrurus*	LC
Andaman Nightjar	*Caprimulgus andamanicus*	LC
Indian Nightjar	*Caprimulgus asiaticus*	LC
Savanna Nightjar	*Caprimulgus affinis*	LC
Hemiprocnidae (Treeswift)		
Crested Treeswift	*Hemiprocne coronata*	LC
Apodidae (Swifts)		
White-rumped Spinetail	*Zoonavena sylvatica*	LC
White-throated Needletail	*Hirundapus caudacutus*	LC
Silver-backed Needletail	*Hirundapus cochinchinensis*	LC
Brown-backed Needletail	*Hirundapus giganteus*	LC
Glossy Swiftlet	*Collocalia esculenta*	LC
Himalayan Swiftlet	*Aerodramus brevirostris*	LC
Indian Swiftlet	*Aerodramus unicolor*	LC
Edible-nest Swiftlet	*Aerodramus fuciphagus*	LC
Asian Palm Swift	*Cypsiurus balasiensis*	LC
Alpine Swift	*Tachymarptis melba*	LC
Dark-rumped Swift	*Apus acuticauda*	VU
Pacific Swift	*Apus pacificus*	LC
Nepal House Swift	*Apus nipalensis*	LC
Indian House Swift	*Apus affinis*	LC
Common Swift	*Apus apus*	LC
Cuculiformes		
Cuculidae (Cuckoos)		
Greater Coucal	*Centropus sinensis*	LC
Andaman Coucal	*Centropus andamanensis*	LC
Lesser Coucal	*Centropus bengalensis*	LC
Sirkeer Malkoha	*Taccocua leschenaultii*	LC
Blue-faced Malkoha	*Phaenicophaeus viridirostris*	LC
Green-billed Malkoha	*Phaenicophaeus tristis*	LC
Pied Cuckoo	*Clamator jacobinus*	LC
Chestnut-winged Cuckoo	*Clamator coromandus*	LC
Asian Koel	*Eudynamys scolopaceus*	LC
Horsfield's Bronze Cuckoo	*Chalcites basalis*	LC
Asian Emerald Cuckoo	*Chrysococcyx maculatus*	LC
Violet Cuckoo	*Chrysococcyx xanthorhynchus*	LC
Banded Bay Cuckoo	*Cacomantis sonneratii*	LC
Plaintive Cuckoo	*Cacomantis merulinus*	LC
Grey-bellied Cuckoo	*Cacomantis passerinus*	LC
Fork-tailed Drongo Cuckoo	*Surniculus dicruroides*	LC

English Name	Scientific Name	IUCN Category
Square-tailed Drongo Cuckoo	Surniculus lugubris	LC
Large Hawk Cuckoo	Hierococcyx sparverioides	LC
Common Hawk Cuckoo	Hierococcyx varius	LC
Whistling Hawk Cuckoo	Hierococcyx nisicolor	LC
Indian Cuckoo	Cuculus micropterus	LC
Common Cuckoo	Cuculus canorus	LC
Himalayan Cuckoo	Cuculus saturatus	LC
Lesser Cuckoo	Cuculus poliocephalus	LC
Gruiformes		
Rallidae (Crakes, Rails)		
Andaman Crake	Rallina canningi	LC
Slaty-legged Crake	Rallina eurizonoides	LC
Western Water Rail	Rallus aquaticus	LC
Eastern Water Rail	Rallus indicus	LC
Slaty-breasted Rail	Lewinia striata	LC
Corncrake	Crex crex	LC
Spotted Crake	Porzana porzana	LC
Ruddy-breasted Crake	Zapornia fusca	LC
Brown Crake	Zapornia akool	LC
Little Crake	Zapornia parva	LC
Baillon's Crake	Zapornia pusilla	LC
Black-tailed Crake	Zapornia bicolor	LC
White-breasted Waterhen	Amaurornis phoenicurus	LC
White-browed Crake	Amaurornis cinerea	LC
Watercock	Gallicrex cinerea	LC
Grey-headed Swamphen	Porphyrio porphyrio	LC
Common Moorhen	Gallinula chloropus	LC
Common Coot	Fulica atra	LC
Heliornithidae		
Masked Finfoot	Heliopais personatus	EN
Gruidae (Cranes)		
Siberian Crane	Leucogeranus leucogeranus	CR
Sarus Crane	Antigone antigone	VU
Demoiselle Crane	Grus virgo	LC
Common Crane	Grus grus	LC
Black-necked Crane	Grus nigricollis	VU
Otidiformes		
Otididae (Bustards)		
Great Indian Bustard	Ardeotis nigriceps	CR
Little Bustard	Tetrax tetrax	NT
Bengal Florican	Houbaropsis bengalensis	CR
Lesser Florican	Sypheotides indicus	EN
MacQueen's Bustard	Chlamydotis macqueenii	VU
Gaviiformes		
Gaviidae (Divers)		
Red-throated Diver	Gavia stellata	LC
Black-throated Diver	Gavia arctica	LC
Procellariiformes		
Oceanitidae (Storm-petrels)		
Wilson's Storm-petrel	Oceanites oceanicus	LC
White-faced Storm-petrel	Pelagodroma marina	LC
Black-bellied Storm-petrel	Fregetta tropica	LC
Hydrobatidae (Storm Petrels)		
Swinhoe's Storm-petrel	Hydrobates monorhis	NT
Procellariidae (Shearwaters and Petrels)		
Barau's Petrel	Pterodroma baraui	EN
Wedge-tailed Shearwater	Ardenna pacifica	LC
Short-tailed Shearwater	Ardenna tenuirostris	LC
Flesh-footed Shearwater	Ardenna carneipes	NT
Streaked Shearwater	Calonectris leucomelas	NT
Cory's Shearwater	Calonectris borealis	LC
Tropical Shearwater	Puffinus bailloni	LC
Persian Shearwater	Puffinus persicus	LC
Jouanin's Petrel	Bulweria fallax	NT
Ciconiiformes		
Ciconiidae (Storks)		
Greater Adjutant	Leptoptilos dubius	EN
Lesser Adjutant	Leptoptilos javanicus	VU
Painted Stork	Mycteria leucocephala	NT
Asian Openbill	Anastomus oscitans	LC
Black Stork	Ciconia nigra	LC

English Name	Scientific Name	IUCN Category
Woolly-necked Stork	*Ciconia episcopus*	VU
European White Stork	*Ciconia ciconia*	LC
Black-necked Stork	*Ephippiorhynchus asiaticus*	NT
Pelecaniformes		
Pelecanidae (Pelicans)		
Great White Pelican	*Pelecanus onocrotalus*	LC
Spot-billed Pelican	*Pelecanus philippensis*	NT
Dalmatian Pelican	*Pelecanus crispus*	NT
Ardeidae (Bitterns, Herons and Egrets)		
Eurasian Bittern	*Botaurus stellaris*	LC
Little Bittern	*Ixobrychus minutus*	LC
Yellow Bittern	*Ixobrychus sinensis*	LC
Cinnamon Bittern	*Ixobrychus cinnamomeus*	LC
Black Bittern	*Ixobrychus flavicollis*	LC
White-eared Night Heron	*Gorsachius magnificus*	EN
Malayan Night Heron	*Gorsachius melanolophus*	LC
Black-crowned Night Heron	*Nycticorax nycticorax*	LC
Striated Heron	*Butorides striata*	LC
Indian Pond Heron	*Ardeola grayii*	LC
Chinese Pond Heron	*Ardeola bacchus*	LC
Javan Pond Heron	*Ardeola speciosa*	LC
Cattle Egret	*Bubulcus ibis*	LC
Grey Heron	*Ardea cinerea*	LC
White-bellied Heron	*Ardea insignis*	CR
Goliath Heron	*Ardea goliath*	LC
Purple Heron	*Ardea purpurea*	LC
Great Egret	*Ardea alba*	LC
Intermediate Egret	*Ardea intermedia*	LC
Little Egret	*Egretta garzetta*	LC
Western Reef Egret	*Egretta gularis*	LC
Pacific Reef Egret	*Egretta sacra*	LC
Chinese Egret	*Egretta eulophotes*	VU
Threskiornithidae (Ibises and Spoonbill)		
Black-headed Ibis	*Threskiornis melanocephalus*	NT
Eurasian Spoonbill	*Platalea leucorodia*	LC
Indian Black Ibis	*Pseudibis papillosa*	LC
Glossy Ibis	*Plegadis falcinellus*	LC
Suliformes		
Fregatidae (Frigatebirds)		
Lesser Frigatebird	*Fregata ariel*	LC
Great Frigatebird	*Fregata minor*	LC
Christmas Island Frigatebird	*Fregata andrewsi*	CR
Sulidae (Boobies)		
Red-footed Booby	*Sula sula*	LC
Brown Booby	*Sula leucogaster*	LC
Masked Booby	*Sula dactylatra*	LC
Phalacrocoracidae (Cormorants)		
Little Cormorant	*Microcarbo niger*	LC
Great Cormorant	*Phalacrocorax carbo*	LC
Indian Cormorant	*Phalacrocorax fuscicollis*	LC
Anhingidae (Darter)		
Oriental Darter	*Anhinga melanogaster*	NT
Charadriiformes		
Burhinidae (Thick-knees)		
Indian Thick-knee	*Burhinus indicus*	LC
Great Thick-knee	*Esacus recurvirostris*	NT
Beach Thick-knee	*Esacus magnirostris*	NT
Haematopodidae (Oystercatchers)		
Eurasian Oystercatcher	*Haematopus ostralegus*	NT
Ibidorhynchidae (Ibisbill)		
Ibisbill	*Ibidorhyncha struthersii*	LC
Recurvirostridae (Stilts and Avocets)		
Pied Avocet	*Recurvirostra avosetta*	LC
Black-winged Stilt	*Himantopus himantopus*	LC
Charadriidae (Plovers and Lapwings)		
Grey Plover	*Pluvialis squatarola*	LC
Eurasian Golden Plover	*Pluvialis apricaria*	LC
Pacific Golden Plover	*Pluvialis fulva*	LC
Common Ringed Plover	*Charadrius hiaticula*	LC
Long-billed Plover	*Charadrius placidus*	LC
Little Ringed Plover	*Charadrius dubius*	LC

English Name	Scientific Name	IUCN Category
Kentish Plover	Charadrius alexandrinus	LC
Lesser Sand Plover	Charadrius mongolus	LC
Greater Sand Plover	Charadrius leschenaultii	LC
Caspian Plover	Charadrius asiaticus	LC
Oriental Plover	Charadrius veredus	LC
Northern Lapwing	Vanellus vanellus	NT
River Lapwing	Vanellus duvaucelii	NT
Yellow-wattled Lapwing	Vanellus malabaricus	LC
Grey-headed Lapwing	Vanellus cinereus	LC
Red-wattled Lapwing	Vanellus indicus	LC
Sociable Lapwing	Vanellus gregarius	CR
White-tailed Lapwing	Vanellus leucurus	LC
Rostratulidae (Painted-snipe)		
Greater Painted-snipe	Rostratula benghalensis	LC
Jacanidae (Jacanas)		
Pheasant-tailed Jacana	Hydrophasianus chirurgus	LC
Bronze-winged Jacana	Metopidius indicus	LC
Scolopacidae (Snipes, Sandpipers and other Waders)		
Whimbrel	Numenius phaeopus	LC
Eurasian Curlew	Numenius arquata	NT
Bar-tailed Godwit	Limosa lapponica	NT
Black-tailed Godwit	Limosa limosa	NT
Ruddy Turnstone	Arenaria interpres	LC
Great Knot	Calidris tenuirostris	EN
Red Knot	Calidris canutus	NT
Ruff	Calidris pugnax	LC
Broad-billed Sandpiper	Calidris falcinellus	LC
Sharp-tailed Sandpiper	Calidris acuminata	LC
Curlew Sandpiper	Calidris ferruginea	NT
Temminck's Stint	Calidris temminckii	LC
Long-toed Stint	Calidris subminuta	LC
Spoon-billed Sandpiper	Calidris pygmaea	CR
Red-necked Stint	Calidris ruficollis	NT
Sanderling	Calidris alba	LC
Dunlin	Calidris alpina	LC
Little Stint	Calidris minuta	LC
Buff-breasted Sandpiper	Calidris subruficollis	NT
Pectoral Sandpiper	Calidris melanotos	LC
Asian Dowitcher	Limnodromus semipalmatus	NT
Long-billed Dowitcher	Limnodromus scolopaceus	LC
Eurasian Woodcock	Scolopax rusticola	LC
Solitary Snipe	Gallinago solitaria	LC
Wood Snipe	Gallinago nemoricola	VU
Pintail Snipe	Gallinago stenura	LC
Swinhoe's Snipe	Gallinago megala	LC
Great Snipe	Gallinago media	NT
Common Snipe	Gallinago gallinago	LC
Jack Snipe	Lymnocryptes minimus	LC
Terek Sandpiper	Xenus cinereus	LC
Common Sandpiper	Actitis hypoleucos	LC
Green Sandpiper	Tringa ochropus	LC
Grey-tailed Tattler	Tringa brevipes	NT
Spotted Redshank	Tringa erythropus	LC
Common Greenshank	Tringa nebularia	LC
Common Redshank	Tringa totanus	LC
Wood Sandpiper	Tringa glareola	LC
Marsh Sandpiper	Tringa stagnatilis	LC
Red-necked Phalarope	Phalaropus lobatus	LC
Red Phalarope	Phalaropus fulicarius	LC
Turnicidae (Buttonquails)		
Small Buttonquail	Turnix sylvaticus	LC
Yellow-legged Buttonquail	Turnix tanki	LC
Barred Buttonquail	Turnix suscitator	LC
Dromadidae (Crab Plover)		
Crab-plover	Dromas ardeola	LC
Glareolidae (Coursers & Pratincoles)		
Jerdon's Courser	Rhinoptilus bitorquatus	CR
Cream-coloured Courser	Cursorius cursor	LC
Indian Courser	Cursorius coromandelicus	LC
Collared Pratincole	Glareola pratincola	LC
Oriental Pratincole	Glareola maldivarum	LC

English Name	Scientific Name	IUCN Category
Little Pratincole	*Glareola lactea*	LC
Stercorariidae (Skuas)		
Long-tailed Skua	*Stercorarius longicaudus*	LC
Arctic Skua	*Stercorarius parasiticus*	LC
Pomarine Skua	*Stercorarius pomarinus*	LC
South Polar Skua	*Stercorarius maccormicki*	LC
Brown Skua	*Stercorarius antarcticus*	LC
Laridae (Gulls, Terns, Noddies & Skimmer)		
Brown Noddy	*Anous stolidus*	LC
Lesser Noddy	*Anous tenuirostris*	LC
Black Noddy	*Anous minutus*	LC
White Tern	*Gygis alba*	LC
Indian Skimmer	*Rynchops albicollis*	VU
Black-legged Kittiwake	*Rissa tridactyla*	VU
Sabine's Gull	*Xema sabini*	LC
Slender-billed Gull	*Chroicocephalus genei*	LC
Brown-headed Gull	*Chroicocephalus brunnicephalus*	LC
Black-headed Gull	*Chroicocephalus ridibundus*	LC
Little Gull	*Hydrocoloeus minutus*	LC
Franklin's Gull	*Leucophaeus pipixcan*	LC
White-eyed Gull	*Ichthyaetus leucophthalmus*	LC
Sooty Gull	*Ichthyaetus hemprichii*	LC
Pallas's Gull	*Ichthyaetus ichthyaetus*	LC
Mew Gull	*Larus canus*	LC
Lesser Black-backed Gull	*Larus fuscus*	LC
Caspian Gull	*Larus cachinnans*	LC
Mongolian Gull	*Larus smithsonianus mongolicus*	LC
Sooty Tern	*Onychoprion fuscatus*	LC
Bridled Tern	*Onychoprion anaethetus*	LC
Little Tern	*Sternula albifrons*	LC
Saunders's Tern	*Sternula saundersi*	LC
Gull-billed Tern	*Gelochelidon nilotica*	LC
Caspian Tern	*Hydroprogne caspia*	LC
Whiskered Tern	*Chlidonias hybrida*	LC
White-winged Tern	*Chlidonias leucopterus*	LC
Black Tern	*Chlidonias niger*	LC
River Tern	*Sterna aurantia*	NT
Roseate Tern	*Sterna dougallii*	LC
Black-naped Tern	*Sterna sumatrana*	LC
Common Tern	*Sterna hirundo*	LC
White-cheeked Tern	*Sterna repressa*	LC
Arctic Tern	*Sterna paradisaea*	LC
Black-bellied Tern	*Sterna acuticauda*	EN
Lesser Crested Tern	*Thalasseus bengalensis*	LC
Sandwich Tern	*Thalasseus sandvicensis*	LC
Greater Crested Tern	*Thalasseus bergii*	LC
Accipitriformes		
Pandionidae (Osprey)		
Osprey	*Pandion haliaetus*	LC
Accipitridae (Hawks, Kites and Eagles)		
Black-winged Kite	*Elanus caeruleus*	LC
Oriental Honey Buzzard	*Pernis ptilorhynchus*	LC
Jerdon's Baza	*Aviceda jerdoni*	LC
Black Baza	*Aviceda leuphotes*	LC
Bearded Vulture	*Gypaetus barbatus*	NT
Egyptian Vulture	*Neophron percnopterus*	EN
Crested Serpent Eagle	*Spilornis cheela*	LC
Nicobar Serpent Eagle	*Spilornis klossi*	NT
Andaman Serpent Eagle	*Spilornis elgini*	VU
Short-toed Snake Eagle	*Circaetus gallicus*	LC
Red-headed Vulture	*Sarcogyps calvus*	CR
Himalayan Vulture	*Gyps himalayensis*	NT
White-rumped Vulture	*Gyps bengalensis*	CR
Indian Vulture	*Gyps indicus*	CR
Slender-billed Vulture	*Gyps tenuirostris*	CR
Griffon Vulture	*Gyps fulvus*	LC
Cinereous Vulture	*Aegypius monachus*	NT
Mountain Hawk Eagle	*Nisaetus nipalensis*	LC
Changeable Hawk Eagle	*Nisaetus cirrhatus*	LC
Rufous-bellied Eagle	*Lophotriorchis kienerii*	NT
Black Eagle	*Ictinaetus malaiensis*	LC

English Name	Scientific Name	IUCN Category
Indian Spotted Eagle	Clanga hastata	VU
Greater Spotted Eagle	Clanga clanga	VU
Tawny Eagle	Aquila rapax	VU
Steppe Eagle	Aquila nipalensis	EN
Eastern Imperial Eagle	Aquila heliaca	VU
Golden Eagle	Aquila chrysaetos	LC
Bonelli's Eagle	Aquila fasciata	LC
Booted Eagle	Hieraaetus pennatus	LC
Western Marsh Harrier	Circus aeruginosus	LC
Eastern Marsh Harrier	Circus spilonotus	LC
Hen Harrier	Circus cyaneus	LC
Pallid Harrier	Circus macrourus	NT
Pied Harrier	Circus melanoleucos	LC
Montagu's Harrier	Circus pygargus	LC
Crested Goshawk	Accipiter trivirgatus	LC
Shikra	Accipiter badius	LC
Nicobar Sparrowhawk	Accipiter butleri	VU
Chinese Sparrowhawk	Accipiter soloensis	LC
Japanese Sparrowhawk	Accipiter gularis	LC
Besra	Accipiter virgatus	LC
Eurasian Sparrowhawk	Accipiter nisus	LC
Northern Goshawk	Accipiter gentilis	LC
White-bellied Sea Eagle	Haliaeetus leucogaster	LC
Pallas's Fish Eagle	Haliaeetus leucoryphus	EN
White-tailed Sea Eagle	Haliaeetus albicilla	LC
Lesser Fish Eagle	Icthyophaga humilis	NT
Grey-headed Fish Eagle	Icthyophaga ichthyaetus	NT
Brahminy Kite	Haliastur indus	LC
Red Kite	Milvus milvus	NT
Black Kite	Milvus migrans	LC
White-eyed Buzzard	Butastur teesa	LC
Grey-faced Buzzard	Butastur indicus	LC
Common Buzzard	Buteo buteo	LC
Himalayan Buzzard	Buteo refectus	LC
Long-legged Buzzard	Buteo rufinus	LC
Upland Buzzard	Buteo hemilasius	LC
Strigiformes		
Tytonidae (Barn Owls)		
Oriental Bay Owl	Phodilus badius	LC
Sri Lanka Bay Owl	Phodilus assimilis	LC
Eastern Grass Owl	Tyto longimembris	LC
Andaman Barn Owl	Tyto deroepstorffi	LC
Common Barn Owl	Tyto alba	LC
Strigidae (Owls)		
Brown Hawk Owl	Ninox scutulata	LC
Hume's Hawk Owl	Ninox obscura	LC
Andaman Hawk Owl	Ninox affinis	LC
Collared Owlet	Glaucidium brodiei	LC
Asian Barred Owlet	Glaucidium cuculoides	LC
Jungle Owlet	Glaucidium radiatum	LC
Spotted Owlet	Athene brama	LC
Little Owl	Athene noctua	LC
Forest Owlet	Heteroglaux blewitti	EN
Boreal Owl	Aegolius funereus	LC
Andaman Scops Owl	Otus balli	LC
Mountain Scops Owl	Otus spilocephalus	LC
Eurasian Scops Owl	Otus scops	LC
Pallid Scops Owl	Otus brucei	LC
Oriental Scops Owl	Otus sunia	LC
Nicobar Scops Owl	Otus alius	NT
Indian Scops Owl	Otus bakkamoena	LC
Collared Scops Owl	Otus lettia	LC
Northern Long-eared Owl	Asio otus	LC
Short-eared Owl	Asio flammeus	LC
Mottled Wood Owl	Strix ocellata	LC
Brown Wood Owl	Strix leptogrammica	LC
Tawny Owl	Strix aluco	LC
Himalayan Owl	Strix nivicolum	LC
Eurasian Eagle Owl	Bubo bubo	LC
Indian Eagle Owl	Bubo bengalensis	LC
Spot-bellied Eagle Owl	Bubo nipalensis	LC

English Name	Scientific Name	IUCN Category
Dusky Eagle Owl	*Bubo coromandus*	LC
Brown Fish Owl	*Ketupa zeylonensis*	LC
Tawny Fish Owl	*Ketupa flavipes*	LC
Buffy Fish Owl	*Ketupa ketupu*	LC
Trogoniformes		
Trogonidae (Trogons)		
Malabar Trogon	*Harpactes fasciatus*	LC
Red-headed Trogon	*Harpactes erythrocephalus*	LC
Ward's Trogon	*Harpactes wardi*	NT
Bucerotiformes		
Bucerotidae (Hornbills)		
Great Hornbill	*Buceros bicornis*	VU
Malabar Pied Hornbill	*Anthracoceros coronatus*	NT
Oriental Pied Hornbill	*Anthracoceros albirostris*	LC
Austen's Brown Hornbill	*Anorrhinus austeni*	NT
Malabar Grey Hornbill	*Ocyceros griseus*	LC
Indian Grey Hornbill	*Ocyceros birostris*	LC
Rufous-necked Hornbill	*Aceros nipalensis*	VU
Narcondam Hornbill	*Rhyticeros narcondami*	EN
Wreathed Hornbill	*Rhyticeros undulatus*	VU
Upupidae (Hoopoe)		
Common Hoopoe	*Upupa epops*	LC
Piciformes		
Indicatoridae (Honeyguide)		
Yellow-rumped Honeyguide	*Indicator xanthonotus*	NT
Picidae (Woodpeckers)		
Eurasian Wryneck	*Jynx torquilla*	LC
White-browed Piculet	*Sasia ochracea*	LC
Speckled Piculet	*Picumnus innominatus*	LC
Heart-spotted Woodpecker	*Hemicircus canente*	LC
Himalayan Golden-backed Woodpecker	*Dinopium shorii*	LC
Common Golden-backed Woodpecker	*Dinopium javanense*	LC
Lesser Golden-backed Woodpecker	*Dinopium benghalense*	LC
Pale-headed Woodpecker	*Gecinulus grantia*	LC
Rufous Woodpecker	*Micropternus brachyurus*	LC
Greater Yellow-naped Woodpecker	*Chrysophlegma flavinucha*	LC
Lesser Yellow-naped Woodpecker	*Picus chlorolophus*	LC
Streak-throated Woodpecker	*Picus xanthopygaeus*	LC
Grey-headed Woodpecker	*Picus canus*	LC
Scaly-bellied Woodpecker	*Picus squamatus*	LC
Great Slaty Woodpecker	*Mulleripicus pulverulentus*	VU
White-bellied Woodpecker	*Dryocopus javensis*	LC
Andaman Woodpecker	*Dryocopus hodgei*	VU
Bay Woodpecker	*Blythipicus pyrrhotis*	LC
Greater Golden-backed Woodpecker	*Chrysocolaptes guttacristatus*	LC
White-naped Woodpecker	*Chrysocolaptes festivus*	LC
Brown-capped Pygmy Woodpecker	*Dendrocopos nanus*	LC
Grey-capped Pygmy Woodpecker	*Dendrocopos canicapillus*	LC
Fulvous-breasted Pied Woodpecker	*Dendrocopos macei*	LC
Spot-breasted Pied Woodpecker	*Dendrocopos analis*	LC
Stripe-breasted Pied Woodpecker	*Dendrocopos atratus*	LC
Brown-fronted Pied Woodpecker	*Dendrocopos auriceps*	LC
Yellow-fronted Pied Woodpecker	*Leiopicus mahrattensis*	LC
Crimson-breasted Pied Woodpecker	*Dendrocopos cathpharius*	LC
Darjeeling Pied Woodpecker	*Dendrocopos darjellensis*	LC
Himalayan Pied Woodpecker	*Dendrocopos himalayensis*	LC
Sind Pied Woodpecker	*Dendrocopos assimilis*	LC
Great Spotted Woodpecker	*Dendrocopos major*	LC
Rufous-bellied Woodpecker	*Dendrocopos hyperythrus*	LC
Megalaimidae (Barbets)		
Great Barbet	*Psilopogon virens*	LC
Brown-headed Barbet	*Psilopogon zeylanicus*	LC
Lineated Barbet	*Psilopogon lineatus*	LC
White-cheeked Barbet	*Psilopogon viridis*	LC
Golden-throated Barbet	*Psilopogon franklinii*	LC
Blue-throated Barbet	*Psilopogon asiaticus*	LC
Blue-eared Barbet	*Psilopogon duvaucelii*	LC
Malabar Barbet	*Psilopogon malabaricus*	LC
Coppersmith Barbet	*Psilopogon haemacephalus*	LC
Coraciiformes		
Meropidae (Bee-eaters)		

English Name	Scientific Name	IUCN Category
Blue-bearded Bee-eater	*Nyctyornis athertoni*	LC
Green Bee-eater	*Merops orientalis*	LC
Chestnut-headed Bee-eater	*Merops leschenaulti*	LC
Blue-throated Bee-eater	*Merops viridis*	LC
Blue-tailed Bee-eater	*Merops philippinus*	LC
Blue-cheeked Bee-eater	*Merops persicus*	LC
European Bee-eater	*Merops apiaster*	LC
Coraciidae (Rollers)		
Indian Roller	*Coracias benghalensis*	LC
European Roller	*Coracias garrulus*	LC
Dollarbird	*Eurystomus orientalis*	LC
Alcedinidae (Kingfishers)		
Oriental Dwarf Kingfisher	*Ceyx erithaca*	LC
Blue-eared Kingfisher	*Alcedo meninting*	LC
Blyth's Kingfisher	*Alcedo hercules*	NT
Common Kingfisher	*Alcedo atthis*	LC
Crested Kingfisher	*Megaceryle lugubris*	LC
Pied Kingfisher	*Ceryle rudis*	LC
Stork-billed Kingfisher	*Pelargopsis capensis*	LC
Brown-winged Kingfisher	*Pelargopsis amauroptera*	NT
Ruddy Kingfisher	*Halcyon coromanda*	LC
White-throated Kingfisher	*Halcyon smyrnensis*	LC
Black-capped Kingfisher	*Halcyon pileata*	LC
Collared Kingfisher	*Todiramphus chloris*	LC
Falconiformes		
Falconidae (Falcons)		
Collared Falconet	*Microhierax caerulescens*	LC
Pied Falconet	*Microhierax melanoleucos*	LC
Lesser Kestrel	*Falco naumanni*	LC
Common Kestrel	*Falco tinnunculus*	LC
Red-necked Falcon	*Falco chicquera*	NT
Red-footed Falcon	*Falco vespertinus*	NT
Amur Falcon	*Falco amurensis*	LC
Merlin	*Falco columbarius*	LC
Eurasian Hobby	*Falco subbuteo*	LC
Oriental Hobby	*Falco severus*	LC
Laggar Falcon	*Falco jugger*	NT
Saker Falcon	*Falco cherrug*	EN
Peregrine Falcon	*Falco peregrinus*	LC
Psittaciformes		
Psittaculidae (Parakeets)		
Grey-headed Parakeet	*Psittacula finschii*	NT
Slaty-headed Parakeet	*Psittacula himalayana*	LC
Blossom-headed Parakeet	*Psittacula roseata*	NT
Plum-headed Parakeet	*Psittacula cyanocephala*	LC
Red-breasted Parakeet	*Psittacula alexandri*	NT
Lord Derby's Parakeet	*Psittacula derbiana*	NT
Long-tailed Parakeet	*Psittacula longicauda*	VU
Malabar Parakeet	*Psittacula columboides*	LC
Alexandrine Parakeet	*Psittacula eupatria*	NT
Rose-ringed Parakeet	*Psittacula krameri*	LC
Nicobar Parakeet	*Psittacula caniceps*	NT
Vernal Hanging Parrot	*Loriculus vernalis*	LC
Pittidae (Pittas)		
Blue-naped Pitta	*Hydrornis nipalensis*	LC
Blue Pitta	*Hydrornis cyaneus*	LC
Indian Pitta	*Pitta brachyura*	LC
Blue-winged Pitta	*Pitta moluccensis*	LC
Mangrove Pitta	*Pitta megarhyncha*	NT
Hooded Pitta	*Pitta sordida*	LC
Eurylaimidae (Broadbills)		
Long-tailed Broadbill	*Psarisomus dalhousiae*	LC
Silver-breasted Broadbill	*Serilophus lunatus*	LC
Campephagidae (Cuckooshrikes, Minivets)		
White-bellied Minivet	*Pericrocotus erythropygius*	LC
Small Minivet	*Pericrocotus cinnamomeus*	LC
Grey-chinned Minivet	*Pericrocotus solaris*	LC
Short-billed Minivet	*Pericrocotus brevirostris*	LC
Long-tailed Minivet	*Pericrocotus ethologus*	LC
Scarlet Minivet	*Pericrocotus flammeus*	LC
Ashy Minivet	*Pericrocotus divaricatus*	LC

English Name	Scientific Name	IUCN Category
Swinhoe's Minivet	*Pericrocotus cantonensis*	LC
Rosy Minivet	*Pericrocotus roseus*	LC
Large Cuckooshrike	*Coracina javensis*	LC
Andaman Cuckooshrike	*Coracina dobsoni*	NT
Pied Triller	*Lalage nigra*	LC
Black-winged Cuckooshrike	*Lalage melaschistos*	LC
Black-headed Cuckooshrike	*Lalage melanoptera*	LC
Pachycephalidae (Whistler)		
Mangrove Whistler	*Pachycephala cinerea*	LC
Vireonidae (Shrike-babblers)		
Black-headed Shrike-babbler	*Pteruthius rufiventer*	LC
Himalayan Shrike-babbler	*Pteruthius ripleyi*	LC
Blyth's Shrike-babbler	*Pteruthius aeralatus*	LC
Green Shrike-babbler	*Pteruthius xanthochlorus*	LC
Black-eared Shrike-babbler	*Pteruthius melanotis*	LC
Clicking Shrike-babbler	*Pteruthius intermedius*	LC
White-bellied Erpornis	*Erpornis zantholeuca*	LC
Oriolidae (Orioles)		
Maroon Oriole	*Oriolus traillii*	LC
Black-hooded Oriole	*Oriolus xanthornus*	LC
Eurasian Golden Oriole	*Oriolus oriolus*	LC
Indian Golden Oriole	*Oriolus kundoo*	LC
Black-naped Oriole	*Oriolus chinensis*	LC
Slender-billed Oriole	*Oriolus tenuirostris*	LC
Artamidae (Woodswallows)		
White-breasted Woodswallow	*Artamus leucoryn*	LC
Ashy Woodswallow	*Artamus fuscus*	LC
Vangidae (Flycatcher-shrike and Woodshrikes)		
Bar-winged Flycatcher-shrike	*Hemipus picatus*	LC
Malabar Woodshrike	*Tephrodornis sylvicola*	LC
Large Woodshrike	*Tephrodornis virgatus*	LC
Common Woodshrike	*Tephrodornis pondicerianus*	LC
Aegithinidae (Ioras)		
Common Iora	*Aegithina tiphia*	LC
Marshall's Iora	*Aegithina nigrolutea*	LC
Dicruridae (Drongos)		
Black Drongo	*Dicrurus macrocercus*	LC
Ashy Drongo	*Dicrurus leucophaeus*	LC
White-bellied Drongo	*Dicrurus caerulescens*	LC
Crow-billed Drongo	*Dicrurus annectens*	LC
Bronzed Drongo	*Dicrurus aeneus*	LC
Lesser Racket-tailed Drongo	*Dicrurus remifer*	LC
Hair-crested Drongo	*Dicrurus hottentottus*	LC
Andaman Drongo	*Dicrurus andamanensis*	LC
Greater Racket-tailed Drongo	*Dicrurus paradiseus*	LC
Rhipiduridae (Fantails)		
White-browed Fantail	*Rhipidura aureola*	LC
White-spotted Fantail	*Rhipidura albogularis*	LC
White-throated Fantail	*Rhipidura albicollis*	LC
Laniidae (Shrikes)		
Brown Shrike	*Lanius cristatus*	LC
Red-backed Shrike	*Lanius collurio*	LC
Red-tailed Shrike	*Lanius phoenicuroides*	LC
Isabelline Shrike	*Lanius isabellinus*	LC
Burmese Shrike	*Lanius collurioides*	LC
Bay-backed Shrike	*Lanius vittatus*	LC
Long-tailed Shrike	*Lanius schach*	LC
Grey-backed Shrike	*Lanius tephronotus*	LC
Lesser Grey Shrike	*Lanius minor*	LC
Great Grey Shrike	*Lanius excubitor*	LC
Woodchat Shrike	*Lanius senator*	LC
Masked Shrike	*Lanius nubicus*	LC
Corvidae (Jays, Magpies, and Crows)		
Rufous Treepie	*Dendrocitta vagabunda*	LC
Grey Treepie	*Dendrocitta formosae*	LC
White-bellied Treepie	*Dendrocitta leucogastra*	LC
Collared Treepie	*Dendrocitta frontalis*	LC
Andaman Treepie	*Dendrocitta bayleii*	VU
Red-billed Chough	*Pyrrhocorax pyrrhocorax*	LC
Yellow-billed Chough	*Pyrrhocorax graculus*	LC
Yellow-billed Blue Magpie	*Urocissa flavirostris*	LC

English Name	Scientific Name	IUCN Category
Red-billed Blue Magpie	Urocissa erythroryncha	LC
Common Green Magpie	Cissa chinensis	LC
Eurasian Jay	Garrulus glandarius	LC
Black-headed Jay	Garrulus lanceolatus	LC
Eurasian Magpie	Pica pica	LC
Spotted Nutcracker	Nucifraga caryocatactes	LC
Large-spotted Nutcracker	Nucifraga multipunctata	LC
Eurasian Jackdaw	Corvus monedula	LC
Rook	Corvus frugilegus	LC
Common Raven	Corvus corax	LC
Pied Crow	Corvus albus	LC
Carrion Crow	Corvus corone	LC
House Crow	Corvus splendens	LC
Large-billed Crow	Corvus macrorhynchos	LC
Monarchidae (Monarch flycatchers)		
Black-naped Monarch	Hypothymis azurea	LC
Amur Paradise-flycatcher	Terpsiphone incei	LC
Blyth's Paradise-flycatcher	Terpsiphone affinis	LC
Indian Paradise-flycatcher	Terpsiphone paradisi	LC
Dicaeidae (Flowerpeckers)		
Yellow-bellied Flowerpecker	Dicaeum melanozanthum	LC
Yellow-vented Flowerpecker	Dicaeum chrysorrheum	LC
Thick-billed Flowerpecker	Dicaeum agile	LC
Pale-billed Flowerpecker	Dicaeum erythrorhynchos	LC
Nilgiri Flowerpecker	Dicaeum concolor	LC
Plain Flowerpecker	Dicaeum minullum	LC
Scarlet-backed Flowerpecker	Dicaeum cruentatum	LC
Fire-breasted Flowerpecker	Dicaeum ignipectus	LC
Nectariniidae (Sunbirds and Spiderhunters)		
Little Spiderhunter	Arachnothera longirostra	LC
Streaked Spiderhunter	Arachnothera magna	LC
Ruby-cheeked Sunbird	Chalcoparia singalensis	LC
Purple-rumped Sunbird	Leptocoma zeylonica	LC
Crimson-backed Sunbird	Leptocoma minima	LC
Van Hasselt's Sunbird	Leptocoma brasiliana	LC
Purple Sunbird	Cinnyris asiaticus	LC
Olive-backed Sunbird	Cinnyris jugularis	LC
Loten's Sunbird	Cinnyris lotenius	LC
Fire-tailed Sunbird	Aethopyga ignicauda	LC
Black-throated Sunbird	Aethopyga saturata	LC
Green-tailed Sunbird	Aethopyga nipalensis	LC
Mrs Gould's Sunbird	Aethopyga gouldiae	LC
Vigors's Sunbird	Aethopyga vigorsii	LC
Crimson Sunbird	Aethopyga siparaja	LC
Irenidae (Fairy Bluebird)		
Asian Fairy-bluebird	Irena puella	LC
Chloropseidae (Leafbirds)		
Golden-fronted Leafbird	Chloropsis aurifrons	LC
Jerdon's Leafbird	Chloropsis jerdoni	LC
Orange-bellied Leafbird	Chloropsis hardwickii	LC
Blue-winged Leafbird	Chloropsis cochinchinensis	LC
Prunellidae (Accentors)		
Altai Accentor	Prunella himalayana	LC
Alpine Accentor	Prunella collaris	LC
Maroon-backed Accentor	Prunella immaculata	LC
Robin Accentor	Prunella rubeculoides	LC
Rufous-breasted Accentor	Prunella strophiata	LC
Brown Accentor	Prunella fulvescens	LC
Black-throated Accentor	Prunella atrogularis	LC
Ploceidae (Weavers)		
Black-breasted Weaver	Ploceus benghalensis	LC
Streaked Weaver	Ploceus manyar	LC
Baya Weaver	Ploceus philippinus	LC
Finn's Weaver	Ploceus megarhynchus	VU
Estrildidae (Avadavats and Munias)		
Red Munia	Amandava amandava	LC
Green Munia	Amandava formosa	VU
Indian Silverbill	Euodice malabarica	LC
White-rumped Munia	Lonchura striata	LC
Scaly-breasted Munia	Lonchura punctulata	LC
Black-throated Munia	Lonchura kelaarti	LC

English Name	Scientific Name	IUCN Category
Tricoloured Munia	*Lonchura malacca*	LC
Chestnut Munia	*Lonchura atricapilla*	LC
Passeridae (Sparrows and Snowfinches)		
House Sparrow	*Passer domesticus*	LC
Spanish Sparrow	*Passer hispaniolensis*	LC
Sind Sparrow	*Passer pyrrhonotus*	LC
Russet Sparrow	*Passer cinnamomeus*	LC
Eurasian Tree Sparrow	*Passer montanus*	LC
Pale Rock Sparrow	*Carpospiza brachydactyla*	LC
Eurasian Rock Sparrow	*Petronia petronia*	LC
Yellow-throated Sparrow	*Gymnoris xanthocollis*	LC
Black-winged Snowfinch	*Montifringilla adamsi*	LC
White-rumped Snowfinch	*Onychostruthus taczanowskii*	LC
Rufous-necked Snowfinch	*Pyrgilauda ruficollis*	LC
Blanford's Snowfinch	*Pyrgilauda blanfordi*	LC
Motacillidae (Wagtails and Pipits)		
Forest Wagtail	*Dendronanthus indicus*	LC
Tree Pipit	*Anthus trivialis*	LC
Olive-backed Pipit	*Anthus hodgsoni*	LC
Red-throated Pipit	*Anthus cervinus*	LC
Rosy Pipit	*Anthus roseatus*	LC
Buff-bellied Pipit	*Anthus rubescens*	LC
Water Pipit	*Anthus spinoletta*	LC
Upland Pipit	*Anthus sylvanus*	LC
Nilgiri Pipit	*Anthus nilghiriensis*	VU
Richard's Pipit	*Anthus richardi*	LC
Paddyfield Pipit	*Anthus rufulus*	LC
Blyth's Pipit	*Anthus godlewskii*	LC
Tawny Pipit	*Anthus campestris*	LC
Long-billed Pipit	*Anthus similis*	LC
Western Yellow Wagtail	*Motacilla flava*	LC
Grey Wagtail	*Motacilla cinerea*	LC
Citrine Wagtail	*Motacilla citreola*	LC
Eastern Yellow Wagtail	*Motacilla tschutschensis*	LC
White-browed Wagtail	*Motacilla maderaspatensis*	LC
White Wagtail	*Motacilla alba*	LC
Fringillidae (Finches)		
Common Chaffinch	*Fringilla coelebs*	LC
Brambling	*Fringilla montifringilla*	LC
Black-and-yellow Grosbeak	*Mycerobas icterioides*	LC
Collared Grosbeak	*Mycerobas affinis*	LC
Spot-winged Grosbeak	*Mycerobas melanozanthos*	LC
White-winged Grosbeak	*Mycerobas carnipes*	LC
Hawfinch	*Coccothraustes coccothraustes*	LC
Common Rosefinch	*Carpodacus erythrinus*	LC
Scarlet Finch	*Carpodacus sipahi*	LC
Pale Rosefinch	*Carpodacus stoliczkae*	LC
Sillem's Rosefinch	*Carpodacus sillemi*	DD
Streaked Rosefinch	*Carpodacus rubicilloides*	LC
Great Rosefinch	*Carpodacus rubicilla*	LC
Red-fronted Rosefinch	*Carpodacus puniceus*	LC
Crimson-browed Finch	*Carpodacus subhimachalus*	LC
Himalayan White-browed Rosefinch	*Carpodacus thura*	LC
Chinese White-browed Rosefinch	*Carpodacus dubius*	LC
Blyth's Rosefinch	*Carpodacus grandis*	LC
Himalayan Beautiful Rosefinch	*Carpodacus pulcherrimus*	LC
Pink-rumped Rosefinch	*Carpodacus waltoni*	LC
Dark-rumped Rosefinch	*Carpodacus edwardsii*	LC
Pink-browed Rosefinch	*Carpodacus rodochroa*	LC
Spot-winged Rosefinch	*Carpodacus rodopeplus*	LC
Vinaceous Rosefinch	*Carpodacus vinaceus*	LC
Brown Bullfinch	*Pyrrhula nipalensis*	LC
Orange Bullfinch	*Pyrrhula aurantiaca*	LC
Red-headed Bullfinch	*Pyrrhula erythrocephala*	LC
Grey-headed Bullfinch	*Pyrrhula erythaca*	LC
Trumpeter Finch	*Bucanetes githagineus*	LC
Mongolian Finch	*Bucanetes mongolicus*	LC
Blanford's Rosefinch	*Agraphospiza rubescens*	LC
Spectacled Finch	*Callacanthis burtoni*	LC
Gold-naped Finch	*Pyrrhoplectes epauletta*	LC
Dark-breasted Rosefinch	*Procarduelis nipalensis*	LC

English Name	Scientific Name	IUCN Category
Plain Mountain Finch	*Leucosticte nemoricola*	LC
Brandt's Mountain Finch	*Leucosticte brandti*	LC
Yellow-breasted Greenfinch	*Chloris spinoides*	LC
Black-headed Greenfinch	*Chloris ambigua*	LC
Twite	*Linaria flavirostris*	LC
Common Linnet	*Linaria cannabina*	LC
Red Crossbill	*Loxia curvirostra*	LC
European Goldfinch	*Carduelis carduelis*	LC
Fire-fronted Serin	*Serinus pusillus*	LC
Tibetan Siskin	*Spinus thibetanus*	LC
Eurasian Siskin	*Spinus spinus*	LC
Emberizidae (Buntings)		
Crested Bunting	*Emberiza lathami*	LC
Black-headed Bunting	*Emberiza melanocephala*	LC
Red-headed Bunting	*Emberiza bruniceps*	LC
Chestnut-eared Bunting	*Emberiza fucata*	LC
Rock Bunting	*Emberiza cia*	LC
Godlewski's Bunting	*Emberiza godlewskii*	LC
Grey-necked Bunting	*Emberiza buchanani*	LC
Ortolan Bunting	*Emberiza hortulana*	LC
White-capped Bunting	*Emberiza stewarti*	LC
Yellowhammer	*Emberiza citrinella*	LC
Pine Bunting	*Emberiza leucocephalos*	LC
Striolated Bunting	*Emberiza striolata*	LC
Eurasian Reed Bunting	*Emberiza schoeniclus*	LC
Yellow-breasted Bunting	*Emberiza aureola*	CR
Black-faced Bunting	*Emberiza spodocephala*	LC
Chestnut Bunting	*Emberiza rutila*	LC
Yellow-browed Bunting	*Emberiza chrysophrys*	LC
Tristram's Bunting	*Emberiza tristrami*	LC
Little Bunting	*Emberiza pusilla*	LC
Stenostiridae (Fantails)		
Yellow-bellied Fairy-fantail	*Chelidorhynx hypoxanthus*	LC
Grey-headed Canary-flycatcher	*Culicicapa ceylonensis*	LC
Paridae (Tits)		
Fire-capped Tit	*Cephalopyrus flammiceps*	LC
Yellow-browed Tit	*Sylviparus modestus*	LC
Sultan Tit	*Melanochlora sultanea*	LC
Coal Tit	*Periparus ater*	LC
Rufous-naped Tit	*Periparus rufonuchalis*	LC
Rufous-vented Tit	*Periparus rubidiventris*	LC
Grey-crested Tit	*Lophophanes dichrous*	LC
Azure Tit	*Cyanistes cyanus*	LC
Ground Tit	*Pseudopodoces humilis*	LC
Green-backed Tit	*Parus monticolus*	LC
Cinereous Tit	*Parus cinereus*	LC
White-naped Tit	*Machlolophus nuchalis*	VU
Black-lored Tit	*Machlolophus xanthogenys*	LC
Yellow-cheeked Tit	*Machlolophus spilonotus*	LC
Remizidae (Penduline Tits)		
White-crowned Penduline Tit	*Remiz coronatus*	LC
Alaudidae (Larks)		
Greater Hoopoe Lark	*Alaemon alaudipes*	LC
Rufous-tailed Lark	*Ammomanes phoenicura*	LC
Desert Lark	*Ammomanes deserti*	LC
Black-crowned Sparrow Lark	*Eremopterix nigriceps*	LC
Ashy-crowned Sparrow Lark	*Eremopterix griseus*	LC
Singing Bushlark	*Mirafra cantillans*	LC
Bengal Bushlark	*Mirafra assamica*	LC
Indian Bushlark	*Mirafra erythroptera*	LC
Jerdon's Bushlark	*Mirafra affinis*	LC
Lesser Short-toed Lark	*Alaudala rufescens*	LC
Sand Lark	*Alaudala raytal*	LC
Bimaculated Lark	*Melanocorypha bimaculata*	LC
Tibetan Lark	*Melanocorypha maxima*	LC
Hume's Short-toed Lark	*Calandrella acutirostris*	LC
Greater Short-toed Lark	*Calandrella brachydactyla*	LC
Sykes's Short-toed Lark	*Calandrella dukhunensis*	LC
Horned Lark	*Eremophila alpestris*	LC
Eurasian Skylark	*Alauda arvensis*	LC
Oriental Skylark	*Alauda gulgula*	LC

English Name	Scientific Name	IUCN Category
Crested Lark	*Galerida cristata*	LC
Malabar Lark	*Galerida malabarica*	LC
Sykes's Lark	*Galerida deva*	LC
Cisticolidae (Cisticolas, Prinias and Tailorbirds)		
Zitting Cisticola	*Cisticola juncidis*	LC
Golden-headed Cisticola	*Cisticola exilis*	LC
Striated Prinia	*Prinia crinigera*	LC
Black-throated Prinia	*Prinia atrogularis*	LC
Hill Prinia	*Prinia superciliaris*	LC
Grey-crowned Prinia	*Prinia cinereocapilla*	VU
Rufous-fronted Prinia	*Prinia buchanani*	LC
Rufescent Prinia	*Prinia rufescens*	LC
Grey-breasted Prinia	*Prinia hodgsonii*	LC
Graceful Prinia	*Prinia gracilis*	LC
Jungle Prinia	*Prinia sylvatica*	LC
Yellow-bellied Prinia	*Prinia flaviventris*	LC
Ashy Prinia	*Prinia socialis*	LC
Plain Prinia	*Prinia inornata*	LC
Common Tailorbird	*Orthotomus sutorius*	LC
Dark-necked Tailorbird	*Orthotomus atrogularis*	LC
Locustellidae (Locustella Warblers and Grassbirds)		
Rusty-rumped Warbler	*Locustella certhiola*	LC
Lanceolated Warbler	*Locustella lanceolata*	LC
Brown Bush Warbler	*Locustella luteoventris*	LC
Chinese Bush Warbler	*Locustella tacsanowskia*	LC
Long-billed Bush Warbler	*Locustella major*	NT
Grasshopper Warbler	*Locustella naevia*	LC
Baikal Bush Warbler	*Locustella davidi*	LC
West Himalayan Bush Warbler	*Locustella kashmirensis*	LC
Spotted Bush Warbler	*Locustella thoracica*	LC
Russet Bush Warbler	*Locustella mandelli*	LC
Striated Grassbird	*Megalurus palustris*	LC
Broad-tailed Grassbird	*Schoenicola platyurus*	VU
Bristled Grassbird	*Chaetornis striata*	VU
Acrocephalidae (Reed Warblers)		
Thick-billed Warbler	*Arundinax aedon*	LC
Booted Warbler	*Iduna caligata*	LC
Sykes's Warbler	*Iduna rama*	LC
Black-browed Reed Warbler	*Acrocephalus bistrigiceps*	LC
Moustached Warbler	*Acrocephalus melanopogon*	LC
Sedge Warbler	*Acrocephalus schoenobaenus*	LC
Large-billed Reed Warbler	*Acrocephalus orinus*	DD
Blyth's Reed Warbler	*Acrocephalus dumetorum*	LC
Paddyfield Warbler	*Acrocephalus agricola*	LC
Blunt-winged Warbler	*Acrocephalus concinens*	LC
Great Reed Warbler	*Acrocephalus arundinaceus*	LC
Oriental Reed Warbler	*Acrocephalus orientalis*	LC
Clamorous Reed Warbler	*Acrocephalus stentoreus*	LC
Pnoepygidae (Wren Babblers)		
Nepal Wren Babbler	*Pnoepyga immaculata*	LC
Pygmy Wren Babbler	*Pnoepyga pusilla*	LC
Scaly-breasted Wren Babbler	*Pnoepyga albiventer*	LC
Hirundinidae (Swallows and Martins)		
Northern House Martin	*Delichon urbicum*	LC
Asian House Martin	*Delichon dasypus*	LC
Nepal House Martin	*Delichon nipalense*	LC
Streak-throated Swallow	*Petrochelidon fluvicola*	LC
Red-rumped Swallow	*Cecropis daurica*	LC
Striated Swallow	*Cecropis striolata*	LC
Pacific Swallow	*Hirundo tahitica*	LC
Wire-tailed Swallow	*Hirundo smithii*	LC
Barn Swallow	*Hirundo rustica*	LC
Eurasian Crag Martin	*Ptyonoprogne rupestris*	LC
Dusky Crag Martin	*Ptyonoprogne concolor*	LC
Grey-throated Martin	*Riparia chinensis*	LC
Sand Martin	*Riparia riparia*	LC
Pale Martin	*Riparia diluta*	LC
Pycnonotidae (Bulbuls)		
White-throated Bulbul	*Alophoixus flaveolus*	LC
Olive Bulbul	*Iole viridescens*	LC
Ashy Bulbul	*Hemixos flavala*	LC

English Name	Scientific Name	IUCN Category
Nicobar Bulbul	Ixos nicobariensis	NT
Mountain Bulbul	Ixos mcclellandii	LC
Square-tailed Bulbul	Hypsipetes ganeesa	LC
Black Bulbul	Hypsipetes leucocephalus	LC
Crested Finchbill	Spizixos canifrons	LC
Striated Bulbul	Pycnonotus striatus	LC
Flame-throated Bulbul	Pycnonotus gularis	LC
Black-crested Bulbul	Pycnonotus flaviventris	LC
Red-whiskered Bulbul	Pycnonotus jocosus	LC
Himalayan Bulbul	Pycnonotus leucogenis	LC
White-eared Bulbul	Pycnonotus leucotis	LC
Red-vented Bulbul	Pycnonotus cafer	LC
Yellow-throated Bulbul	Pycnonotus xantholaemus	VU
Flavescent Bulbul	Pycnonotus flavescens	LC
White-browed Bulbul	Pycnonotus luteolus	LC
Black-headed Bulbul	Brachypodius atriceps	LC
Andaman Bulbul	Brachypodius fuscoflavescens	LC
Grey-headed Bulbul	Brachypodius priocephalus	NT
Yellow-browed Bulbul	Acritillas indica	LC
Phylloscopidae (Leaf Warblers)		
Wood Warbler	Phylloscopus sibilatrix	LC
Chinese Leaf Warbler	Phylloscopus yunnanensis	LC
Brooks's Leaf Warbler	Phylloscopus subviridis	LC
Yellow-browed Warbler	Phylloscopus inornatus	LC
Hume's Leaf Warbler	Phylloscopus humei	LC
Lemon-rumped Warbler	Phylloscopus chloronotus	LC
Buff-barred Warbler	Phylloscopus pulcher	LC
Ashy-throated Warbler	Phylloscopus maculipennis	LC
Dusky Warbler	Phylloscopus fuscatus	LC
Smoky Warbler	Phylloscopus fuligiventer	LC
Buff-throated Warbler	Phylloscopus subaffinis	LC
Common Chiffchaff	Phylloscopus collybita	LC
Kashmir Chiffchaff	Phylloscopus sindianus	LC
Plain Leaf Warbler	Phylloscopus neglectus	LC
Tytler's Leaf Warbler	Phylloscopus tytleri	NT
Sulphur-bellied Warbler	Phylloscopus griseolus	LC
Tickell's Leaf Warbler	Phylloscopus affinis	LC
White-spectacled Warbler	Phylloscopus intermedius	LC
Grey-cheeked Warbler	Phylloscopus poliogenys	LC
Green-crowned Warbler	Phylloscopus burkii	LC
Grey-crowned Warbler	Phylloscopus tephrocephalus	LC
Whistler's Warbler	Phylloscopus whistleri	LC
Chestnut-crowned Warbler	Phylloscopus castaniceps	LC
Green Leaf Warbler	Phylloscopus nitidus	LC
Greenish Leaf Warbler	Phylloscopus trochiloides	LC
Two-barred Leaf Warbler	Phylloscopus plumbeitarsus	LC
Arctic Warbler	Phylloscopus borealis	LC
Pale-legged Leaf Warbler	Phylloscopus tenellipes	LC
Large-billed Leaf Warbler	Phylloscopus magnirostris	LC
Yellow-vented Leaf Warbler	Phylloscopus cantator	LC
Claudia's Leaf Warbler	Phylloscopus claudiae	LC
Blyth's Leaf Warbler	Phylloscopus reguloides	LC
Western Crowned Leaf Warbler	Phylloscopus occipitalis	LC
Grey-hooded Leaf Warbler	Phylloscopus xanthoschistos	LC
Scotocercidae (Tesias and Bush Warblers)		
Slaty-bellied Tesia	Tesia olivea	LC
Grey-bellied Tesia	Tesia cyaniventer	LC
Chestnut-crowned Bush Warbler	Cettia major	LC
Grey-sided Bush Warbler	Cettia brunnifrons	LC
Chestnut-headed Tesia	Cettia castaneocoronata	LC
Cetti's Warbler	Cettia cetti	LC
Pale-footed Bush Warbler	Hemitesia pallidipes	LC
Asian Stubtail	Urosphena squameiceps	LC
Yellow-bellied Warbler	Abroscopus superciliaris	LC
Rufous-faced Warbler	Abroscopus albogularis	LC
Black-faced Warbler	Abroscopus schisticeps	LC
Mountain Tailorbird	Phyllergates cucullatus	LC
Broad-billed Warbler	Tickellia hodgsoni	LC
Brownish-flanked Bush Warbler	Horornis fortipes	LC
Hume's Bush Warbler	Horornis brunnescens	LC
Aberrant Bush Warbler	Horornis flavolivaceus	LC

English Name	Scientific Name	IUCN Category
Manchurian Bush Warbler	*Horornis canturians*	LC
Aegithalidae (Tits)		
White-browed Tit Warbler	*Leptopoecile sophiae*	LC
Crested Tit Warbler	*Leptopoecile elegans*	LC
Black-throated Tit	*Aegithalos concinnus*	LC
White-cheeked Tit	*Aegithalos leucogenys*	LC
White-throated Tit	*Aegithalos niveogularis*	LC
Black-browed Tit	*Aegithalos iouschistos*	LC
Sylviidae (Sylvia Warblers)		
Garden Warbler	*Sylvia borin*	LC
Asian Desert Warbler	*Sylvia nana*	LC
Barred Warbler	*Sylvia nisoria*	LC
Eastern Orphean Warbler	*Sylvia crassirostris*	LC
Lesser Whitethroat	*Sylvia curruca*	LC
Common Whitethroat	*Sylvia communis*	LC
Fire-tailed Myzornis	*Myzornis pyrrhoura*	LC
Golden-breasted Fulvetta	*Lioparus chrysotis*	LC
Yellow-eyed Babbler	*Chrysomma sinense*	LC
Jerdon's Babbler	*Chrysomma altirostre*	VU
White-browed Fulvetta	*Fulvetta vinipectus*	LC
Ludlow's Fulvetta	*Fulvetta ludlowi*	LC
Manipur Fulvetta	*Fulvetta manipurensis*	LC
Black-breasted Parrotbill	*Paradoxornis flavirostris*	VU
Spot-breasted Parrotbill	*Paradoxornis guttaticollis*	LC
White-breasted Parrotbill	*Psittiparus ruficeps*	LC
Rufous-headed Parrotbill	*Psittiparus bakeri*	LC
Grey-headed Parrotbill	*Psittiparus gularis*	LC
Great Parrotbill	*Conostoma aemodium*	LC
Brown Parrotbill	*Cholornis unicolor*	LC
Fulvous Parrotbill	*Suthora fulvifrons*	LC
Black-throated Parrotbill	*Suthora nipalensis*	LC
Lesser Rufous-headed Parrotbill	*Chleuasicus atrosuperciliaris*	LC
Zosteropidae (White-eyes)		
Striated Yuhina	*Yuhina castaniceps*	LC
Black-chinned Yuhina	*Yuhina nigrimenta*	LC
Stripe-throated Yuhina	*Yuhina gularis*	LC
Whiskered Yuhina	*Yuhina flavicollis*	LC
Rufous-vented Yuhina	*Yuhina occipitalis*	LC
White-naped Yuhina	*Yuhina bakeri*	LC
Chestnut-flanked White-eye	*Zosterops erythropleurus*	LC
Oriental White-eye	*Zosterops palpebrosus*	LC
Timaliidae (Babblers)		
Rufous-throated Wren Babbler	*Spelaeornis caudatus*	NT
Mishmi Wren Babbler	*Spelaeornis badeigularis*	VU
Bar-winged Wren Babbler	*Spelaeornis troglodytoides*	LC
Naga Wren Babbler	*Spelaeornis chocolatinus*	NT
Chin Hills Wren Babbler	*Spelaeornis oatesi*	LC
Grey-bellied Wren Babbler	*Spelaeornis reptatus*	LC
Tawny-breasted Wren Babbler	*Spelaeornis longicaudatus*	VU
Red-billed Scimitar Babbler	*Pomatorhinus ochraceiceps*	LC
Coral-billed Scimitar Babbler	*Pomatorhinus ferruginosus*	LC
Slender-billed Scimitar Babbler	*Pomatorhinus superciliaris*	LC
Indian Scimitar Babbler	*Pomatorhinus horsfieldii*	LC
White-browed Scimitar Babbler	*Pomatorhinus schisticeps*	LC
Streak-breasted Scimitar Babbler	*Pomatorhinus ruficollis*	LC
Large Scimitar Babbler	*Erythrogenys hypoleucos*	LC
Rusty-cheeked Scimitar Babbler	*Erythrogenys erythrogenys*	LC
Spot-breasted Scimitar Babbler	*Erythrogenys mcclellandi*	LC
Grey-throated Babbler	*Stachyris nigriceps*	LC
Sikkim Wedge-billed Babbler	*Stachyris humei*	NT
Cachar Wedge-billed Babbler	*Stachyris roberti*	NT
Snowy-throated Babbler	*Stachyris oglei*	VU
Tawny-bellied Babbler	*Dumetia hyperythra*	LC
Dark-fronted Babbler	*Rhopocichla atriceps*	LC
Chestnut-capped Babbler	*Timalia pileata*	LC
Pin-striped Tit Babbler	*Mixornis gularis*	LC
Golden Babbler	*Cyanoderma chrysaeum*	LC
Black-chinned Babbler	*Cyanoderma pyrrhops*	LC
Rufous-capped Babbler	*Cyanoderma ruficeps*	LC
Buff-chested Babbler	*Cyanoderma ambiguum*	LC
Pellorneidae (Babblers and Fulvettas)		

English Name	Scientific Name	IUCN Category
White-hooded Babbler	*Gampsorhynchus rufulus*	LC
Rusty-capped Fulvetta	*Schoeniparus dubius*	LC
Rufous-throated Fulvetta	*Schoeniparus rufogularis*	LC
Yellow-throated Fulvetta	*Schoeniparus cinereus*	LC
Rufous-winged Fulvetta	*Schoeniparus castaneceps*	LC
Rufous-vented Grass Babbler	*Laticilla burnesii*	NT
Swamp Grass Babbler	*Laticilla cinerascens*	EN
Puff-throated Babbler	*Pellorneum ruficeps*	LC
Marsh Babbler	*Pellorneum palustre*	VU
Spot-throated Babbler	*Pellorneum albiventre*	LC
Buff-breasted Babbler	*Trichastoma tickelli*	LC
Abbott's Babbler	*Malacocincla abbotti*	LC
Streaked Wren Babbler	*Turdinus brevicaudatus*	LC
Eyebrowed Wren Babbler	*Napothera epilepidota*	LC
Long-billed Wren Babbler	*Rimator malacoptilus*	LC
Indian Grass Babbler	*Graminicola bengalensis*	NT
Leiothrichidae (Babblers and Laughingthrushs)		
Quaker Tit Babbler	*Alcippe poioicephala*	LC
Nepal Tit Babbler	*Alcippe nipalensis*	LC
Striated Laughingthrush	*Grammatoptila striata*	LC
Himalayan Cutia	*Cutia nipalensis*	LC
Large Grey Babbler	*Argya malcolmi*	LC
Rufous Babbler	*Argya subrufa*	LC
Striated Babbler	*Argya earlei*	LC
Common Babbler	*Argya caudata*	LC
Slender-billed Babbler	*Chatarrhaea longirostris*	VU
Jungle Babbler	*Turdoides striata*	LC
Yellow-billed Babbler	*Turdoides affinis*	LC
Spot-breasted Laughingthrush	*Garrulax merulinus*	LC
Lesser Necklaced Laughingthrush	*Garrulax monileger*	LC
White-crested Laughingthrush	*Garrulax leucolophus*	LC
Spotted Laughingthrush	*Garrulax ocellatus*	LC
Moustached Laughingthrush	*Garrulax cineraceus*	LC
Rufous-chinned Laughingthrush	*Garrulax rufogularis*	LC
White-browed Laughingthrush	*Garrulax sannio*	LC
Chestnut-backed Laughingthrush	*Garrulax nuchalis*	NT
Greater Necklaced Laughingthrush	*Garrulax pectoralis*	LC
Chinese Babax	*Garrulax lanceolatus*	LC
White-throated Laughingthrush	*Garrulax albogularis*	LC
Grey-sided Laughingthrush	*Garrulax caerulatus*	LC
Rufous-necked Laughingthrush	*Garrulax ruficollis*	LC
Yellow-throated Laughingthrush	*Garrulax galbanus*	LC
Wayanad Laughingthrush	*Garrulax delesserti*	LC
Rufous-vented Laughingthrush	*Garrulax gularis*	LC
Scaly Laughingthrush	*Trochalopteron subunicolor*	LC
Brown-capped Laughingthrush	*Trochalopteron austeni*	LC
Blue-winged Laughingthrush	*Trochalopteron squamatum*	LC
Streaked Laughingthrush	*Trochalopteron lineatum*	LC
Bhutan Laughingthrush	*Trochalopteron imbricatum*	LC
Striped Laughingthrush	*Trochalopteron virgatum*	LC
Variegated Laughingthrush	*Trochalopteron variegatum*	LC
Black-faced Laughingthrush	*Trochalopteron affine*	LC
Elliot's Laughingthrush	*Trochalopteron elliotii*	LC
Chestnut-crowned Laughingthrush	*Trochalopteron erythrocephalum*	LC
Assam Laughingthrush	*Trochalopteron chrysopterum*	LC
Banasura Laughingthrush	*Montecincla jerdoni*	EN
Nilgiri Laughingthrush	*Montecincla cachinnans*	EN
Palani Laughingthrush	*Montecincla fairbanki*	NT
Ashambu Laughingthrush	*Montecincla meridionalis*	VU
Long-tailed Sibia	*Heterophasia picaoides*	LC
Beautiful Sibia	*Heterophasia pulchella*	LC
Rufous Sibia	*Heterophasia capistrata*	LC
Grey Sibia	*Heterophasia gracilis*	LC
Silver-eared Mesia	*Leiothrix argentauris*	LC
Red-billed Leiothrix	*Leiothrix lutea*	LC
Rufous-backed Sibia	*Leioptila annectens*	LC
Red-tailed Minla	*Minla ignotincta*	LC
Red-faced Liocichla	*Liocichla phoenicea*	LC
Bugun Liocichla	*Liocichla bugunorum*	CR
Hoary-throated Barwing	*Sibia nipalensis*	LC
Streak-throated Barwing	*Sibia waldeni*	LC

English Name	Scientific Name	IUCN Category
Blue-winged Minla	Siva cyanouroptera	LC
Chestnut-tailed Minla	Chrysominla strigula	LC
Rusty-fronted Barwing	Actinodura egertoni	LC
Regulidae (Goldcrest)		
Goldcrest	Regulus regulus	LC
Elachuridae (Elachuras)		
Spotted Elachura	Elachura formosa	LC
Bombycillidae (Waxwing)		
Bohemian Waxwing	Bombycilla garrulus	LC
Hypocoliidae (Hypocolius)		
Grey Hypocolius	Hypocolius ampelinus	LC
Certhiidae (Treecreepers)		
Rusty-flanked Treecreeper	Certhia nipalensis	LC
Sikkim Treecreeper	Certhia discolor	LC
Manipur Treecreeper	Certhia manipurensis	LC
Bar-tailed Treecreeper	Certhia himalayana	LC
Hodgson's Treecreeper	Certhia hodgsoni	LC
Sittidae (Nuthatches)		
Chestnut-vented Nuthatch	Sitta nagaensis	LC
Kashmir Nuthatch	Sitta cashmirensis	LC
Indian Nuthatch	Sitta castanea	LC
Chestnut-bellied Nuthatch	Sitta cinnamoventris	LC
White-tailed Nuthatch	Sitta himalayensis	LC
Yunnan Nuthatch	Sitta yunnanensis	NT
White-cheeked Nuthatch	Sitta leucopsis	LC
Velvet-fronted Nuthatch	Sitta frontalis	LC
Beautiful Nuthatch	Sitta formosa	VU
Indian Spotted Creeper	Salpornis spilonota	LC
Wallcreeper	Tichodroma muraria	LC
Troglodytidae (Wren)		
Eurasian Wren	Troglodytes troglodytes	LC
Sturnidae (Starlings and Mynas)		
Common Starling	Sturnus vulgaris	LC
Rosy Starling	Pastor roseus	LC
Purple-backed Starling	Agropsar sturninus	LC
Chestnut-cheeked Starling	Agropsar philippensis	LC
Asian Pied Starling	Gracupica contra	LC
Brahminy Starling	Sturnia pagodarum	LC
Chestnut-tailed Starling	Sturnia malabarica	LC
White-headed Starling	Sturnia erythropygia	LC
White-cheeked Starling	Spodiopsar cineraceus	LC
Common Myna	Acridotheres tristis	LC
Bank Myna	Acridotheres ginginianus	LC
Jungle Myna	Acridotheres fuscus	LC
Collared Myna	Acridotheres albocinctus	LC
Great Myna	Acridotheres grandis	LC
Spot-winged Starling	Saroglossa spilopterus	LC
Southern Hill Myna	Gracula indica	LC
Common Hill Myna	Gracula religiosa	LC
Golden-crested Myna	Ampeliceps coronatus	LC
Asian Glossy Starling	Aplonis panayensis	LC
Cinclidae (Dippers)		
White-throated Dipper	Cinclus cinclus	LC
Brown Dipper	Cinclus pallasii	LC
Muscicapidae (Flycatchers)		
Rufous-tailed Scrub Robin	Cercotrichas galactotes	LC
Indian Robin	Saxicoloides fulicatus	LC
Oriental Magpie Robin	Copsychus saularis	LC
White-rumped Shama	Kittacincla malabarica	LC
Andaman Shama	Kittacincla albiventris	LC
Spotted Flycatcher	Muscicapa striata	LC
Dark-sided Flycatcher	Muscicapa sibirica	LC
Asian Brown Flycatcher	Muscicapa dauurica	LC
Brown-breasted Flycatcher	Muscicapa muttui	LC
Ferruginous Flycatcher	Muscicapa ferruginea	LC
Nilgiri Sholakili	Sholicola major	EN
White-bellied Sholakili	Sholicola albiventris	VU
Pale Blue Flycatcher	Cyornis unicolor	LC
White-bellied Blue Flycatcher	Cyornis pallidipes	LC
Pale-chinned Flycatcher	Cyornis poliogenys	LC
Large Blue Flycatcher	Cyornis magnirostris	LC

Checklist of Birds of India

English Name	Scientific Name	IUCN Category
Hill Blue Flycatcher	Cyornis banyumas	LC
Tickell's Blue Flycatcher	Cyornis tickelliae	LC
Blue-throated Flycatcher	Cyornis rubeculoides	LC
White-tailed Blue Flycatcher	Cyornis concretus	LC
Nicobar Jungle Flycatcher	Cyornis nicobaricus	NT
White-gorgeted Flycatcher	Anthipes monileger	LC
Rufous-bellied Niltava	Niltava sundara	LC
Vivid Niltava	Niltava vivida	LC
Large Niltava	Niltava grandis	LC
Small Niltava	Niltava macgrigoriae	LC
Blue-and-white Flycatcher	Cyanoptila cyanomelana	LC
Zappey's Flycatcher	Cyanoptila cumatilis	NT
Verditer Flycatcher	Eumyias thalassinus	LC
Nilgiri Flycatcher	Eumyias albicaudatus	LC
White-browed Shortwing	Brachypteryx montana	LC
Lesser Shortwing	Brachypteryx leucophris	LC
Rusty-bellied Shortwing	Brachypteryx hyperythra	NT
Gould's Shortwing	Heteroxenicus stellatus	LC
Indian Blue Robin	Larvivora brunnea	LC
Siberian Blue Robin	Larvivora cyane	LC
Bluethroat	Luscinia svecica	LC
Hodgson's Blue Robin	Luscinia phaenicuroides	LC
Little Forktail	Enicurus scouleri	LC
Black-backed Forktail	Enicurus immaculatus	LC
Slaty-backed Forktail	Enicurus schistaceus	LC
White-crowned Forktail	Enicurus leschenaulti	LC
Spotted Forktail	Enicurus maculatus	LC
Blue-fronted Robin	Cinclidium frontale	LC
Malabar Whistling Thrush	Myophonus horsfieldii	LC
Blue Whistling Thrush	Myophonus caeruleus	LC
Firethroat	Calliope pectardens	NT
Himalayan Rubythroat	Calliope pectoralis	LC
Chinese Rubythroat	Calliope tschebaiewi	LC
Siberian Rubythroat	Calliope calliope	LC
White-tailed Robin	Myiomela leucura	LC
White-browed Bush Robin	Tarsiger indicus	LC
Golden Bush Robin	Tarsiger chrysaeus	LC
Red-flanked Bush Robin	Tarsiger cyanurus	LC
Himalayan Bush Robin	Tarsiger rufilatus	LC
Rufous-breasted Bush Robin	Tarsiger hyperythrus	LC
Rusty-tailed Flycatcher	Ficedula ruficauda	LC
Kashmir Flycatcher	Ficedula subrubra	VU
Red-breasted Flycatcher	Ficedula parva	LC
Taiga Flycatcher	Ficedula albicilla	LC
Snowy-browed Flycatcher	Ficedula hyperythra	LC
Rufous-gorgeted Flycatcher	Ficedula strophiata	LC
Ultramarine Flycatcher	Ficedula superciliaris	LC
Little Pied Flycatcher	Ficedula westermanni	LC
Mugimaki Flycatcher	Ficedula mugimaki	LC
Yellow-rumped Flycatcher	Ficedula zanthopygia	LC
Slaty-blue Flycatcher	Ficedula tricolor	LC
Black-and-orange Flycatcher	Ficedula nigrorufa	LC
Pygmy Blue Flycatcher	Ficedula hodgsoni	LC
Slaty-backed Flycatcher	Ficedula sordida	LC
Sapphire Flycatcher	Ficedula sapphira	LC
Eversmann's Redstart	Phoenicurus erythronotus	LC
Blue-fronted Redstart	Phoenicurus frontalis	LC
Blue-capped Redstart	Phoenicurus coeruleocephala	LC
White-throated Redstart	Phoenicurus schisticeps	LC
White-capped Water Redstart	Phoenicurus leucocephalus	LC
Plumbeous Water Redstart	Phoenicurus fuliginosus	LC
Black Redstart	Phoenicurus ochruros	LC
Common Redstart	Phoenicurus phoenicurus	LC
Daurian Redstart	Phoenicurus auroreus	LC
White-winged Redstart	Phoenicurus erythrogastrus	LC
Hodgson's Redstart	Phoenicurus hodgsoni	LC
Blue-capped Rock Thrush	Monticola cinclorhyncha	LC
Chestnut-bellied Rock Thrush	Monticola rufiventris	LC
Rufous-tailed Rock Thrush	Monticola saxatilis	LC
Blue Rock Thrush	Monticola solitarius	LC
Stoliczka's Bushchat	Saxicola macrorhynchus	VU

English Name	Scientific Name	IUCN Category
Hodgson's Bushchat	*Saxicola insignis*	VU
Siberian Stonechat	*Saxicola maurus*	LC
White-tailed Stonechat	*Saxicola leucurus*	LC
Pied Bushchat	*Saxicola caprata*	LC
Jerdon's Bushchat	*Saxicola jerdoni*	LC
Grey Bushchat	*Saxicola ferreus*	LC
Northern Wheatear	*Oenanthe oenanthe*	LC
Isabelline Wheatear	*Oenanthe isabellina*	LC
Desert Wheatear	*Oenanthe deserti*	LC
Pied Wheatear	*Oenanthe pleschanka*	LC
Brown Rock Chat	*Oenanthe fusca*	LC
Variable Wheatear	*Oenanthe picata*	LC
Hume's Wheatear	*Oenanthe albonigra*	LC
Red-tailed Wheatear	*Oenanthe chrysopygia*	LC
Turdidae (Flycatchers)		
Grandala	*Grandala coelicolor*	LC
Long-tailed Thrush	*Zoothera dixoni*	LC
Alpine Thrush	*Zoothera mollissima*	LC
Himalayan Forest Thrush	*Zoothera salimalii*	LC
Dark-sided Thrush	*Zoothera marginata*	LC
Long-billed Thrush	*Zoothera monticola*	LC
Scaly Thrush	*Zoothera dauma*	LC
Purple Cochoa	*Cochoa purpurea*	LC
Green Cochoa	*Cochoa viridis*	LC
Siberian Thrush	*Geokichla sibirica*	LC
Pied Thrush	*Geokichla wardii*	LC
Orange-headed Thrush	*Geokichla citrina*	LC
Chinese Thrush	*Otocichla mupinensis*	LC
Mistle Thrush	*Turdus viscivorus*	LC
Song Thrush	*Turdus philomelos*	LC
Grey-winged Blackbird	*Turdus boulboul*	LC
Indian Blackbird	*Turdus simillimus*	LC
Black-breasted Thrush	*Turdus dissimilis*	LC
Tickell's Thrush	*Turdus unicolor*	LC
Eyebrowed Thrush	*Turdus obscurus*	LC
Grey-sided Thrush	*Turdus feae*	VU
Kessler's Thrush	*Turdus kessleri*	LC
Tibetan Blackbird	*Turdus maximus*	LC
Fieldfare	*Turdus pilaris*	LC
White-collared Blackbird	*Turdus albocinctus*	LC
Chestnut Thrush	*Turdus rubrocanus*	LC
Naumann's Thrush	*Turdus naumanni*	LC
Dusky Thrush	*Turdus eunomus*	LC
Black-throated Thrush	*Turdus atrogularis*	LC
Red-throated Thrush	*Turdus ruficollis*	LC

Index

Aberrant Bush Warbler *Horornis flavolivaceus* 187
Alexandrine Parakeet *Psittacula eupatria* 14, 20
Alpine Accentor *Prunella collaris* 45, 228, 230
Alpine Swift *Tachymarptis melba* 108, 118, 224
Altai Accentor *Prunella himalayana* 228, 230
Amur Falcon *Falco amurensis* 189
Andaman Barn Owl *Tyto deroepstorffi* 17, 23
Andaman Bulbul *Brachypodius fuscoflavescens* 16, 17, 25
Andaman Coucal *Centropus andamanensis* 18, 19, 20, 22, 24, 25
Andaman Crake *Rallina canningi* 12, 16, 17, 18, 20, 22, 25
Andaman Cuckoo Dove *Macropygia rufipennis* 12, 17, 18, 19, 22, 25
Andaman Cuckooshrike *Coracina dobsoni* 22
Andaman Drongo *Dicrurus andamanensis* 12, 18, 20, 22, 24, 25
Andaman Hawk Owl *Ninox affinis* 12, 16, 18, 19, 22, 25
Andaman Nightjar *Caprimulgus andamanicus* 16, 17
Andaman Scops Owl *Otus balli* 17, 18, 19, 20, 22
Andaman Serpent Eagle *Spilornis elgini* 12, 17, 18, 19, 21, 22
Andaman Shama *Kittacincla albiventris* 17, 18, 19
Andaman Teal *Anas albogularis* 16, 18, 22, 23
Andaman Treepie *Dendrocitta bayleii* 16, 17, 18, 20, 21, 22, 25
Andaman Wood Pigeon *Columba palumboides* 12, 17, 18, 22, 25
Andaman Woodpecker *Dryocopus hodgei* 12, 18, 19, 21, 22, 25
Arctic Skua *Stercorarius parasiticus* 144, 146
Ashy Bulbul *Hemixos flavala* 187, 255
Ashy Drongo *Dicrurus leucophaeus* 30, 68, 94, 256
Ashy Minivet *Pericrocotus divaricatus* 120
Ashy Prinia *Prinia socialis* 27, 28, 156, 166, 204, 216
Ashy Wood Pigeon *Columba pulchricollis* 176, 252, 265
Ashy Woodswallow *Artamus fuscus* 27, 70
Ashy-crowned Sparrow Lark *Eremopterix griseus* 30, 128, 130, 168, 216
Ashy-throated Warbler *Phylloscopus maculipennis* 187
Asian Barred Owlet *Glaucidium cuculoides* 68
Asian Brown Flycatcher *Muscicapa dauurica* 24, 174
Asian Dowitcher *Limnodromus semipalmatus* 200, 203, 234
Asian Emerald Cuckoo *Chrysococcyx maculatus* 59, 71, 184, 266
Asian Emerald Dove *Chalcophaps indica* 21, 46, 51, 78, 171, 173, 181, 266
Asian Fairy-bluebird *Irena puella* 19, 20, 51,78, 172, 173, 175,
Asian Glossy Starling *Aplonis panayensis* 20
Asian Koel *Eudynamys scolopaceus* 37, 78, 80, 123, 200, 248
Asian Openbill *Anastomus oscitans* 28, 32, 36, 37, 57, 70, 71, 78,132, 133, 138, 162, 166, 175, 199, 201, 246
Asian Palm Swift *Cypsiurus balasiensis* 28, 70, 138
Asian Pied Starling *Gracupica contra* 70
Austen's Brown Hornbill *Anorrhinus austeni* 52,178, 194
Baer's Pochard *Aythya baeri* 12, 46, 56, 57, 58, 59, 64, 66, 87, 94, 98, 150, 177, 197, 198, 202, 203, 214, 225, 251, 261, 270
Baikal Teal *Sibirionetta formosa* 67, 150, 151, 177
Baillon's Crake *Zapornia pusilla* 86,

Bank Myna *Acridotheres ginginianus* 204, 216
Bar-headed Goose *Anser indicus* 10, 30, 32, 67, 118, 119, 166, 175, 200, 218, 248
Bar-tailed Godwit *Limosa lapponica* 32, 146
Bar-tailed Treecreeper *Certhia himalayana* 106, 114
Bar-winged Flycatcher-shrike *Hemipus picatus* 78
Bar-winged Wren Babbler *Spelaeornis troglodytoides* 45
Barn Swallow *Hirundo rustica* 20, 28, 29, 174, 218
Barred Buttonquail *Turnix suscitator* 128, 152
Barred Cuckoo Dove *Macropygia unchall* 46, 176, 224
Bay Woodpecker *Blythipicus pyrrhotis* 69, 186, 194, 224, 226, 265, 266
Bay-backed Shrike *Lanius vittatus* 154
Baya Weaver *Ploceus philippinus* 243,
Beach Thick-knee *Esacus magnirostris* 16, 17
Bean Goose *Anser fabalis* 207
Bearded Vulture *Gypaetus barbatus* 102, 108, 109, 114, 116, 230, 252
Beautiful Nuthatch *Sitta formosa* 12, 41, 42, 43, 44, 46, 47, 52, 54, 55, 184, 262, 264, 268 184,
Beautiful Sibia *Heterophasia pulchella* 42, 44, 45, 46, 52
Bengal Bushlark *Mirafra assamica* 261
Bengal Florican *Houbaropsis bengalensis* 8, 12, 57, 58, 60, 61, 62, 64, 66, 185, 244, 245, 261, 265, 266
Besra *Accipiter virgatus* 114, 236
Bimaculated Lark *Melanocorypha bimaculata* 84, 92, 218, 219
Black Baza *Aviceda leuphotes* 21, 56, 58, 262, 263
Black Bittern *Ixobrychus flavicollis* 136
Black Bulbul *Hypsipetes leucocephalus* 186, 187, 224, 256
Black Drongo *Dicrurus macrocercus* 28, 30, 36, 68, 70, 78, 158, 168, 174, 202, 212, 224, 248, 256
Black Eagle *Ictinaetus malaiensis* 48, 78, 187, 224, 237, 258, 263, 265
Black Francolin *Francolinus francolinus* 152, 153, 159, 186, 248, 253, 258
Black Kite *Milvus migrans* 34, 70, 80, 112, 152, 174
Black Noddy *Anous minutus* 144, 146
Black Redstart *Phoenicurus ochruros* 98, 104, 128, 167, 218
Black Stork *Ciconia nigra* 166, 242, 247
Black-and-orange Flycatcher *Ficedula nigrorufa* 11, 121, 136, 142, 236, 238, 240
Black-and-yellow Grosbeak *Mycerobas icterioides* 103, 250
Black-backed Forktail *Enicurus immaculatus* 62, 186
Black-bellied Sandgrouse *Pterocles orientalis* 210, 221
Black-bellied Storm-petrel *Fregetta tropica* 146
Black-bellied Tern *Sterna acuticauda* 12, 58, 66, 150, 157, 198, 202, 206, 214, 216, 249, 270, 271
Black-breasted Parrotbill *Paradoxornis flavirostris* 55, 57, 58, 59, 60, 66, 67, 228, 262, 264, 266, 267
Black-breasted Thrush *Turdus dissimilis* 53
Black-breasted Weaver *Ploceus benghalensis* 153, 244
Black-browed Reed Warbler *Acrocephalus bistrigiceps* 44, 50, 52, 260
Black-browed Tit *Aegithalos iouschistos* 54, 55
Black-capped Kingfisher *Halcyon pileata* 36, 74, 82, 83
Black-crowned Night Heron *Nycticorax nycticorax* 28, 36, 77, 82, 112, 247
Black-eared Shrike-babbler *Pteruthius melanotis* 182
Black-faced Laughingthrush *Trochalopteron affine* 45, 49, 68, 226
Black-headed Cuckooshrike *Lalage melanoptera* 130,

295

168, 216
Black-headed Greenfinch *Chloris ambigua* 47, 54
Black-headed Gull *Chroicocephalus ridibundus* 94, 105, 199
Black-headed Ibis *Threskiornis melanocephalus* 12, 28, 32, 36, 50, 58, 82, 86, 88, 94, 95, 124, 132, 133, 134, 138, 149, 150, 158, 156, 160, 162, 198, 208, 214, 216, 234, 246, 254, 266
Black-headed Jay *Garrulus lanceolatus* 106, 107, 256, 259
Black-headed Shrike-babbler *Pteruthius rufiventer* 265
Black-hooded Oriole *Oriolus xanthornus* 68, 137, 173, 246, 261
Black-lored Tit *Machlolophus xanthogenys*
Black-naped Monarch *Hypothymis azurea* 19, 126, 127, 140, 166, 167, 171, 172, 174, 255
Black-naped Oriole *Oriolus chinensis* 20
Black-necked Crane *Grus nigricollis* 10, 45, 110, 111, 117, 118, 119, 262
Black-necked Grebe *Podiceps nigricollis* 179, 225
Black-necked Stork *Ephippiorhynchus asiaticus* 12, 57, 58, 60, 64, 66, 86, 88, 94, 149, 150, 151, 154, 160, 162, 179, 198, 208, 214, 216, 234, 243, 246, 249, 254, 255, 261, 270, 271
Black-tailed Crake *Zapornia bicolor* 40, 45, 61
Black-tailed Godwit *Limosa limosa* 26, 32, 196, 201, 203
Black-throated Accentor *Prunella atrogularis* 102, 228
Black-throated Munia *Lonchura kelaarti* 72, 76, 126, 239
Black-throated Parrotbill *Suthora nipalensis* 182, 263, 268
Black-throated Sunbird *Aethopyga saturata* 103
Black-throated Thrush *Turdus atrogularis* 102
Black-throated Tit *Aegithalos concinnus* 258
Black-winged Cuckooshrike *Lalage melaschistos* 186, 187, 266
Black-winged Kite *Elanus caeruleus* 34, 112, 152, 174, 202, 212, 240
Black-winged Stilt *Himantopus himantopus* 28, 36, 153, 233, 92
Blanford's Rosefinch *Agraphospiza rubescens* 44
Blanford's Snowfinch *Pyrgilauda blanfordi* 118
Blood Pheasant *Ithaginis cruentus* 45, 46, 222, 229, 230, 261
Blue Pitta *Hydrornis cyaneus* 176, 182, 183
Blue Rock Thrush *Monticola solitarius* 98, 158, 182
Blue Whistling Thrush *Myophonus caeruleus* 104, 224
Blue-bearded Bee-eater *Nyctyornis athertoni* 82, 158
Blue-breasted Quail *Synoicus chinensis* 120
Blue-capped Redstart *Phoenicurus coeruleocephala* 102, 106
Blue-cheeked Bee-eater *Merops persicus* 106, 152
Blue-eared Kingfisher *Alcedo meninting* 68, 72, 73, 74, 75, 172, 173
Blue-faced Malkoha *Phaenicophaeus viridirostris* 30, 126, 140
Blue-fronted Robin *Cinclidium frontale* 43
Blue-naped Pitta *Hydrornis nipalensis* 51, 53, 58, 59, 184
Blue-tailed Bee-eater *Merops philippinus* 20, 28, 209, 255
Blue-throated Barbet *Psilopogon asiaticus* 186, 194, 205, 224, 256, 257, 258, 261
Bluethroat *Luscinia svecica* 138, 158, 167, 201
Blyth's Kingfisher *Alcedo hercules* 40, 41, 50, 51, 58, 64, 66, 182, 183, 187
Blyth's Leaf Warbler *Phylloscopus reguloides* 45
Blyth's Reed Warbler *Acrocephalus dumetorum* 29,

240
Blyth's Rosefinch *Carpodacus grandis* 118
Blyth's Tragopan *Tragopan blythii* 12,41, 42, 43, 44, 46, 49, 52, 53, 177, 181, 182, 188, 189, 190, 192, 226
Bonelli's Eagle *Aquila fasciata* 148, 166, 168, 175, 236
Booted Eagle *Hieraaetus pennatus* 102, 181
Brahminy Kite *Haliastur indus* 33, 36, 80, 82, 132, 133, 139, 172, 174, 175, 201, 202
Brahminy Starling *Sturnia pagodarum* 98, 138, 156, 166, 168, 175, 198, 204, 216
Brambling *Fringilla montifringilla* 100
Brandt's Mountain Finch *Leucosticte brandti* 45
Bridled Tern *Onychoprion anaethetus* 145, 146, 147, 232
Bristled Grassbird *Chaetornis striata* 60, 64, 135, 150, 151, 197, 208, 262
Broad-billed Sandpiper *Calidris falcinellus* 232
Broad-billed Warbler *Tickellia hodgsoni* 44, 47, 50, 195, 226
Broad-tailed Grassbird *Schoenicola platyurus* 11, 121, 135, 136, 140, 142, 143, 233, 234, 236, 238
Bronze-winged Jacana *Metopidius indicus* 80, 149, 166, 246, 247, 261
Bronzed Drongo *Dicrurus aeneus* 68
Brown Accentor *Prunella fulvescens* 116
Brown Booby *Sula leucogaster* 120
Brown Bullfinch *Pyrrhula nipalensis* 43, 268
Brown Bush Warbler *Locustella luteoventris* 44
Brown Crake *Zapornia akool* 200
Brown Dipper *Cinclus pallasii* 102
Brown Fish Owl *Ketupa zeylonensis* 24, 76, 121, 128, 140, 255
Brown Hawk Owl *Ninox scutulata* 80, 236, 249
Brown Noddy *Anous stolidus* 145, 146
Brown Parrotbill *Cholornis unicolor* 45, 260
Brown Rock Chat *Oenanthe fusca* 166, 216
Brown Shrike *Lanius cristatus* 18, 209, 261
Brown Skua *Stercorarius antarcticus* 145
Brown Wood Owl *Strix leptogrammica* 256, 257, 265
Brown-backed Needletail *Hirundapus giganteus* 20, 22, 120, 238, 239, 240
Brown-capped Laughingthrush *Trochalopteron austeni* 188, 189, 190, 191, 192
Brown-capped Pygmy Woodpecker *Dendrocopos nanus* 246
Brown-fronted Pied Woodpecker *Dendrocopos auriceps* 258
Brown-headed Barbet *Psilopogon zeylanicus* 24, 38, 166, 202, 205, 246
Brown-headed Gull *Chroicocephalus brunnicephalus* 32, 36, 81, 116, 175, 200, 203, 208
Brown-winged Kingfisher *Pelargopsis amauroptera* 270, 271
Brownish-flanked Bush Warbler *Horornis fortipes* 44
Buff-barred Warbler *Phylloscopus pulcher* 187
Buffy Fish Owl *Ketupa ketupu* 68, 261, 270, 271
Bugun Liocichla *Liocichla bugunorum* 40, 42
Cachar Wedge-billed Babbler *Stachyris roberti* 189, 190
Caspian Gull *Larus cachinnans* 202
Caspian Plover *Charadrius asiaticus* 134
Caspian Tern *Hydroprogne caspia* 32, 138, 146, 147
Cattle Egret *Bubulcus ibis* 34, 36, 68, 80, 104, 152, 158, 168, 174, 200, 212, 246
Changeable Hawk Eagle *Nisaetus cirrhatus* 20, 21, 34, 122, 123, 132, 133, 140, 166, 236
Cheer Pheasant *Catreus wallichii* 10, 100, 101, 102, 106, 107, 250, 251, 252, 258
Chestnut Thrush *Turdus rubrocanus* 68, 230, 264
Chestnut-bellied Nuthatch *Sitta cinnamoventris* 205

Chestnut-bellied Sandgrouse *Pterocles exustus* 30, 218, 248
Chestnut-breasted Hill Partridge *Arborophila mandellii* 12, 47, 226, 228, 230, 262, 268
Chestnut-crowned Laughingthrush *Trochalopteron erythrocephalum* 258
Chestnut-crowned Warbler *Phylloscopus castaniceps* 224, 252
Chestnut-headed Bee-eater *Merops leschenaulti* 20, 21, 156
Chestnut-headed Tesia *Cettia castaneocoronata* 43
Chestnut-tailed Minla *Chrysominla strigula* 263,
Chestnut-tailed Starling *Sturnia malabarica* 138, 158, 168, 204, 261
Chestnut-vented Nuthatch *Sitta nagaensis* 188, 189, 190, 191, 192
Chestnut-winged Cuckoo *Clamator coromandus* 134, 267
Chinese Francolin *Francolinus pintadeanus* 176
Chinese Pond Heron *Ardeola bacchus* 40, 48, 180
Chukar Partridge *Alectoris chukar* 102, 108, 116, 118, 230
Cinereous Tit *Parus cinereus* 68, 112, 122, 167, 175
Cinereous Vulture *Aegypius monachus* 58, 60, 66, 88, 162, 212, 213, 214, 216, 218, 220, 254
Cinnamon Bittern *Ixobrychus cinnamomeus* 74, 139, 200, 202, 262
Citrine Wagtail *Motacilla citreola* 112
Clamorous Reed Warbler *Acrocephalus stentoreus* 112
Coal Tit *Periparus ater* 252
Collared Falconet *Microhierax caerulescens* 52, 205, 263
Collared Kingfisher *Todiramphus chloris* 20, 82, 83, 271
Collared Owlet *Glaucidium brodiei* 114, 115
Collared Pratincole *Glareola pratincola* 90, 203
Collared Scops Owl *Otus lettia* 24
Collared Treepie *Dendrocitta frontalis* 53
Comb Duck *Sarkidiornis melanotos* 167, 248
Common Babbler *Argya caudata* 174
Common Buzzard *Buteo buteo* 42, 45, 102, 213
Common Chaffinch *Fringilla coelebs* 206
Common Chiffchaff *Phylloscopus collybita* 246
Common Coot *Fulica atra* 28, 112, 200, 240, 247, 248, 263
Common Crane *Grus grus* 90, 92, 93, 152, 214
Common Cuckoo *Cuculus canorus* 112, 258
Common Golden-backed Woodpecker *Dinopium javanense*
Common Goldeneye *Bucephala clangula* 252
Common Green Magpie *Cissa chinensis* 259
Common Greenshank *Tringa nebularia* 32, 83, 104, 153
Common Hawk Cuckoo *Hierococcyx varius* 138, 207, 255
Common Hill Myna *Gracula religiosa* 68, 71
Common Hill Partridge *Arborophila torqueola* 43, 102, 186, 250, 252, 268, 269
Common Hoopoe *Upupa epops* 24, 30, 38, 106, 152, 174, 202, 240, 246
Common Iora *Aegithina tiphia* 30, 31, 68, 174, 175
Common Kestrel *Falco tinnunculus* 34, 45, 154, 166, 175, 212, 240, 163
Common Kingfisher *Alcedo atthis* 82, 112, 121, 122, 168, 202, 212, 255
Common Moorhen *Gallinula chloropus* 28, 87, 138, 160, 181, 248
Common Myna *Acridotheres tristis* 70, 138, 174, 212, 224
Common Pochard *Aythya ferina* 105, 160, 155, 157, 215, 248, 263
Common Quail *Coturnix coturnix* 248
Common Raven *Corvus corax* 206
Common Redshank *Tringa totanus* 28, 36, 37, 138
Common Rosefinch *Carpodacus erythrinus* 39, 240
Common Sandpiper *Actitis hypoleucos* 105, 128, 146, 218
Common Shelduck *Tadorna tadorna* 150, 200, 248
Common Snipe *Gallinago gallinago* 33, 112, 173
Common Starling *Sturnus vulgaris* 112
Common Tailorbird *Orthotomus sutorius* 68, 142, 175, 246
Common Teal *Anas crecca* 28, 67, 92, 105, 112, 153, 158, 168, 175, 215, 248, 262, 263
Common Whitethroat *Sylvia communis* 93
Common Woodshrike *Tephrodornis pondicerianus* 130, 131, 168, 205, 209, 216
Coppersmith Barbet *Psilopogon haemacephalus* 106, 121, 175, 246
Coral-billed Scimitar Babbler *Pomatorhinus ferruginosus* 43, 263
Crab-plover *Dromas ardeola* 84
Cream-coloured Courser *Cursorius cursor* 10, 90, 91, 210, 221
Crested Finchbill *Spizixos canifrons* 182, 183, 188, 191
Crested Goshawk *Accipiter trivirgatus* 68, 140, 173, 177, 236, 252
Crested Kingfisher *Megaceryle lugubris* 256, 257
Crested Lark *Galerida cristata* 90, 92, 96
Crested Serpent Eagle *Spilornis cheela* 20, 57, 68, 78, 98, 130, 158, 174, 166, 168, 236, 240, 263
Crested Treeswift *Hemiprocne coronata* 46, 78, 137, 166, 174
Crimson Sunbird *Aethopyga siparaja* 228
Crimson-backed Sunbird *Leptocoma minima* 75, 76, 122, 126, 136, 140, 142, 171, 172, 233, 236, 238, 239
Crimson-breasted Pied Woodpecker *Dendrocopos cathpharius* 225
Crimson-browed Finch *Carpodacus subhimachalus* 45
Crow-billed Drongo *Dicrurus annectens* 51, 62, 68
Curlew Sandpiper *Calidris ferruginea* 32, 135
Dalmatian Pelican *Pelecanus crispus* 85, 86, 88, 90, 93, 94, 98, 150, 151, 214
Darjeeling Pied Woodpecker *Dendrocopos darjellensis* 49, 184, 264
Dark-fronted Babbler *Rhopocichla atriceps* 76, 78, 126, 127, 140, 142
Dark-rumped Swift *Apus acuticauda* 177, 184, 185, 187, 189, 190, 192
Dark-sided Thrush *Zoothera marginata* 47
Daurian Redstart *Phoenicurus auroreus* 48
Demoiselle Crane *Grus virgo* 30, 86, 96, 175, 216, 218
Desert Lark *Ammomanes deserti* 220
Desert Wheatear *Oenanthe deserti* 104, 218
Dollarbird *Eurystomus orientalis* 19, 20, 22, 106, 267
Dunlin *Calidris alpina* 200
Dusky Crag Martin *Ptyonoprogne concolor* 30, 174
Dusky Eagle Owl *Bubo coromandus* 148
Eastern Grass Owl *Tyto longimembris* 120
Eastern Imperial Eagle *Aquila heliaca* 11, 58, 62, 88, 96, 97, 98, 114, 121, 162, 208, 212, 214, 216, 218, 254, 261, 264
Eastern Orphean Warbler *Sylvia crassirostris* 175
Eastern Water Rail *Rallus indicus* 180, 200
Edible-nest Swiftlet *Aerodramus fuciphagus* 12, 142, 236
Egyptian Vulture *Neophron percnopterus* 212, 157, 160
Eurasian Collared Dove *Streptopelia decaocto* 29, 30, 70, 92, 152, 218

Eurasian Crag Martin *Ptyonoprogne rupestris* 172
Eurasian Curlew *Numenius arquata* 67, 82, 145, 146
Eurasian Hobby *Falco subbuteo* 202, 209, 263
Eurasian Jackdaw *Corvus monedula* 110, 112
Eurasian Jay *Garrulus glandarius* 256
Eurasian Oystercatcher *Haematopus ostralegus* 84
Eurasian Sparrowhawk *Accipiter nisus* 102, 128, 166, 202, 209
Eurasian Spoonbill *Platalea leucorodia* 90, 132, 133, 153, 216, 234, 247, 255
Eurasian Tree Sparrow *Passer montanus* 268
Eurasian Wigeon *Mareca penelope* 67, 90, 112, 113, 135, 200, 201, 203, 206, 248, 262, 263
Eurasian Woodcock *Scolopax rusticola* 102, 188, 229
Eurasian Wren *Troglodytes troglodytes* 114
Eurasian Wryneck *Jynx torquilla* 45
European Bee-eater *Merops apiaster* 112
European Goldfinch *Carduelis carduelis* 108, 112, 252
European Nightjar *Caprimulgus europaeus* 93
European White Stork *Ciconia ciconia* 86, 88, 89, 90, 92, 178
Eyebrowed Thrush *Turdus obscurus* 189
Eyebrowed Wren Babbler *Napothera epilepidota* 53
Falcated Duck *Mareca falcata* 56, 58,179, 200
Ferruginous Duck *Aythya nyroca* 12, 46, 67, 87, 135, 160, 179, 200, 248
Finn's Weaver *Ploceus megarhynchus* 56, 58, 60, 61, 64, 66, 71, 150, 242, 262
Fire-breasted Flowerpecker *Dicaeum ignipectus* 43, 224, 265
Fire-capped Tit *Cephalopyrus flammiceps* 100, 114, 118, 225, 252
Fire-tailed Myzornis *Myzornis pyrrhoura* 43, 45, 47, 49, 50, 227, 228, 268
Fire-tailed Sunbird *Aethopyga ignicauda* 265, 268
Flame-throated Bulbul *Pycnonotus gularis* 126
Flavescent Bulbul *Pycnonotus flavescens* 182, 183, 188
Flesh-footed Shearwater *Ardenna carneipes* 135, 144, 146
Forest Owlet *Heteroglaux blewitti* 11, 168, 169
Forest Wagtail *Dendronanthus indicus* 78, 132, 133, 167
Fulvous Parrotbill *Suthora fulvifrons* 268
Fulvous Whistling Duck *Dendrocygna bicolor* 28, 57, 200, 201
Fulvous-breasted Pied Woodpecker *Dendrocopos macei* 261
Gadwall *Mareca strepera* 28, 32, 46, 64, 153, 203, 248, 263
Garganey *Spatula querquedula* 28, 32, 71, 81, 112, 113, 200, 201, 203
Glossy Ibis *Plegadis falcinellus* 28, 37, 67, 174
Godlewski's Bunting *Emberiza godlewskii* 55
Gold-naped Finch *Pyrrhoplectes epauletta* 108
Golden Bush Robin *Tarsiger chrysaeus* 45, 102, 187
Golden Eagle *Aquila chrysaetos* 102, 108, 116, 118, 119, 222, 230, 252, 268
Golden-breasted Fulvetta *Lioparus chrysotis* 222, 265, 268
Golden-crested Myna *Ampeliceps coronatus* 52
Golden-fronted Leafbird *Chloropsis aurifrons* 39
Golden-throated Barbet *Psilopogon franklinii* 184, 187, 194, 265
Goliath Heron *Ardea goliath* 56, 70, 71, 270, 271
Gould's Shortwing *Heteroxenicus stellatus* 44, 47, 226, 228, 264
Grandala *Grandala coelicolor* 45, 114, 228, 230
Great Barbet *Psilopogon virens* 106, 224, 256, 258
Great Cormorant *Phalacrocorax carbo* 36, 57
Great Crested Grebe *Podiceps cristatus* 11, 57, 94, 118, 119, 209
Great Eared Nightjar *Lyncornis macrotis* 46,134, 135, 137, 238, 239
Great Egret *Ardea alba* 28, 30, 32, 80
Great Frigatebird *Fregata minor* 135, 144, 146, 232
Great Hornbill *Buceros bicornis* 40, 42, 48, 50, 52, 58, 60, 66, 121, 140, 178, 194, 238, 239, 266, 267
Great Indian Bustard *Ardeotis nigriceps* 10, 11, 27, 30, 85, 92, 93, 165, 210, 211, 220, 221
Great Knot *Calidris tenuirostris* 58, 134, 144, 146, 200
Great Parrotbill *Conostoma aemodium* 228
Great Rosefinch *Carpodacus rubicilla* 108, 116, 118
Great Slaty Woodpecker *Mulleripicus pulverulentus* 70, 71, 242, 254, 255
Great Snipe *Gallinago media* 232
Great Thick-knee *Esacus recurvirostris* 132, 133, 158
Great White Pelican *Pelecanus onocrotalus* 85, 153, 200
Greater Adjutant *Leptoptilos dubius* 12, 56, 58, 60, 64, 66, 70, 71, 177, 185, 186, 261, 270
Greater Coucal *Centropus sinensis* 28, 35, 36, 80, 138, 174, 218
Greater Crested Tern *Thalasseus bergii* 82, 146
Greater Golden-backed Woodpecker *Chrysocolaptes guttacristatus* 68, 227
Greater Hoopoe Lark *Alaemon alaudipes* 10, 92, 220
Greater Necklaced Laughingthrush *Garrulax pectoralis* 186, 224
Greater Painted-snipe *Rostratula benghalensis* 138
Greater Racket-tailed Drongo *Dicrurus paradiseus* 18, 20, 71, 75, 239
Greater Sand Plover *Charadrius leschenaultii* 147
Greater Scaup *Aythya marila* 242
Greater Short-toed Lark *Calandrella brachydactyla* 30
Greater Spotted Eagle *Clanga clanga* 11, 24, 38, 52, 58, 60, 62, 66, 67, 71, 80, 86, 88, 92, 94, 96, 97, 98, 121, 124, 132, 136, 149, 150, 151, 155, 162, 166, 170, 171, 175, 177, 178, 185, 198, 204, 208, 214, 220, 226, 228, 244, 246, 247, 248, 254, 268, 270, 271
Greater White-fronted Goose *Anser albifrons* 100
Green Bee-eater *Merops orientalis* 28, 36, 38, 152, 168, 172, 174, 202
Green Cochoa *Cochoa viridis* 51, 53, 68, 70, 71, 193, 266
Green Imperial Pigeon *Ducula aenea* 18, 20, 21, 76, 140
Green Peafowl *Pavo muticus* 177, 178, 180, 182
Green Sandpiper *Tringa ochropus* 28, 105
Green Shrike-babbler *Pteruthius xanthochlorus* 43, 55
Green-backed Tit *Parus monticolus* 106, 114, 187, 224, 256, 267
Green-billed Malkoha *Phaenicophaeus tristis* 28, 59, 76, 124, 204, 205, 236
Greenish Leaf Warbler *Phylloscopus trochiloides* 258
Grey Bushchat *Saxicola ferreus* 158
Grey Francolin *Francolinus pondicerianus* 30, 34, 99, 128, 152, 153, 200
Grey Heron *Ardea cinerea* 28, 32, 36, 90, 112, 234
Grey Hypocolius *Hypocolius ampelinus* 84, 85, 92, 93
Grey Junglefowl *Gallus sonneratii* 34, 78, 120, 122, 123, 239, 240
Grey Nightjar *Caprimulgus jotaka* 46, 242, 258
Grey Peacock Pheasant *Polyplectron bicalcaratum* 42, 53, 62, 68, 69, 176, 186, 194, 230
Grey Plover *Pluvialis squatarola* 146, 200
Grey Sibia *Heterophasia gracilis* 52, 190, 191, 194, 268
Grey Treepie *Dendrocitta formosae* 186, 194, 258, 253
Grey Wagtail *Motacilla cinerea* 92, 268
Grey-backed Shrike *Lanius tephronotus* 68, 116, 182, 212

Grey-bellied Cuckoo *Cacomantis passerinus* 174
Grey-capped Pygmy Woodpecker *Dendrocopos canicapillus* 58, 106
Grey-chinned Minivet *Pericrocotus solaris* 186, 225
Grey-crested Tit *Lophophanes dichrous* 225
Grey-crowned Prinia *Prinia cinereocapilla* 60, 62, 228, 244, 250, 254, 259, 262
Grey-crowned Warbler *Phylloscopus tephrocephalus* 192
Grey-fronted Green Pigeon *Treron affinis* 75, 122, 181,
Grey-headed Bulbul *Brachypodius priocephalus* 11, 73, 74, 75, 76, 121, 122, 136, 140, 142, 233, 238
Grey-headed Bullfinch *Pyrrhula erythaca* 47
Grey-headed Canary-flycatcher *Culicicapa ceylonensis* 224, 256
Grey-headed Fish Eagle *Icthyophaga ichthyaetus* 57, 58, 59, 62, 64, 66, 121, 124, 140, 149, 150, 156, 158, 162, 166, 214, 242, 254
Grey-headed Lapwing *Vanellus cinereus* 67
Grey-headed Parakeet *Psittacula finschii* 186
Grey-headed Parrotbill *Psittiparus gularis* 43
Grey-headed Woodpecker *Picus canus* 101
Grey-hooded Leaf Warbler *Phylloscopus xanthoschistos* 187
Grey-necked Bunting *Emberiza buchanani* 128, 175
Grey-sided Thrush *Turdus feae* 185, 192
Grey-throated Babbler *Stachyris nigriceps* 194
Grey-winged Blackbird *Turdus boulboul* 106, 224, 264
Greylag Goose *Anser anser* 57, 90, 112, 248
Griffon Vulture *Gyps fulvus* 57, 212
Ground Tit *Pseudopodoces humilis* 110, 117
Hair-crested Drongo *Dicrurus hottentottus* 224
Heart-spotted Woodpecker *Hemicircus canente* 122, 126, 164, 173, 238, 239, 240
Hen Harrier *Circus cyaneus* 105, 209, 268
Hill Blue Flycatcher *Cyornis banyumas* 48
Hill Pigeon *Columba rupestris* 229
Hill Prinia *Prinia superciliosa* 187
Himalayan Beautiful Rosefinch *Carpodacus pulcherrimus* 225, 252, 268
Himalayan Bulbul *Pycnonotus leucogenis* 106, 253, 256, 266
Himalayan Buzzard *Buteo refectus* 45, 180, 230
Himalayan Cuckoo *Cuculus saturatus* 222
Himalayan Cutia *Cutia nipalensis* 181, 263
Himalayan Golden-backed Woodpecker *Dinopium shorii* 62
Himalayan Monal *Lophophorus impejanus* 4, 45, 46, 100, 102, 114, 229, 230, 250, 252, 253, 268
Himalayan Pied Woodpecker *Dendrocopos himalayensis* 114, 115, 256, 258
Himalayan Quail *Ophrysia superciliosa* 251
Himalayan Rubythroat *Calliope pectoralis* 111, 114
Himalayan Snowcock *Tetraogallus himalayensis* 10, 100, 102, 108, 116, 222, 229
Himalayan Swiftlet *Aerodramus brevirostris* 45, 224, 252
Himalayan Vulture *Gyps himalayensis* 102, 108, 114, 116, 212, 213, 224, 228, 230, 252, 259
Himalayan White-browed Rosefinch *Carpodacus thura* 45, 106,
Hoary-throated Barwing *Sibia nipalensis* 42, 44, 50, 226, 228
Hodgson's Bushchat *Saxicola insignis* 12, 58, 60, 62, 243, 254, 255
Hodgson's Frogmouth *Batrachostomus hodgsoni* 40, 42, 44
Hodgson's Redstart *Phoenicurus hodgsoni* 45, 226
Hodgson's Treecreeper *Certhia hodgsoni* 258

Hooded Pitta *Pitta sordida* 70, 71, 181
Horned Lark *Eremophila alpestris* 10, 231
House Crow *Corvus splendens* 70, 218, 224, 240
House Sparrow *Passer domesticus* 70, 118, 212, 224, 240
Hume's Bush Warbler *Horornis brunnescens* 45
Hume's Hawk Owl *Ninox obscura* 19, 21
Hume's Leaf Warbler *Phylloscopus humei* 258
Ibisbill *Ibidorhyncha struthersii* 51, 53, 63, 116, 222, 254
Indian Black Ibis *Pseudibis papillosa* 34, 152, 166, 168
Indian Blackbird *Turdus simillimus* 98, 175, 240
Indian Blue Robin *Larvivora brunnea* 45, 98, 141, 264
Indian Bushlark *Mirafra erythroptera* 30, 204, 216
Indian Cormorant *Phalacrocorax fuscicollis* 36, 153
Indian Courser *Cursorius coromandelicus* 30, 128, 148, 168, 212, 216
Indian Cuckoo *Cuculus micropterus* 20, 224
Indian Eagle Owl *Bubo bengalensis* 92, 97, 128, 230
Indian Golden Oriole *Oriolus kundoo* 36, 82, 112, 124, 132, 202
Indian Grey Hornbill *Ocyceros birostris* 30, 124, 154, 156, 158, 168, 246, 262
Indian Nightjar *Caprimulgus asiaticus* 130, 166, 168, 215, 216
Indian Nuthatch *Sitta castanea* 246
Indian Paradise-flycatcher *Terpsiphone paradisi* 88, 98, 130, 132, 133, 174, 197, 255
Indian Peafowl *Pavo cristatus* 80, 124, 128, 152, 155, 158, 166, 167, 168, 174, 204, 216, 246, 248
Indian Pitta *Pitta brachyura* 24, 78, 132, 133, 154, 162, 166, 172, 173, 197, 240
Indian Pond Heron *Ardeola grayii* 28, 34, 36, 68, 80
Indian Robin *Saxicoloides fulicatus* 128, 130, 138, 156, 166, 167, 168, 175, 216
Indian Roller *Coracias benghalensis* 70, 80, 106, 153, 158, 167, 168, 202, 212, 218
Indian Scimitar Babbler *Pomatorhinus horsfieldii* 74, 76, 122, 124, 140, 142, 170, 171, 204
Indian Scops Owl *Otus bakkamoena* 154, 219
Indian Silverbill *Euodice malabarica* 14, 27, 218
Indian Skimmer *Rynchops albicollis* 58, 66, 86, 88, 94, 98, 149, 150, 151, 170, 196, 198, 202, 203, 207, 208, 209, 211, 214, 216, 248, 249, 262
Indian Spot-billed Duck *Anas poecilorhyncha* 28, 56, 81, 160, 203, 248
Indian Spotted Creeper *Salpornis spilonota* 210, 211, 219
Indian Spotted Eagle *Clanga hastata* 90, 91, 157, 215
Indian Swiftlet *Aerodramus unicolor* 122, 134, 239, 238
Indian Thick-knee *Burhinus indicus* 166, 248
Indian Vulture *Gyps indicus* 174, 211, 212
Intermediate Egret *Ardea intermedia* 80, 138
Jack Snipe *Lymnocryptes minimus* 184
Japanese Quail *Coturnix japonica* 176, 179
Japanese Sparrowhawk *Accipiter gularis* 22
Jerdon's Babbler *Chrysomma altirostre* 12, 57, 58, 60, 66, 67, 262, 266, 267
Jerdon's Baza *Aviceda jerdoni* 26, 51, 205, 236, 265
Jerdon's Bushchat *Saxicola jerdoni* 56, 67
Jerdon's Bushlark *Mirafra affinis* 67, 122, 128
Jerdon's Courser *Rhinoptilus bitorquatus* 11, 26, 27, 34, 35
Jerdon's Leafbird *Chloropsis jerdoni* 167
Jerdon's Nightjar *Caprimulgus atripennis* 74, 76, 81, 120, 124, 164
Jouanin's Petrel *Bulweria fallax* 135
Jungle Babbler *Turdoides striata* 30, 38, 78, 130, 156, 158, 166, 168, 204, 216
Jungle Bush Quail *Perdicula asiatica* 30, 34, 128, 129, 130, 166, 167, 168, 174, 200, 204, 216, 248

299

Jungle Myna *Acridotheres fuscus* 58, 78, 80, 98, 138, 158
Jungle Nightjar *Caprimulgus indicus* 73
Jungle Prinia *Prinia sylvatica* 166, 168, 216
Kalij Pheasant *Lophura leucomelanos* 43, 46, 68, 102, 186, 225, 227, 228, 252, 253, 254, 256, 257, 268
Kashmir Flycatcher *Ficedula subrubra* 101, 111, 112, 114, 208, 240
Kashmir Nuthatch *Sitta cashmirensis* 101, 110, 111
Kentish Plover *Charadrius alexandrinus* 200
Kessler's Thrush *Turdus kessleri* 44
Koklass Pheasant *Pucrasia macrolopha* 10, 100, 102, 106, 114, 252, 259
Laggar Falcon *Falco jugger* 84, 92, 213, 218, 220
Large Cuckooshrike *Coracina javensis* 18
Large Grey Babbler *Argya malcolmi* 30, 130, 156, 216, 218
Large Hawk Cuckoo *Hierococcyx sparverioides* 224
Large Niltava *Niltava grandis* 68, 193
Large Scimitar Babbler *Erythrogenys hypoleucos* 53
Large Woodshrike *Tephrodornis virgatus* 68
Large-billed Crow *Corvus macrorhynchos* 70, 218
Large-billed Reed Warbler *Acrocephalus orinus* 239
Large-tailed Nightjar *Caprimulgus macrurus* 173
Laughing Dove *Streptopelia senegalensis* 174, 248
Lesser Adjutant *Leptoptilos javanicus* 11, 57, 58, 60, 62, 64, 66, 68, 71, 80, 82, 96, 97, 98, 122, 124, 149, 150, 151, 155, 156,157, 158, 159, 160, 166, 167, 170, 178, 197, 198, 202, 203, 208, 214, 216, 233, 236, 244, 254, 262, 266, 267, 270, 271
Lesser Coucal *Centropus bengalensis* 196, 248
Lesser Crested Tern *Thalasseus bengalensis* 32, 82, 145, 146
Lesser Cuckoo *Cuculus poliocephalus* 265
Lesser Fish Eagle *Icthyophaga humilis* 242, 254
Lesser Flamingo *Phoeniconaias minor* 10, 32, 36, 84, 85, 86, 88, 90, 94, 96, 153, 165, 170, 200, 214, 216, 234, 235, 248
Lesser Florican *Sypheotides indicus* 11, 85, 90, 96, 98, 155, 158, 159, 210, 211, 243, 244
Lesser Frigatebird *Fregata ariel* 146
Lesser Golden-backed Woodpecker *Dinopium benghalense* 74, 130, 156, 166, 168
Lesser Kestrel *Falco naumanni* 11, 24, 30, 38, 58, 60, 62, 88, 106, 135, 185, 162, 168, 198, 233, 264
Lesser Necklaced Laughingthrush *Garrulax monileger* 186
Lesser Noddy *Anous tenuirostris* 32, 144, 232
Lesser Rufous-headed Parrotbill *Chleuasicus atrosuperciliaris* 43
Lesser Sand Plover *Charadrius mongolus* 22, 145, 214
Lesser Shortwing *Brachypteryx leucophris* 225
Lesser Whistling Duck *Dendrocygna javanica* 30, 81, 138, 159, 179, 214, 263
Lesser White-fronted Goose *Anser erythropus* 58, 59, 202, 203, 214
Lesser Yellow-naped Woodpecker *Picus chlorolophus* 68, 224, 257
Little Bittern *Ixobrychus minutus* 112
Little Bunting *Emberiza pusilla* 268
Little Cormorant *Microcarbo niger* 28, 29, 32, 36, 70, 128, 132, 138, 153, 172, 246
Little Crake *Zapornia parva* 110
Little Egret *Egretta garzetta* 28, 32, 36, 78, 112, 200
Little Grebe *Tachybaptus ruficollis* 90, 112, 159, 160, 167, 168, 200, 248
Little Owl *Athene noctua* 110, 230, 231
Little Pied Flycatcher *Ficedula westermanni* 65
Little Spiderhunter *Arachnothera longirostra* 79
Little Stint *Calidris minuta* 36

Little Tern *Sternula albifrons* 26, 33, 202
Long-billed Bush Warbler *Locustella major* 114
Long-billed Dowitcher *Limnodromus scolopaceus* 134, 164
Long-billed Pipit *Anthus similis* 218, 242
Long-billed Plover *Charadrius placidus* 45, 51
Long-billed Thrush *Zoothera monticola* 252, 264, 266
Long-legged Buzzard *Buteo rufinus* 92, 105, 213, 218
Long-tailed Broadbill *Psarisomus dalhousiae* 43, 48, 256
Long-tailed Duck *Clangula hyemalis* 112, 206
Long-tailed Minivet *Pericrocotus ethologus* 68, 195, 258, 262
Long-tailed Parakeet *Psittacula longicauda* 16, 18, 19, 20
Long-tailed Shrike *Lanius schach* 70, 112, 174, 224, 258
Long-tailed Sibia *Heterophasia picaoides* 46, 48
Long-toed Stint *Calidris subminuta* 57
Loten's Sunbird *Cinnyris lotenius* 74, 76, 124, 125, 126, 128, 170, 171, 204, 205
MacQueen's Bustard *Chlamydotis macqueenii* 10, 84, 90, 91, 93, 220
Malabar Barbet *Psilopogon malabaricus* 72, 74, 75, 76, 236
Malabar Grey Hornbill *Ocyceros griseus* 11, 74, 75, 76, 82, 121, 122, 124, 126, 136, 140 142, 172, 173, 233, 236, 238, 239
Malabar Lark *Galerida malabarica* 122
Malabar Parakeet *Psittacula columboides* 11, 72, 73, 74, 75, 76, 121, 122, 124, 126, 136, 140, 142, 233, 239, 238
Malabar Pied Hornbill *Anthracoceros coronatus* 73, 74, 75, 76, 126, 140, 156, 158, 162, 204, 205, 236
Malabar Trogon *Harpactes fasciatus* 72, 73, 74, 76, 78, 122, 126, 127, 140, 142, 170, 171, 204, 205, 236, 239
Malabar Whistling Thrush *Myophonus horsfieldii* 72, 76, 79, 98, 122, 124, 126, 142, 170, 204, 205, 236
Malayan Night Heron *Gorsachius melanolophus* 76, 77
Mallard *Anas platyrhynchos* 46, 64, 105, 112
Mangrove Pitta *Pitta megarhyncha* 196, 197, 198, 260, 270
Mangrove Whistler *Pachycephala cinerea* 260, 271
Manipur Bush Quail *Perdicula manipurensis* 176
Manipur Fulvetta *Fulvetta manipurensis* 177
Maroon Oriole *Oriolus traillii* 68, 186, 185, 194, 225, 252
Maroon-backed Accentor *Prunella immaculata* 265, 268
Marsh Babbler *Pellorneum palustre* 57, 58, 60, 62, 66, 71, 185
Marsh Sandpiper *Tringa stagnatilis* 32, 200
Marshall's Iora *Aegithina nigrolutea* 92, 98, 216
Masked Booby *Sula dactylatra* 164
Masked Finfoot *Heliopais personatus* 62, 66, 68, 260, 261, 270, 271
Merlin *Falco columbarius* 93
Mishmi Wren Babbler *Spelaeornis badeigularis* 40, 47
Montagu's Harrier *Circus pygargus* 97, 128, 175, 218
Mottled Wood Owl *Strix ocellata* 99, 122, 157, 159, 236, 237
Mountain Bamboo Partridge *Bambusicola fytchii* 68, 178, 180, 186, 187, 188, 190, 191
Mountain Bulbul *Ixos mcclellandii* 50
Mountain Hawk Eagle *Nisaetus nipalensis* 134, 224, 225, 230, 252, 263, 265, 268
Mountain Imperial Pigeon *Ducula badia* 51, 73, 75, 76, 238, 239, 265
Mountain Scops Owl *Otus spilocephalus* 180, 188, 191

Mrs Gould's Sunbird *Aethopyga gouldiae* 45, 183
Mrs Hume's Pheasant *Syrmaticus humiae* 12, 52, 177, 180, 181, 182, 183
Naga Wren Babbler *Spelaeornis chocolatinus* 188, 189, 190
Narcondam Hornbill *Rhyticeros narcondami* 12, 16, 17
Nepal House Martin *Delichon nipalense* 186
Nepal Wren Babbler *Pnoepyga immaculata* 252
Nicobar Bulbul *Ixos nicobariensis* 22
Nicobar Megapode *Megapodius nicobariensis* 12, 22
Nicobar Sparrowhawk *Accipiter butleri* 22
Nilgiri Flowerpecker *Dicaeum concolor* 78, 174
Nilgiri Flycatcher *Eumyias albicaudatus* 121, 136, 142, 233, 238, 240, 241
Nilgiri Laughingthrush *Montecincla cachinnans* 135, 233, 240
Nilgiri Pipit *Anthus nilghiriensis* 11, 136, 142, 143, 233, 238
Nilgiri Wood Pigeon *Columba elphinstonii* 11, 72, 73, 74, 75, 76, 78, 121, 122, 124, 126, 130, 135, 131, 136, 142, 170, 171, 172, 173, 233, 236, 238, 240
Northern Goshawk *Accipiter gentilis* 114, 263
Northern House Martin *Delichon urbicum* 178
Northern Pintail *Anas acuta* 28, 64, 67, 71, 81, 105, 153, 160, 201, 203, 216, 248, 263
Northern Shoveler *Spatula clypeata* 28, 30, 32, 71, 81, 90, 92, 96, 112, 153, 175, 201, 246, 263
Olive-backed Pipit *Anthus hodgsoni* 268
Olive-backed Sunbird *Cinnyris jugularis* 18
Orange Bullfinch *Pyrrhula aurantiaca* 101, 102, 110, 111, 114, 115, 251
Orange-bellied Leafbird *Chloropsis hardwickii* 186, 187
Orange-headed Thrush *Geokichla citrina* 78, 104, 130, 138, 157, 158, 166, 174, 255
Oriental Bay Owl *Phodilus badius* 69
Oriental Darter *Anhinga melanogaster* 12, 50, 58, 64, 66, 80, 86, 88, 94, 124, 132, 133, 138, 140, 149, 150, 151, 156, 160, 162, 197, 198, 200, 202, 208, 214, 216, 234, 238, 246, 247, 254, 260, 262, 266, 270, 271
Oriental Dwarf Kingfisher *Ceyx erithaca* 72, 78, 79, 172, 173
Oriental Hobby *Falco severus* 62, 182, 183
Oriental Honey Buzzard *Pernis ptilorhynchus* 71, 171, 166, 236, 240, 263
Oriental Magpie Robin *Copsychus saularis* 20, 98, 138, 158, 175, 240, 246
Oriental Pied Hornbill *Anthracoceros albirostris* 51, 68, 178, 181, 255
Oriental Pratincole *Glareola maldivarum* 152
Oriental Scops Owl *Otus sunia* 24
Oriental Skylark *Alauda gulgula* 30, 70
Oriental Turtle Dove *Streptopelia orientalis* 46, 102, 138, 256, 258
Oriental White-eye *Zosterops palpebrosus* 20, 24, 174, 246, 258
Osprey *Pandion haliaetus* 77, 80, 81, 82, 105, 139, 175, 263
Pacific Golden Plover *Pluvialis fulva* 138, 144, 146, 164
Pacific Reef Egret *Egretta sacra* 22
Paddyfield Pipit *Anthus rufulus* 31, 153
Painted Bush Quail *Perdicula erythrorhyncha* 124, 135, 136, 204, 206, 216, 224, 240, 241
Painted Francolin *Francolinus pictus* 24, 34, 120, 160, 161, 164, 166, 168, 175, 216
Painted Sandgrouse *Pterocles indicus* 93, 128, 167
Painted Spurfowl *Galloperdix lunulata* 26, 34, 128, 129, 130, 135, 154, 157, 160, 204, 216
Painted Stork *Mycteria leucocephala* 32, 34, 36, 37, 86, 88, 90, 94, 95, 96, 98, 132, 133, 150, 158, 160, 162, 175, 198, 208, 214, 216, 234, 235, 246, 254
Palani Laughingthrush *Montecincla fairbanki* 142, 238

Pale-capped Pigeon *Columba punicea* 11, 38, 46, 47, 57, 58, 66, 67, 68, 155, 196, 197, 204, 264
Pale-headed Woodpecker *Gecinulus grantia* 40, 68, 186, 194
Pallas's Fish Eagle *Haliaeetus leucoryphus* 12, 58, 62, 64, 65, 66, 67, 71, 88, 105, 150, 170, 177, 185, 198, 202, 203, 208, 214, 228, 244, 245, 247, 248, 254, 270, 271
Pallas's Gull *Ichthyaetus ichthyaetus* 138
Pallid Harrier *Circus macrourus* 24, 26, 30, 38, 58, 60, 92, 106, 121, 140, 158, 169, 202, 234, 238, 208, 214, 266, 267
Peregrine Falcon *Falco peregrinus* 82, 102, 171, 201
Pheasant-tailed Jacana *Hydrophasianus chirurgus* 71, 80, 86, 113, 138, 139
Pied Avocet *Recurvirostra avosetta* 90, 200
Pied Bushchat *Saxicola caprata* 98, 138, 240
Pied Cuckoo *Clamator jacobinus* 246, 255
Pied Falconet *Microhierax melanoleucos* 51, 68, 184
Pied Harrier *Circus melanoleucos* 71, 97, 262
Pied Imperial Pigeon *Ducula bicolor* 16, 18, 24, 25
Pied Kingfisher *Ceryle rudis* 30, 36, 106
Pied Thrush *Geokichla wardii* 136, 252
Pied Wheatear *Oenanthe pleschanka* 108
Pin-tailed Green Pigeon *Treron apicauda* 46, 48
Pink-browed Rosefinch *Carpodacus rodochroa* 226
Pintail Snipe *Gallinago stenura* 200
Plain Flowerpecker *Dicaeum minullum* 18, 45
Plain Mountain Finch *Leucosticte nemoricola* 250
Plain Prinia *Prinia inornata* 24, 30, 38, 128
Plum-headed Parakeet *Psittacula cyanocephala* 124, 130, 157, 161, 166, 168, 204
Pomarine Skua *Stercorarius pomarinus* 196
Puff-throated Babbler *Pellorneum ruficeps* 130, 171, 175
Purple Cochoa *Cochoa purpurea* 47, 266
Purple Heron *Ardea purpurea* 36, 68, 133, 138, 246, 162, 167, 175, 200
Purple Sunbird *Cinnyris asiaticus* 28, 36, 37, 78, 80, 153
Purple-rumped Sunbird *Leptocoma zeylonica* 133
Pygmy Blue Flycatcher *Ficedula hodgsoni* 45
Rain Quail *Coturnix coromandelica* 26, 30, 166, 168, 171, 175, 204, 205, 216, 248
Red Collared Dove *Streptopelia tranquebarica* 30, 96, 152
Red Crossbill *Loxia curvirostra* 44
Red Junglefowl *Gallus gallus* 10, 46, 107, 186, 230
Red Knot *Calidris canutus* 200, 232
Red Spurfowl *Galloperdix spadicea* 73, 78, 122, 126, 128, 134, 136, 140, 141, 238, 239
Red-backed Shrike *Lanius collurio* 93
Red-billed Blue Magpie *Urocissa erythroryncha* 115, 256
Red-billed Chough *Pyrrhocorax pyrrhocorax* 116
Red-billed Leiothrix *Leiothrix lutea* 107, 224
Red-billed Tropicbird *Phaethon aethereus* 120, 144, 146, 196
Red-breasted Flycatcher *Ficedula parva* 104
Red-breasted Parakeet *Psittacula alexandri* 58, 59, 182
Red-crested Pochard *Netta rufina* 36, 37, 112, 113, 179, 200, 201, 248
Red-faced Liocichla *Liocichla phoenicea* 43, 42, 55
Red-footed Booby *Sula sula* 196
Red-fronted Rosefinch *Carpodacus puniceus* 114, 228
Red-headed Bullfinch *Pyrrhula erythrocephala* 52, 253
Red-headed Bunting *Emberiza bruniceps* 14, 242
Red-headed Trogon *Harpactes erythrocephalus* 48, 51, 68, 71, 267
Red-headed Vulture *Sarcogyps calvus* 58, 60, 64, 88,

Index

94, 106, 124, 130, 135, 154, 158, 160, 161, 162, 204, 205, 208, 214, 216, 218, 220, 230, 251, 266
Red-necked Falcon *Falco chicquera* 202, 204, 205
Red-necked Grebe *Podiceps grisegena* 100, 105
Red-necked Phalarope *Phalaropus lobatus* 90
Red-rumped Swallow *Cecropis daurica* 82, 256
Red-tailed Minla *Minla ignotincta* 226, 263
Red-tailed Shrike *Lanius phoenicuroides* 93
Red-throated Thrush *Turdus ruficollis* 206
Red-vented Bulbul *Pycnonotus cafer* 36, 38, 68, 70, 80, 174, 212, 224
Red-wattled Lapwing *Vanellus indicus* 28, 30, 32, 36, 80, 152, 166
Red-whiskered Bulbul *Pycnonotus jocosus* 18, 68, 82, 240
Richard's Pipit *Anthus richardi* 261
River Lapwing *Vanellus duvaucelii* 151, 160, 200
River Tern *Sterna aurantia* 132, 133, 202, 203
Robin Accentor *Prunella rubeculoides* 116, 230
Rock Bunting *Emberiza cia* 118
Rock Bush Quail *Perdicula argoondah* 129
Rose-ringed Parakeet *Psittacula krameri* 28, 68, 157, 218
Rosy Minivet *Pericrocotus roseus* 254
Rosy Pipit *Anthus roseatus* 228, 268
Rosy Starling *Pastor roseus* 132, 133, 138, 212
Ruby-cheeked Sunbird *Chalcoparia singalensis* 57, 182
Ruddy Kingfisher *Halcyon coromanda* 50, 51, 63, 181, 197, 260, 261
Ruddy Shelduck *Tadorna ferruginea* 45, 46, 64, 67, 86, 88, 89, 112, 118, 175, 179, 199, 201 216, 229, 248
Ruddy Turnstone *Arenaria interpres* 32, 100, 145, 146
Ruddy-breasted Crake *Zapornia fusca* 18, 138, 152, 200
Ruff *Calidris pugnax* 86, 153, 197, 200, 201
Rufous Babbler *Argya subrufa* 122, 136, 140, 142, 233, 236, 238, 239
Rufous Sibia *Heterophasia capistrata* 106, 256
Rufous Treepie *Dendrocitta vagabunda* 82, 154, 174, 216, 218
Rufous Woodpecker *Micropternus brachyurus* 24, 38, 263
Rufous-bellied Eagle *Lophotriorchis kienerii* 180, 230, 236
Rufous-bellied Niltava *Niltava sundara* 71, 104, 259
Rufous-bellied Woodpecker *Dendrocopos hyperythrus* 106, 252, 258, 259
Rufous-breasted Accentor *Prunella strophiata* 238, 230
Rufous-capped Babbler *Cyanoderma ruficeps* 194
Rufous-chinned Laughingthrush *Garrulax rufogularis* 256, 166
Rufous-fronted Prinia *Prinia buchanani* 216
Rufous-gorgeted Flycatcher *Ficedula strophiata* 67, 187
Rufous-headed Parrotbill *Psittiparus bakeri* 43, 48
Rufous-naped Tit *Periparus rufonuchalis* 114
Rufous-necked Hornbill *Aceros nipalensis* 40, 42, 43, 44, 47, 50, 51, 52, 53, 54, 55, 60, 62, 71, 177, 178, 185, 194, 226, 228, 262, 263, 264, 266, 267
Rufous-necked Laughingthrush *Garrulax ruficollis* 62
Rufous-necked Snowfinch *Pyrgilauda ruficollis* 228
Rufous-tailed Lark *Ammomanes phoenicura* 30, 96, 128, 157
Rufous-tailed Scrub Robin *Cercotrichas galactotes* 92, 93
Rufous-throated Fulvetta *Schoeniparus rufogularis* 53
Rufous-throated Hill Partridge *Arborophila rufogularis* 46, 176, 257, 265
Rufous-throated Wren Babbler *Spelaeornis caudatus*

44, 50, 62, 226
Rufous-vented Laughingthrush *Garrulax gularis* 53
Rufous-vented Tit *Periparus rubidiventris* 45
Rufous-vented Yuhina *Yuhina occipitalis* 182, 193
Russet Bush Warbler *Locustella mandelli* 45
Russet Sparrow *Passer cinnamomeus* 224, 256
Rusty-bellied Shortwing *Brachypteryx hyperythra* 42, 43, 44, 49, 52, 53, 54, 55, 226, 227, 264, 266, 267, 268
Rusty-capped Fulvetta *Schoeniparus dubius* 189, 190, 191, 192
Rusty-cheeked Scimitar Babbler *Erythrogenys erythrogenys* 105, 227, 255
Rusty-flanked Treecreeper *Certhia nipalensis* 252
Rusty-fronted Barwing *Actinodura egertoni* 45, 224
Rusty-tailed Flycatcher *Ficedula ruficauda* 98, 114, 136, 140, 142
Saker Falcon *Falco cherrug* 210, 212
Sanderling *Calidris alba* 200
Sandwich Tern *Thalasseus sandvicensis*
Sapphire Flycatcher *Ficedula sapphira* 68
Sarus Crane *Antigone antigone* 11, 31, 66, 85, 86, 88, 94, 96, 98, 148, 149, 150, 151, 155, 158, 156, 163, 166, 178, 179, 214, 216, 242, 244, 246, 247, 248, 249, 254
Satyr Tragopan *Tragopan satyra* 42, 222, 225, 226, 227, 230, 264, 265, 268, 269
Saunders's Tern *Sternula saundersi* 196, 202
Savanna Nightjar *Caprimulgus affinis* 74, 75, 196
Scaly Laughingthrush *Trochalopteron subunicolor* 226
Scaly Thrush *Zoothera dauma* 68
Scaly-breasted Munia *Lonchura punctulata* 14, 174
Scaly-breasted Wren Babbler *Pnoepyga albiventer* 102, 226, 259
Scarlet Finch *Carpodacus sipahi* 47, 224, 250, 252, 253
Scarlet Minivet *Pericrocotus flammeus* 18, 39, 68, 140, 224
Scarlet-backed Flowerpecker *Dicaeum cruentatum* 187
Sclater's Monal *Lophophorus sclateri* 10, 12, 40, 46, 47, 49
Shikra *Accipiter badius* 33, 34, 125, 166, 174
Short-billed Minivet *Pericrocotus brevirostris* 186
Short-eared Owl *Asio flammeus* 85, 92
Short-toed Snake Eagle *Circaetus gallicus* 26, 30, 34, 92, 166, 169, 175, 212, 218, 240
Siberian Rubythroat *Calliope calliope* 70, 71
Sikkim Wedge-billed Babbler *Stachyris humei* 43, 52, 226
Silver-backed Needletail *Hirundapus cochinchinensis* 222
Silver-breasted Broadbill *Serilophus lunatus* 58, 64
Silver-eared Mesia *Leiothrix argentauris* 176
Sind Sparrow *Passer pyrrhonotus* 148
Sirkeer Malkoha *Taccocua leschenaultii* 124, 128, 130, 205, 216
Slaty-backed Forktail *Enicurus schistaceus* 48, 263
Slaty-bellied Tesia *Tesia olivea* 186
Slaty-breasted Rail *Lewinia striata* 18, 22, 25
Slaty-headed Parakeet *Psittacula himalayana* 106, 258
Slaty-legged Crake *Rallina eurizonoides* 134, 138
Slender-billed Babbler *Chatarrhaea longirostris* 58, 60, 62, 185, 228
Slender-billed Oriole *Oriolus tenuirostris* 45
Slender-billed Scimitar Babbler *Pomatorhinus superciliaris* 43, 49
Slender-billed Vulture *Gyps tenuirostris* 52, 58, 60, 62, 64, 66, 104, 185, 208, 243, 244, 254, 261, 262, 266, 268
Small Minivet *Pericrocotus cinnamomeus* 20, 80, 124, 130, 156, 168, 172, 205
Small Niltava *Niltava macgrigoriae* 68, 192, 224, 263

Snow Partridge *Lerwa lerwa* 45, 46, 102, 103, 110, 228, 229, 252, 253
Snow Pigeon *Columba leuconota* 45, 102, 108, 116, 222, 229, 230
Snowy-throated Babbler *Stachyris oglei* 53, 54, 55, 152
Sociable Lapwing *Vanellus gregarius* 11, 85, 90, 100, 150, 211, 214, 254
Solitary Snipe *Gallinago solitaria* 45, 102, 114, 176, 228
Sooty Tern *Onychoprion fuscatus* 138, 144, 146, 147
South Polar Skua *Stercorarius maccormicki* 232
Spanish Sparrow *Passer hispaniolensis* 206
Speckled Piculet *Picumnus innominatus* 24, 38, 253
Speckled Wood Pigeon *Columba hodgsonii* 45, 46, 102, 176, 225, 264
Spectacled Finch *Callacanthis burtoni* 101, 102, 111, 251
Spoon-billed Sandpiper *Calidris pygmaea* 26, 27, 36, 202, 203, 232, 234, 271
Spot-bellied Eagle Owl *Bubo nipalensis* 122, 140, 251, 254
Spot-billed Pelican *Pelecanus philippensis* 11, 28, 29, 32, 37, 57, 58, 60, 64, 66, 67, 71, 88, 89, 98, 121, 132, 133, 138, 149, 150, 159, 177, 178, 197, 198, 200, 202, 203, 234, 245, 254,
Spot-breasted Laughingthrush *Garrulax merulinus* 182, 183
Spot-breasted Parrotbill *Paradoxornis guttaticollis* 54, 189, 192
Spot-breasted Pied Woodpecker *Dendrocopos analis* 18, 20
Spot-breasted Scimitar Babbler *Erythrogenys mcclellandi* 193
Spot-throated Babbler *Pellorneum albiventre* 48
Spot-winged Grosbeak *Mycerobas melanozanthos* 230, 252
Spot-winged Rosefinch *Carpodacus rodopeplus* 252
Spot-winged Starling *Saroglossa spilopterus* 252, 257
Spotted Dove *Streptopelia chinensis* 68, 70, 78, 80, 82, 148, 224, 240, 248, 256, 258
Spotted Flycatcher *Muscicapa striata* 84, 92, 93
Spotted Forktail *Enicurus maculatus* 255
Spotted Laughingthrush *Garrulax ocellatus* 43, 225
Spotted Owlet *Athene brama* 80, 152, 167, 202, 212
Spotted Sandgrouse *Pterocles senegallus* 90, 210, 220
Sri Lanka Bay Owl *Phodilus assimilis* 73, 74, 134, 140, 239
Sri Lanka Frogmouth *Batrachostomus moniliger* 72, 73, 75, 126, 140, 239
Steppe Eagle *Aquila nipalensis* 139, 159, 175, 212
Stoliczka's Bushchat *Saxicola macrorhynchus* 11, 84, 85, 92, 93, 96, 116, 152, 211, 214, 216
Stork-billed Kingfisher *Pelargopsis capensis* 20, 74, 82, 159, 197, 261
Streak-breasted Scimitar Babbler *Pomatorhinus ruficollis* 265
Streak-throated Barwing *Sibia waldeni* 42, 44, 52
Streak-throated Woodpecker *Picus xanthopygaeus* 140, 258
Streaked Laughingthrush *Trochalopteron lineatum* 114, 256
Streaked Rosefinch *Carpodacus rubicilloides* 116
Streaked Shearwater *Calonectris leucomelas* 135
Streaked Spiderhunter *Arachnothera magna* 186, 195
Striated Babbler *Argya earlei* 71, 244
Striated Bulbul *Pycnonotus striatus* 45, 224, 225, 260, 265
Striated Grassbird *Megalurus palustris* 209
Striated Heron *Butorides striata* 82, 138
Striated Laughingthrush *Grammatoptila striata* 190, 225, 226
Striated Swallow *Cecropis striolata* 68

Striolated Bunting *Emberiza striolata* 220
Striped Laughingthrush *Trochalopteron virgatum* 188, 190, 192
Sulphur-bellied Warbler *Phylloscopus griseolus* 209
Sultan Tit *Melanochlora sultanea* 43, 48, 61, 62, 68, 186, 267
Swamp Francolin *Francolinus gularis* 12, 58, 59, 60, 62, 64, 66, 67, 71, 185, 186, 230, 243, 244, 245, 262, 270
Swinhoe's Minivet *Pericrocotus cantonensis* 267
Swinhoe's Storm-petrel *Hydrobates monorhis* 135, 144, 146
Sykes's Lark *Galerida deva* 26, 30, 90
Sykes's Nightjar *Caprimulgus mahrattensis* 92, 220
Sykes's Short-toed Lark *Calandrella dukhunensis* 30
Tawny Eagle *Aquila rapax* 34, 35, 92, 105, 213
Tawny Fish Owl *Ketupa flavipes* 255
Tawny Owl *Strix aluco* 114, 184
Tawny Pipit *Anthus campestris* 97, 218
Tawny-bellied Babbler *Dumetia hyperythra* 130, 204
Tawny-breasted Wren Babbler *Spelaeornis longicaudatus* 185, 187, 189, 192
Temminck's Stint *Calidris temminckii* 105, 200
Temminck's Tragopan *Tragopan temminckii* 10, 40, 43, 44, 45, 46, 47, 48
Terek Sandpiper *Xenus cinereus* 22, 82, 230, 200,
Thick-billed Flowerpecker *Dicaeum agile* 174
Thick-billed Warbler *Arundinax aedon* 19
Tibetan Blackbird *Turdus maximus* 45
Tibetan Partridge *Perdix hodgsoniae* 110, 116, 230
Tibetan Sandgrouse *Syrrhaptes tibetanus* 10, 116
Tibetan Siskin *Spinus thibetanus* 260
Tibetan Snowcock *Tetraogallus tibetanus* 46, 118, 222, 230
Tickell's Blue Flycatcher *Cyornis tickelliae* 38, 130, 132, 133, 266
Tickell's Leaf Warbler *Phylloscopus affinis* 240
Tickell's Thrush *Turdus unicolor* 104, 112, 118, 226, 256
Tropical Shearwater *Puffinus bailloni* 164
Trumpeter Finch *Bucanetes githagineus* 220, 221
Tufted Duck *Aythya fuligula* 160, 201, 203, 209
Twite *Linaria flavirostris* 100, 108
Tytler's Leaf Warbler *Phylloscopus tytleri* 101, 111, 114, 251
Ultramarine Flycatcher *Ficedula superciliaris* 24, 78, 98, 114, 158, 174
Upland Buzzard *Buteo hemilasius* 230, 252
Variable Wheatear *Oenanthe picata* 212
Variegated Laughingthrush *Trochalopteron variegatum* 102, 114, 252
Velvet-fronted Nuthatch *Sitta frontalis* 240
Verditer Flycatcher *Eumyias thalassinus* 24, 158, 224
Vernal Hanging Parrot *Loriculus vernalis* 19, 73, 75
Vigors's Sunbird *Aethopyga vigorsii* 72
Violet Cuckoo *Chrysococcyx xanthorhynchus* 19, 59, 68
Vivid Niltava *Niltava vivida* 43
Wallcreeper *Tichodroma muraria* 42, 117, 118, 254
Ward's Trogon *Harpactes wardi* 41, 42, 43, 44, 46, 47, 49, 52, 53, 55, 228, 264
Water Pipit *Anthus spinoletta* 218
Watercock *Gallicrex cinerea* 209
Wayanad Laughingthrush *Garrulax delesserti* 11, 121, 136, 140, 238, 239
Wedge-tailed Green Pigeon *Treron sphenurus* 46, 224, 258
West Himalayan Bush Warbler *Locustella kashmirensis* 251
Western Crowned Leaf Warbler *Phylloscopus occipitalis* 114, 136

Western Marsh Harrier *Circus aeruginosus* 96, 97, 139, 171
Western Reef Egret *Egretta gularis* 30, 32, 82, 92, 145
Western Tragopan *Tragopan melanocephalus* 100, 101, 102, 251
Western Yellow Wagtail *Motacilla flava* 153, 204
Whimbrel *Numenius phaeopus* 146, 200
Whiskered Tern *Chlidonias hybrida* 36, 139, 201
Whiskered Yuhina *Yuhina flavicollis* 258, 263
White Wagtail *Motacilla alba* 112, 153
White-bellied Blue Flycatcher *Cyornis pallidipes* 73, 76, 122, 126, 140, 142, 233, 236, 239
White-bellied Drongo *Dicrurus caerulescens* 122, 156, 166, 168, 174, 175, 204, 217
White-bellied Erpornis *Erpornis zantholeuca* 71, 268
White-bellied Heron *Ardea insignis* 52, 53, 58, 62, 66, 68, 161
White-bellied Minivet *Pericrocotus erythropygius* 92, 122, 148, 210, 212, 218
White-bellied Sea Eagle *Haliaeetus leucogaster* 20, 22, 32, 80, 82, 171, 199, 201
White-bellied Treepie *Dendrocitta leucogastra* 11, 74, 75, 76, 122, 124, 126, 136, 140, 142, 233, 236, 239
White-bellied Woodpecker *Dryocopus javensis* 73, 75, 120, 140
White-breasted Waterhen *Amaurornis phoenicurus* 20, 80, 138, 149, 246
White-breasted Woodswallow *Artamus leucoryn* 16
White-browed Bulbul *Pycnonotus luteolus* 128, 130, 143
White-browed Bush Robin *Tarsiger indicus* 45, 252, 265
White-browed Fantail *Rhipidura aureola* 156, 168, 209, 216, 218
White-browed Fulvetta *Fulvetta vinipectus* 231
White-browed Piculet *Sasia ochracea* 48, 58, 70, 71
White-browed Shortwing *Brachypteryx montana* 43
White-browed Tit Warbler *Leptopoecile sophiae* 45
White-capped Water Redstart *Phoenicurus leucocephalus* 104
White-cheeked Barbet *Psilopogon viridis* 74, 76, 124, 136, 142, 174, 236
White-cheeked Hill Partridge *Arborophila atrogularis* 51, 52, 54, 68, 182, 184, 186
White-cheeked Tern *Sterna repressa* 144, 146
White-cheeked Tit *Aegithalos leucogenys* 102, 111, 251
White-collared Blackbird *Turdus albocinctus* 42, 226, 264
White-crested Laughingthrush *Garrulax leucolophus* 257
White-crowned Forktail *Enicurus leschenaulti* 68
White-crowned Penduline Tit *Remiz coronatus* 209
White-eared Bulbul *Pycnonotus leucotis* 212, 218, 220, 221
White-eyed Buzzard *Butastur teesa* 30, 34, 39, 154, 156, 159, 166, 168, 204, 205
White-gorgeted Flycatcher *Anthipes monileger* 43
White-headed Duck *Oxyura leucocephala* 208, 243
White-headed Starling *Sturnia erythropygia* 12, 17, 18, 20, 21, 24, 25
White-hooded Babbler *Gampsorhynchus rufulus* 53, 130
White-naped Tit *Machlolophus nuchalis* 84, 92, 93
White-naped Woodpecker *Chrysocolaptes festivus* 157, 216
White-naped Yuhina *Yuhina bakeri* 42, 44, 50, 52, 190, 226
White-rumped Shama *Kittacincla malabarica* 24, 38, 78, 127, 158
White-rumped Snowfinch *Onychostruthus taczanowskii* 230, 231

White-rumped Vulture *Gyps bengalensis* 60, 64, 66, 86, 90, 98, 104, 106, 122, 124, 130, 132, 136, 158, 156, 160, 162, 166, 168, 170, 172, 198, 202, 204, 208, 209, 212, 214, 216, 218, 220, 230, 236, 242, 244, 245, 254, 262, 266
White-spotted Fantail *Rhipidura albogularis* 82, 174
White-tailed Lapwing *Vanellus leucurus* 86, 153
White-tailed Robin *Myiomela leucura* 264, 265
White-tailed Sea Eagle *Haliaeetus albicilla* 58, 62, 66, 214, 254
White-tailed Stonechat *Saxicola leucurus* 209, 244
White-tailed Tropicbird *Phaethon lepturus* 232
White-throated Bulbul *Alophoixus flaveolus* 48, 68, 186, 194, 266
White-throated Dipper *Cinclus cinclus* 103, 118, 230
White-throated Fantail *Rhipidura albicollis* 224
White-throated Kingfisher *Halcyon smyrnensis* 30, 152, 175, 246
White-throated Needletail *Hirundapus caudacutus* 164, 224
White-throated Redstart *Phoenicurus schisticeps* 45, 268
White-throated Tit *Aegithalos niveogularis* 101, 102, 111
White-winged Grosbeak *Mycerobas carnipes* 45, 230, 268
White-winged Redstart *Phoenicurus erythrogastrus* 118
White-winged Tern *Chlidonias leucopterus* 206, 209
White-winged Wood Duck *Asarcornis scutulata* 12, 50, 51, 52, 53, 56, 57, 62, 66, 69, 177, 185, 186, 194
Whooper Swan *Cygnus cygnus* 100
Wilson's Storm-petrel *Oceanites oceanicus* 146
Wire-tailed Swallow *Hirundo smithii* 82
Wood Sandpiper *Tringa glareola* 105
Wood Snipe *Gallinago nemoricola* 54, 55, 135, 136, 164, 185, 228, 229, 233, 244, 254, 262, 268
Woolly-necked Stork *Ciconia episcopus* 64, 82, 87, 200
Wreathed Hornbill *Rhyticeros undulatus* 50, 53, 58, 178, 184
Yellow Bittern *Ixobrychus sinensis* 138, 200
Yellow-billed Babbler *Turdoides affinis* 128, 140
Yellow-billed Blue Magpie *Urocissa flavirostris* 45, 108, 250, 253
Yellow-breasted Bunting *Emberiza aureola* 178
Yellow-breasted Greenfinch *Chloris spinoides* 224
Yellow-browed Bulbul *Acritillas indica* 76, 78, 122, 124, 126, 142, 236, 238, 239
Yellow-browed Tit *Sylviparus modestus* 265
Yellow-eyed Pigeon *Columba eversmanni* 206, 209, 210, 211, 219
Yellow-legged Green Pigeon *Treron phoenicopterus* 124, 156, 166, 204, 216
Yellow-rumped Honeyguide *Indicator xanthonotus* 43, 45, 54, 55, 106, 225, 231, 253
Yellow-throated Bulbul *Pycnonotus xantholaemus* 121, 122, 128, 129, 130, 131, 236
Yellow-throated Laughingthrush *Garrulax galbanus* 188, 192
Yellow-vented Leaf Warbler *Phylloscopus cantator* 44
Yellow-wattled Lapwing *Vanellus malabaricus* 30, 152, 168, 204, 216
Yunnan Nuthatch *Sitta yunnanensis* 54, 55
Zitting Cisticola *Cisticola juncidis* 30, 151